Environmental
PRINCIPLES AND ETHICS

Environmental
PRINCIPLES AND ETHICS

Ming H Wong
Frank W K Lee
Martin K F Fung
Hong Kong Baptist University

World Scientific
NEW JERSEY • LONDON • SINGAPORE • BEIJING • SHANGHAI • HONG KONG • TAIPEI • CHENNAI

Published by

World Scientific Publishing Co. Pte. Ltd.
5 Toh Tuck Link, Singapore 596224
USA office: 27 Warren Street, Suite 401-402, Hackensack, NJ 07601
UK office: 57 Shelton Street, Covent Garden, London WC2H 9HE

British Library Cataloguing-in-Publication Data
A catalogue record for this book is available from the British Library.

ENVIRONMENTAL PRINCIPLES AND ETHICS
(with Field Trip Guide)

ISBN 981-256-838-7 (pbk)

Desk Editor: Tjan Kwang Wei

Acknowledgements

The authors would like to thank all colleagues in the Croucher Institute for Environmental Sciences and the Department of Biology, Hong Kong Baptist University especially Ms. Doris Ng, Ms. Anna Leung, Mr. Dennis Kong, Mr. Eric Yuen, Ms. Janet Chan, Ms. Kitty Siu, Mr. Samuel Sit and Ms. Zoe Lee for their assistance. Financial support from the Teaching Development Grants (Project No.: HKBU-5 for 2001-04 Triennium) of University Grants Committee is gratefully acknowledged.

Authors

M.H. Wong, F.W.K. Lee and M.K.F. Fung

Environmental Principles and Ethics: Textbook
環境法則與倫理：教材本

About the Authors ix

Foreword xi

Introduction 引言 xiii

Basic Principles 基本理論

1. **Ecological Principles** 生態學原理 3
2. **Biodiversity and Conservation** 生物多樣性及保育 14
3. **Human Impact on the Environment** 人類對環境的影響 29
4. **Hong Kong Environment and Precious Species** 香港環境及珍貴物種 40
5. **Environmental Ethics** 環境倫理 53
6. **Ecological Footprint** 生態足印 65
7. **Sustainable Development** 可持續發展 76
8. **Ecotourism** 生態旅遊 91

Advanced Topics 進階題目

9. **Persistent Toxic Substances** 持續性有毒物質 109
10. **Genetically Modified Food** 基因改造食物 123
11. **Solid Waste Management** 固體廢物處理 143
12. **Sewage Treatment** 污水處理 155
13. **E-waste** 電子廢物 165
14. **Environmental Impact Assessment** 環境影響評估 180
15. **Environmental Management System** 環境管理系統 192

Reference Materials 參考資料 204

Index 211

Note 記事 214

About the Authors

Professor Ming H. Wong

After graduation from the Chinese University of Hong Kong with a BSc in Biology, Professor Wong obtained his MSc, PhD and DSc from the University of Durham and also MBA and DSc from the University of Strathclyde. Professor Wong served the Biology Department of the Chinese University of Hong Kong as Lecturer and Senior Lecturer from 1973-85, following which he became Head of the Biology Department of Hong Kong Baptist University (1986-2002) and was promoted to Chair Professor in 1990. He currently serves as Director of the Croucher Institute for Environmental Sciences, Hong Kong.

Professor Wong's research work centers on restoration of derelict land and pollution ecology, especially heavy metals in earlier years, and persistent toxic substances more recently. He serves as regional co-ordinator of Central and North East Asia for the project "Regionally based assessment of persistent toxic substances" sponsored by the Global Environmental Facility (GEF) and implemented by the United Nations Environment Programme (UNEP).

Professor Wong has over 200 papers published in international scientific journals and edited several books. He serves on editorial boards of eight scientific journals related to environmental science, including Editor-in-Chief of the journal "Environmental Geochemistry and Health (Kluwer Academic Press)" and is a Visiting Professor of several major institutes in Mainland China such as Nanjing Institute of Soil Science of the Chinese Academy of Science, Zheijian University, Wuhan University, and Zhongshan University, and also Middlesex University in the UK.

Frank W.K. Lee and Martin K.F. Fung

Mr. Frank Lee obtained his bachelor and master degrees from the Hong Kong University of Science and Technology, whereas Mr. Martin Fung obtained his from the Hong Kong Baptist University. As staff of Croucher Institute for Environmental Sciences, both Mr. Lee and Mr. Fung are project officers for the production of this set of teaching package.

Other Contributors

Name	*Post*	*Institution/Department/Unit*
Dr. G.Y.S. Chan	Lecturer	Department of Applied Biology and Chemical Technology, HKPU
Prof. K.M. Chan	Associate Professor	Department of Biochemistry & Environmental Science Programme, CUHK
Dr. K.C. Cheung	Assistant Professor	Department of Biology, HKBU
Dr. S.S. Chung	Assistant Professor	Croucher Institute for Environmental Sciences, HKBU
Prof. C.Y. Lan	Associate Dean	School of Life Science, Zhongshan University
Dr. Y. Liang	Assistant Professor	Croucher Institute for Environmental Sciences, HKBU
Dr. C.N. Ng	Associate Professor	Department of Geography, HKU
Dr. M.L. So	Associate Professor	Department of Biology, HKBU
Mr. Roy Tam	Chairman	*Green Sense* (NGO)
Dr. C.K.C. Wong	Associate Professor	Department of Biology, HKBU
Prof. K.K.K. Wong	Professor	Department of Geography, HKBU
Prof. R.N.S. Wong	Professor	Department of Biology, HKBU
Dr. M.S.M. Yang	Associate Professor	Department of Biology, HKBU
Prof. Z.H. Ye	Professor	School of Life Science, Zhongshan University, China
Prof. K.K.L. Yung	Professor	Department of Biology, HKBU
Prof. J.H. Zhang	Professor	Department of Biology, HKBU

Environmental Science is a relatively new subject and its development in Asia far lags behind the Western World. It is not easy for local teachers to gather sufficient teaching materials on environmental science since most of the teaching materials, reference books, case studies and field trip manuals are produced by foreign scientists. Moreover, they may not be suitable for the study of the local environment of HK and China. This tailor-made teaching package will help fill the gap. Instead of adopting materials produced in foreign countries, teachers can make use of the local information in the packages to conduct high quality educational activities for their students.

South China has been undergoing an economic take-off, however, its environmental aspects at the same time has deteriorated rapidly. Meanwhile, China's entry into the World Trade Organization will provide a historic opportunity for China to adjust its economic structure and facilitate the building of a more environmentally-friendly economy. It is time to alert the general public of the seriousness of environmental deterioration and raise their level of awareness on environmental conservation. Since the materials for teaching environmental issues of Hong Kong and the Pearl River Delta region are very limited, this multimedia education package will help enhance teaching and the learning quality of environmental education.

The package will not only assist the development of environmental education in formal curriculum, but will also allow students to study independently through interactive CD-ROM. They will also be useful for teaching other relevant subjects of the Departments of Biology and Chemistry of the Hong Kong Baptist University.

I would like to thank the University Grants Committee for the financial support to this project. I also acknowledge the colleagues and staff of CIES and Hong Kong Baptist University who contributed and made this project possible.

Ming Hung Wong

Chair Professor of Biology and Director of CIES,
Hong Kong Baptist University

December 2005

About Croucher Institute for Environmental Sciences

This publication (along with the education package entitled ***Environmental Principles and Ethics***) has been proposed and implemented by the **Croucher Institute for Environmental Sciences (CIES)**, Hong Kong Baptist University.

Hong Kong Baptist University

Led by **Prof. Ming Hung Wong**, (Chair Professor of the Biology Department of Hong Kong Baptist University and Director of CIES), CIES has been involved in research and consultancy projects related to environmental impact assessment, revegetation of derelict land, and treatment and recycling of industrial, agricultural and domestic waste, with emphasis on the use of relatively low cost technology. Through its many distinguished research projects and consultancy services, CIES has, over the years, acquired an international reputation and established international, national and local links.

Since April 2004, the Croucher Institute for Environmental Sciences has been named after the **Croucher Foundation** as the only awarded institution within the Hong Kong Baptist University. CIES was formerly known as the Institute for Natural Resources and Environmental Management (INREM, 1997-2004) and the Centre for Waste Recycling and Environmental Biotechnology (1990-1997) which was inaugurated in 1990. The honourable naming by the Croucher Foundation has allowed the Institute to strengthen in our existing research and consultancy services.

The mission of CIES is to promote interdisciplinary research into the management of our natural resources including air, soil, water and biological resources. Managed by professional staff and equipped with the-state-of-the-art facilities, CIES has an impressive track record in attracting research and consultancy projects in natural resources and environmental studies.

Publications of CIES

CIES' vision is to become a technology transfer center for providing information and assistance in the practical application of new and developing technologies pertinent to the sustainable use and maintenance of natural resources, and waste treatment and disposal and recycling of waste in Hong Kong, the Pearl River Delta, and the Southeast Asian Region.

For more information about CIES, please visit www.cieshk.org.

Importance of Environmental Education

This education package aims to foster environmental literacy of non-science students and to train them to look at environment-related issues from a critical perspective. With this package, students will obtain knowledge on fundamental environmental principles and ethics. Furthermore, they will also be in a position to apply these ethical principles in the debate of local and global environment-related issues.

The main objectives of the textbook are to understand fundamental environmental principles and ethics and to foster critical thinking in environment-related issues. Issues covered include natural science, resources management, food safety, public health, sustainable development, ecotourism, nature conservation and ecological footprint, as well as other current environment-related issues in Hong Kong and South China.

The first part of this book focuses on **basic environmental principles**, including ecology, biodiversity and conservation, human impact, environmental ethics, sustainable development, ecotourism and ecological footprint.

These chapters explain fundamental environmental knowledge with detailed background and illustrations with practical examples on specific issues.

Human Impact on the Environment (Chapter 3)

The problem of E-waste (Chapter 13)

The second part of this book introduces some of the selected **topics of a more advanced level**, with emphasis on updated information and practices of Hong Kong and the Pearl River Delta region.

The topics include persistent toxic substances, electronic waste, sewage treatment, solid waste management, genetically modified food, environmental impact assessment and environmental management system.

This book could be used as a reference material by academic institutions, non-profit organizations and government departments. It provides practical educational materials on the topic of life science and aids in increasing the knowledge on the environment.

關於裘槎環科所

本刊物（連同名為「環境法則與倫理」的教材套）由香港浸會大學　裘槎環科所的員工籌備及製作。

由香港浸會大學生物系黃銘洪講座教授帶領，裘槎環科所是集科研、教育、學術交流、資訊傳遞和人才培訓於一體的多功能環境科學中心，竭誠為香港、珠江三角洲乃至整個東南亞地區提供嶄新的資源管理理念和環境諮詢服務。其所涉及的研究和諮詢領域涵蓋空氣、土壤、水、生物等自然資源以及各種無機和有機污染物的環境行為及其風險評價。

裘槎環科所辦事處

裘槎環科所前身為香港浸會大學「自然資源與環境管理研究所」。二〇〇四年為了慶祝基金會二十五周年志禧，裘槎基金會擬在全港範圍內甄選6間頂級研究所，冠以「裘槎」之名，本所原有幸獲此殊榮，被冠名為「裘槎環科所」。是次甄選由基金會專門聘請國際知名學者組成特別委員會，根據香港各個研究所的科研成果和社會貢獻、以及屬下研究人員的學術成就等多項綜合指標嚴格評審後確定的，具有很高的權威性。長期以來，裘槎基金會對環科所的各項科學研究和國際交流活動都給予了大力的支持，這次獲選正是基於對環科所以往工作成就的高度肯定和認同。

裘槎環科所是集科研、教育、學術交流、資訊傳遞和人才培訓於一體的多功能環境科學中心，竭誠為香港、珠江三角洲乃至整個東南亞地區提供嶄新的資源管理理念和環境諮詢服務。其所涉及的研究和諮詢領域涵蓋空氣、土壤、水、生物等自然資源以及各種無機和有機污染物的環境行為及其風險評價。

詳情請瀏覽網址：www.cieshk.org

裘槎環科所經常舉辦國際性學術研討會

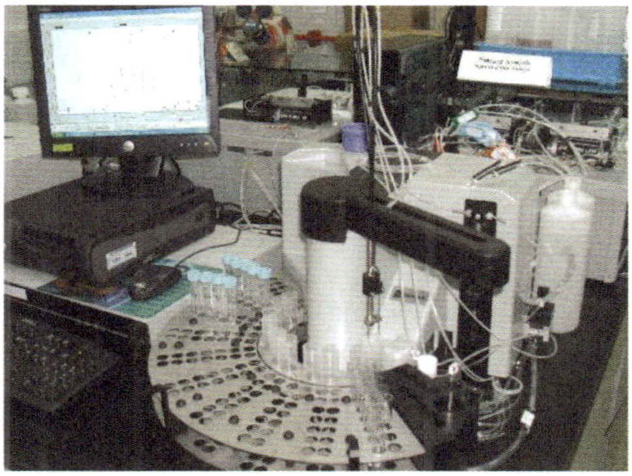
利用最先進實驗室設備及技術，開發科研

環境教育的重要性

本教材套旨在給非理科學生灌輸環境學識，鍛鍊他們以批判的角度去探討有關環境的事項。透過本教材套，學生不但汲取基本環境倫理的知識及其法則，還可以應用於本地及世界的環境議題，作出獨立思考和分析，融會貫通。

透過閱讀本教材本，讀者可以了解各基本環境法則和倫理，以及加強對環境議題的批判性思考。內容全面亦深入淺出，範圍包括自然科學、資源管理、食物安全、公共衛生、可持續發展、生態旅遊、自然保育和生態足印等，當中亦提及本港和南中國最新的環境資訊。

本書的第一部分著重基本環境理論，當中包括生態學、生物多樣性及保育、人類對環境的影響、環境倫理、可持續發展、生態旅遊及生態足印。

這些章節嘗試以不同理論教授基本環境知識，並引用個別題材的實用例子作解釋。

中華白海豚：香港珍貴物種（第四章）

廢物分類回收桶：固體廢物處理（第十一章）

第二部分則選緝一些有關環境教育的進階題目作介紹，並提供本港及珠江三角洲的最新資訊和實用例子作參考。

本部分題目包括持續性有毒物質、電子廢物、污水處理、固體廢物處理、基因改造食物、環境影響評估及環境管理系統。

本刊物可供本港各大學、非牟利團體及政府部門作參考材料，當中包含有關生物學和環境科學的實用教學材料，有助加深對自然環境的認識。

Basic Principles
基本理論

Ecological Principles
生態學原理

Biodiversity and Conservation
生物多樣性及保育

Human Impact on the Environment
人類對環境的影響

Hong Kong Environment and Precious Species
香港環境及珍貴物種

Environmental Ethics
環境倫理

Ecological Footprint
生態足印

Sustainable Development
可持續發展

Ecotourism
生態旅遊

Chapter 1. Ecological Principles

Although this teaching package is designed for students of non-science background, some basic ecological principles are delivered here in order to strengthen their scientific knowledge. These basic concepts may help them to have a better understanding of the preceding chapters as well as a critical perspective on the principles to be taught.

← Nature © Frank Lee

Life first appeared on Earth about 4 billion years ago. It started with a small living cell that fixes and uses energy in order to grow and reproduce. Eventually life involved many systems of energy management (or metabolism), together with repeated replications, and subsequently evolved into more and more complex forms.

But how do we define life? Generally speaking, we may distinguish living from non-living things based on these characteristics (Beeby and Brennan, 1997; Jones, 1997):

- having a complex, organized structure based on organic (carbon) compounds;
- acquiring material and energy from their environment and converting these into different forms;
- actively maintaining their complex structure;
- responding to environmental stimuli;
- growing and reproducing using a molecular blueprint (DNA); and
- having the capacity to evolve.

I. Basic Principles

Laws, Theories & Principles

A **law** is a hypothesis that has survived repeated examination by many investigators over a long time and has central importance to an area of science, e.g. the biogenetic law. A **theory** is a broadly written statement that covers large bodies of scientific knowledge. It is well-thought out and has lots of evidence for its support, e.g. theory of evolution. **Principles** are generally true statements, can be referred to as or base on laws or theories (Bunce, 1993).

Scientific laws with ecological relevance:

Example 1: The first & second law of thermodynamics (Jones, 1997)

1st law	energy can neither be created nor destroyed; it can only be changed from one form into another
2nd law	there is no loss of total energy, but there is a loss of useful energy

Example 2: Limiting Factor Principle (Hanks, 1996)

- Too much or too little of any abiotic factor can limit or prevent growth of a population, even if all other factors are at or near the optimum range of tolerance.

In Greek, the term "Ecology" means "house" or "place to live". **Ecology** is the study of how organisms interact with one another and with their physical and chemical environment (Reiss and Chapman, 2000; Dickinson and Murphy, 1998). It is concerned with the biology of groups of organisms and with functional processes on the land, in the oceans and freshwater, and in the air (Buell and Girard, 1994).

Realms of Ecology (Townsend, Begon and Harper, 2003):

Ecosphere (Biosphere)	- all of the Earth's ecosystems together
↑ Ecosystems	- community of different species interacting with one another and with their nonliving environment of matter and energy
↑ Communities	- populations of all the different species occupying a particular place
↑ Populations	- group of individuals of the same species occupying a given area at the same time
↑ Organisms	- any form of life

Species is a group of organisms that resemble one another in appearance, behavior, chemistry, and genetic structure (Townsend, Begon and Harper, 2003). Each species can breed with one another to produce fertile offspring under natural conditions. Scientists estimate that between 5,000,000 and 100,000,000 species exist (1,400,000 have been identified). A **habitat** is the place where an organism or population lives (Jones, 1997).

Ecosystem – woodland
© Frank Lee

The word '**ecosystem**' was introduced in 1935 by a British ecologist, Sir Arthur Tansley, referring to the whole community of organisms and their environment taken as a single functioning unit (Reiss and Chapman, 2000).

Today, the term means the basic functional unit in ecology. It contains organisms, populations, and communities, each having unique properties and influence on each other.

Ecosystem also represents an order of systems below the so-called biosphere, and that a collection of similar ecosystems makes up the conceptual notion of a biome (see illustration of "Realms of Ecology" above).

The **environment** of an organism relates to the features (including physical, chemical and biological) that are significant to its existence. They are (Dickinson and Murphy, 1998):

- abiotic characteristics such as atmospheric gases, temperature, fire and wind, and components such as mineral nutrients and water;
- food, or nutrients required for the organism;
- living habitat to rest, sleep, hide and reproduce;
- or even other organisms of the same species and of other species.

A **Cycle** deals with matters that circulate, transform and re-circulate in the biosphere from the environment to organisms and back to the environment. It applies to chemical elements: water, carbon, nitrogen, sulphur, phosphorous, etc.

Just as the living world depends on water, so too does it influence how water moves through the **Water Cycle**. ➔

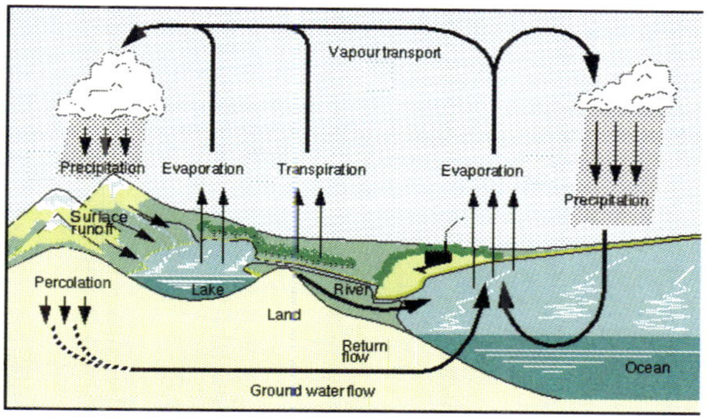

(modified from Dickinson and Murphy, 1998)

Community, Competition and Symbiosis

A community is any assemblage of populations living in a prescribed area (Reiss and Chapman, 2000). In a community, organisms compete for limited resources. Symbiosis means "living together". It is defined as a close interaction between organisms of different species for an extended time.

🡐 Corals and coral fish: example of mutualism
© Frank Lee

Interactions among organisms (Jeffries, 1997; Beeby and Brennan, 1997):

Type of interaction	*Effect on organism A*	*Effect on organism B*
Competition between A & B	Harms	Harms
Predation by A on B	Benefits	Harms
Symbiosis a. Parasitism by A on B b. Commensalism of A with B c. Mutualism between A & B	 Benefits Benefits Benefits	 Harms No effect Benefits

Carrying capacity

Every ecosystem has limits in terms of the size of various populations that it can support. Every species has certain needs that the community must provide in order for it to survive and continue to exist (Reiss and Chapman, 2000).

Total demand on resources = population x average demand per individual

The term **"ecological niche"** refers to the organism's functional role in the community. It could be the species' status in terms of its activities, its rate of metabolism and growth, its effect on other organisms with which it has contact, or its ability to modify important operations in the ecosystem (Hanks, 1996).

Limiting Factors
In the natural world, limiting factors (i.e. the availability of food, water, shelter and space) can affect the populations of animals and plants (Hanks, 1996). In fact, a species may have a wide range of tolerance for one factor but a narrow range of tolerance for another. For example, species of:

- Terrestrial Ecosystems– limiting factors are temperature, water, light, nutrients, etc.
- Aquatic Ecosystems – limiting factors are salinity, temperature, sunlight, dissolved oxygen, nutrients, etc.

Energy Flow in Ecosystems (Townsend, Begon and Harper, 2003)
The **Food Chain** is the transfer of food energy from its source in plants through a series of organisms where eating and being eaten is repeated a number of times. **Food Web** refers to a network of feeding relationships in an ecosystem. **Trophic Level** basically means "feeding level". The exact level depends on whether an organism is a producer or a consumer and what it eats or decomposes.

- 1st Level = producers
- 2nd Level = primary consumers
- 3rd Level = secondary consumers
- 4th Level = tertiary consumers
- detritivores process detritus from all levels

Ecological Pyramid

- Pyramid of Numbers is the number of organisms at each trophic level.
- Pyramid of Biomass is biomass (weight) of all organic matter contained in the organisms at each trophic level. The biomass is transferred from one level to another in the form of chemical energy.

Trophic levels in the biosphere

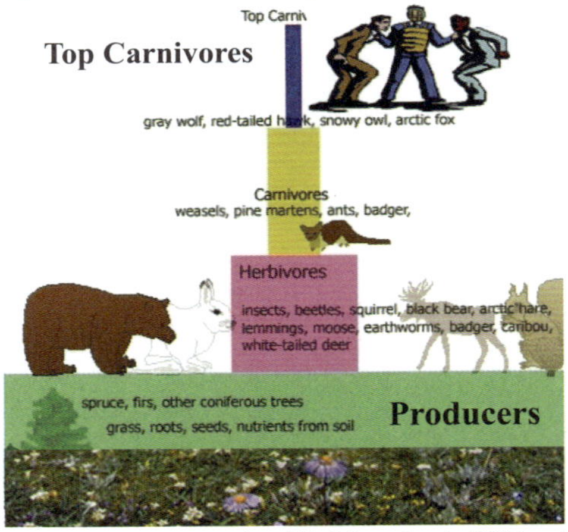

(modified from Jarvis, 2000, p.45)

Pyramid of Energy Flow is the percent of usable energy transferred from one level to another (Hanks, 1996). Only about 5% to 20% of the energy is transferred (depends on the type of organisms in the ecosystem). This explains why food webs rarely have more than 4 trophic levels.

Productivity
Gross Primary Productivity (GPP) is the rate at which an ecosystem's producers capture and store chemical energy as biomass. Net Primary Productivity (NPP) is GPP minus the rate at which producers use stored chemical energy in their biomass. This is the basic food source for all consumers. It is estimated that humans waste 27% of the world's NPP (Jones, 1997).

II. Living Organisms

Types of organisms based on cell type (Beeby and Brennan, 1997):

- **Prokaryotic** organisms lack a distinct nucleus. They also lack internal parts surrounded by a membrane. They all belong to the Kingdom Monera.
- **Eukaryotic** organisms contain a nucleus surrounded by a membrane. They also have other internal parts surrounded by a membrane.

Kingdoms of Organisms:

Monera
- single celled and microscopic
- all are prokaryotic
- e.g. bacteria

Protista
- most primitive of eukaryotic species
- mainly single celled, some are colonial
- basically a catch-all group
- e.g. algae, protozoa

Fungi
- are mainly multicellular.
- are all non-photosynthetic autotrophs.
- obtain energy by decomposing dead and dying organisms and absorbing their nutrients from those organisms

© Frank Lee

Plantae
- are mainly multicellular.
- are all photosynthetic autotrophs.
- are important sources of oxygen, food, and clothing/construction materials, as well as pigments, spices, dyes, and drugs

© Frank Lee

Animalia
- are mainly multicellular
- are all heterotrophic by ingestion
- provide food, clothing, fats, scents and labour

Birds © Zoe Lee

Butterfly © Zoe Lee

Fish © Frank Lee

Turtle © Frank Lee

Monkey © Frank Lee

Types of organisms based on function (Townsend, Begon and Harper, 2003): **Producers/Autotrophs** can make the nutrients they need to survive from compounds in their environment. Photosynthesis is the biological process using sunlight to make complex nutrient compounds.

$$CO_2 + H_2O + \text{solar energy} \rightarrow C_6H_{12}O_6 + O_2$$

Chemosynthesis is the conversion of simple compounds from their environment into more complex nutrient compounds without sunlight. **Consumers/Heterotrophs** get their organic nutrients by feeding on the tissues of producers and other consumers.

As primary consumers, **herbivores** feed directly on producers. Feeding only on other consumers, **carnivores** are secondary, tertiary, and quartinary consumers. **Omnivores** eat both plant and animal tissues. **Detritivores** live off detritus (part of dead organisms and cast-off fragments and wastes of living organisms). **Decomposers** digest complex organic molecules in detritus into simpler inorganic compounds and absorb the soluble nutrients. **Detritus feeders** extract nutrients from partly decomposed organic matter.

Aerobic Respiration is the process by which energy is released to sustain life processes of producers and consumers.

$$C_6H_{12}O_6 + O_2 \rightarrow CO_2 + H_2O + \text{solar energy}$$

Survival of any individual organism depends on the flow of matter and energy through its body. The community of organisms in an ecosystem survives because of a combination of matter recycling and one-way energy flow.

Conclusion

This chapter introduces some of the basic ecological principles, from knowing the definitions of law, theory and principles, realms of ecology, interactions among species, ecological pyramid, to classifications of living organisms. The important message is that there are different interactions among plants, animals and microorganisms – all being applied by the principles mentioned above.

Questions for Further Thinking

1. It was mentioned that every species has its niches in the ecosystem. Name one species and its niche for the following habitats:

Ecosystem	*Species*	*Niche*
- River/stream		
- Rocky shore		
- Forest		

2. Besides the Water Cycle, name one more natural cycle that is essential in the ecosystem.

第一章　生態學原理

雖然這個教材套是為非修讀理科的學生而設計，不過這裡提供一些基本生態學原理，以加強他們的科學知識。這些基本概念也許幫助他們理解以下的章節及一些重要的原則。

大約四十億年前，**生命**首次在地球上出現。它以一個細胞開始，攝取及利用能量來增長和繁殖。最終，生命包括了很多能量管理系統(或新陳代謝)，配合複製，而演變較複雜的形態。

但我們怎樣定義生命？一般而言，我們可以根據以下特徵去區分生命與非生命 (Beeby and Brennan, 1997; Jones, 1997)：

- 由有機物質組成的複雜及有組織的結構；
- 從外界攝取材料和能量，轉換成不同的形式；
- 維持固定的型態；
- 對外界刺激作出反應；
- 利用DNA生長和複製；及
- 有進化的能力。

I. 基礎原理

定律、學說和定理

定律是很多研究者經過長時間反覆驗證假設，並對科學的一個範疇十分重要，例如：生物遺傳定律。學說是涵蓋廣泛的科學知識，而且是一個很好的意念和有許多證據作為支持，例如進化學說。原理是一般事實的陳述，可被稱為定律或學說的基礎 (Bunce, 1993)。

與生態學相關的科學定律：

例子一：熱力學第一及第二定律 (Jones, 1997)

第一定律	能量不能被創造及消滅；能量只可以由一個型態轉變成另一個。
第二定律	總能量是沒有減少的，減少的是有用的能量。

例子二：限制因子定理 (Hanks, 1996)

- 儘管其他因子接近或達到可容忍範圍的最高值，其中一個非生物因子過多或過少可能限制或防礙族群的生長。

在希臘文，「生態學」是指「房子」或「居住的地方」。**生態學**是研究生物與其物理和化學環境的關係 (Reiss and Chapman, 2000; Dickinson and Murphy, 1998)。它與在陸地，海洋，淡水和空氣的生物體和功能過程有關 (Buell and Girard 1994)。

生態學的範疇 (Townsend, Begon and Harper, 2003)：

生物圈	- 所有地球的生態系
↑ 生態系	- 不同物種的群落與環境和群落之間的相互作用
↑ 群落	- 一個地區不同種的放群
↑ 族群	- 一個地區同種的個體
↑ 生物	- 任何生命的形式

物種是一群有類似的形態、行為、化學和基因結構的生物 (Townsend, Begon and Harper, 2003)。在自然的環境下，物種可互相交配產生可育的下一代。科學家估計有5,000,000 至 100,000,000 個物種存在 (有1,400,000個物種被確認)。生境是指生物或族群居住的地方 (Jones, 1997)。

「**生態系**」一詞是由英國生態學家，Sir Arthur Tansley 提出，指整個群落與其環境被視為一個功能單位 (Reiss and Chapman, 2000)。現在，「生態系」一詞是指生態學的基本功能單位。它包括生物、族群和群落，彼此皆有獨特的特徵及互相影響。生態系是生物圈的下一層，而相似的生態系構成生物群落 (參見以上的「生態學的範疇」)。

環境對生物的生存有重大關係，包括物理性，化學性和生物性，包括 (Dickinson and Murphy, 1998)：

- 非生物的特徵，例如大氣氣體、溫度、火、風、礦物質和水；
- 生物需要的食物,或營養；
- 休息、睡覺、躲藏和繁殖的棲息地；
- 同種和其他物動的生物。

循環是指物質從環境到生物和歸還環境之間的流動和轉換的過程。循環的化學元素有：水、碳、氮氣、硫磺、磷等。

群落、競爭和共生

群落是一個地區的族群 (Reiss and Chapman, 2000)。在群落裏，生物爭取有限的資源。共生是指「一起生活」。它的定義是在一段持續的時間，不同物種之間的相互作用。

生物間的相互作用 (Jeffries, 1997; Beeby and Brennan, 1997)：

相互作用的種類	對生物*A*的影響	對生物*B*的影響
競爭	受害	受害
捕食	獲益	受害
共生 a. 寄生 b. 片利共生 c. 互利共生	 獲益 獲益 獲益	 受害 沒有影響 獲益

容納量
每個生態系皆有極限，可支持不同大小的族群。每個物種必須從環境得到其所需，讓它繼續生存 (Reiss and Chapman, 2000)。

資源的總需求量 = 族群 × 個體的平均需求量

「**生態區位**」是指生物在生態系的功能。這是根據物種的活動、代謝和生長速率，對其他物種的影響，或改善生態系的能力 (Hanks, 1996)。

限制因子
在自然界，限制因子(即是食物、水、庇護所和空間)能影響動物和植物的族群 (Hanks, 1996)。實際上，物種也許對一個因素的容忍度較大，但對另一個因素的容忍度較低。例如：
- 陸地生態系 — 限制因素是溫度、水、光、養份等
- 水生生態系 — 限制因素是鹽分、溫度、陽光、溶解氧、養份等

生態系的能量流 (Townsend, Begon and Harper, 2003)
食物鏈是食物能量由植物傳至進食和被攝食的動物的途徑。食物網是指生態系中的進食關係網。食性層次基本上是「進食的層次」。食性層次取決於生物是生產者或消費者，並且它吃什麼或分解什麼。

- 第一層次 = 生產者
- 第二層次 = 初級消費者
- 第三層次 = 次級消費者
- 第四層次 = 三級消費者
- 食碎屑動物處理所有層次的腐屑

(modified from Jarvis, 2000, p.45)

生態金字塔
- 數目金字塔是一個食性層次的生物數目。
- 生物量金字塔是一個食性層次的全部有機物質的生物量 (重量)。生物量以化學能量的形式從一個層次轉移到另一個層次。

能量流程金字塔是從一個層次轉移到另一個層次的可用能量的百分比 (Hanks, 1996)。只有約5% 到20% 能量可被轉移 (取決於生物的種類)。這解釋為什麼食物網很少有超過四個食性層次。

生產力
初級總生產量 (GPP) 是生產者吸取和儲存化學能為生物量的速率。
淨初級生產量 (NPP) 是 GPP 減去生產者儲存化學能為生物量的速率。這是所有消費者的基本食物來源。現估計人類浪費整個世界的NPP百份之二十七 (Jones, 1997)。

II. 生物

以細胞類型劃分的生物種類 (Beeby and Brennan, 1997)：

- 原核生物沒有細胞核。它們沒有由膜包圍著的細胞器。它們全都屬於原核生物界
- 真核生物有由膜包圍的細胞核和細胞器。

生物界：

原核生物界
- 單細胞和微小
- 屬於原核生物
- 例子：細菌

原生生物界
- 原始的真核生物
- 主要是單細胞，有些是菌落
- 例子：藻類，原生動物

真菌界
- 主要是多細胞
- 自營生物，不能作光合作用.
- 分解已死和垂危的生物，攝取和吸收能量和營養

動物界
- 主要是多細胞
- 靠攝取食物的異養生物
- 提供食物、衣物、油脂、香水和勞動

植物界
- 主要是多細胞
- 自營生物，可作光合作用
- 是氧氣、食物、衣物 / 建築材料、顏料、香料、染料和藥物的重要來源

以功能劃分的生物種類 (Townsend, Begon and Harper, 2003)：

生產者 / 自營生物可以從環境取得材料，製造它們需要的養份。光合作用是指生物吸收陽光，製造複雜養份的過程。

$$CO_2 + H_2O + \text{太陽能} \rightarrow C_6H_{12}O_6 + O_2$$

化學合成是指在沒有陽光的情況下，生物從環境中取得簡單的化合物，轉換成複雜的化合物。**消費者 / 異養生物**透過進食生產者和其它消費者，獲得有機養份。

作為初級消費者，**草食性動物**直接進食生產者。因為進食其它消費者，**食肉性動物**是次級，三級和四級消費者。**雜食性動物**進食植物和動物組織。**食腐質動物**進食腐質(屍體，被遺棄和浪費的生物碎片)。**分解者**把複雜有機分子消化為簡單的無機化合物，以吸收溶解的營養。**食腐質生物**從部分被分解的有機物質攝取營養。

需氧呼吸是能量釋放的過程，以維持生產者和消費者的生命。

$$C_6H_{12}O_6 + O_2 \rightarrow CO_2 + H_2O + \text{太陽能}$$

任何一個生物個體的生存取決於體內的物質和能量流動。由於物質和能量的循環，生態系的群落才得以生存。

結論

這章節介紹一些基本的生態學原理，如：定律、學說和定理，生態的範疇，物種間的相互作用，生態學金字塔，生物的分類。重要訊息是以上的原則可被應用到植物、動物和微生物之間的相互作用。

思考問題

1. 每個物種在生態系有其生庇區位。在以下生境中，明名一個物種和其生態區位：

生態系	物種	生態區位
- 河流 / 溪流		
- 岩岸		
- 森林		

2. 除水循環外，試列舉另一個對生態系重要的循環。

References 參考資料

Beeby, A. and Brennan, A. (1997), *First Ecology*, Chapman & Hall, New York, p.23, 26, 86–95.

Buell, P. and Girard, J. (1994), *Chemistry: An Environmental Perspective*, Prentice Hall, Englewood Cliffs, N.J.

Bunce, N.J. (1993), *Introduction to Environmental Chemistry*, Wuerz Publishing Ltd., Winnipeg, Canada.

Dickinson, G. and Murphy K. (1998), *Ecosystems: A Functional Approach*, Routledge, London, p.2, 12, 71.

Hanks S. (1996), *Ecology and the Biosphere: Principles and Problems*, St. Lucie Press, Delray Beach, p.63-68, 85.

Jarvis, P.J. (2000), *Ecological Principles and Environmental Issues*, Pearson Education Ltd., Harlow, England, p.35–122.

Jeffries, M. (1997), *Biodiversity and Conservation*, Routledge, London, p.59.

Jones, A. (1997), *Environmental Biology*, Routledge, London, p.3, 58, 163, 169.

Reiss, M. and Chapman, J. (2000), *Environmental Biology*, Cambridge University Press, Cambridge, p.1–4.

Townsend, C., Begon, M. and Harper, J. (2003), *Essentials of Ecology*, Blackwell Pub., Malden, MA, p.8, 55, 96, 317.

Chapter 2. Biodiversity and Conservation

Lotus © Zoe Lee

For billions of years life on earth has increased in diversity to the richness we see today. To date, some 1.7 million species have been described, but the vast majority is still unknown. The term "Biodiversity" – short for biological diversity – describes the richness and variety of the natural world.

I. What is Biodiversity?

Biodiversity accounts for the number, variety and population sizes of living species in their living habitats. It also refers to the existence of a wide variety of kinds of life on earth, including diversity of individuals, diversity of genes, and diversity of species with the following meanings (Pietra, 2002; Melchias, 2001):

- The composition of all living organisms depends on genes inherited from their parents. For example, rice contains thousands of distinct genetic varieties while cheetah contains very few. **Genetic diversity** represents the variation in characteristics of species, as represented by their genes and chromosomes, and as spread by reproduction and recombination. An application would be the plant breeders and farmers who rely on genetic diversity to improve crop yields and improve resistance to pests and diseases.

- A species is a group of organisms, such as horses or dogs, that are so genetically similar that they can interbreed and produce fertile offspring. **Species diversity** refers to the total of different living species currently on Earth and in particular habitats. The more species there are, the wider the range of natural resources available.

Dog © Frank Lee

- An ecosystem consists of groups of plants, animals, micro-organisms, and non-living elements such as soil, water and minerals. For example, insects like bees and butterflies obtain food from flowers and pollinate them at the same time, enabling plants to reproduce. **Ecosystem (community) diversity** is the relative abundance of species at different trophic levels within its own ecosystem, as described in Part III later on.

Butterfly and flower © Frank Lee

Remember the **Five Kingdoms** of living organisms as mentioned in Chapter 1:
- Monera (e.g. bacteria),
- Protista (e.g. algae, protozoa),
- Fungi (e.g. mushrooms),
- Plantae (e.g. flowering plants), and
- Animalia (e.g. birds, fishes, mammals).

More than 75% of plant species are angiosperms © Frank Lee

- Most animals are insects (>80%)
- More species of mammal belong to the rodents than to any other group (>40%)

The overall number of species is unknown, as only about 1.7-1.8 million have been scientifically described. However, it might reasonably be supposed that these are about 14% of the total, most of the rest being insects and micro-organisms such as viruses, protozoa and bacteria (Cox, 1997).

In terms of numbers of species, biodiversity has an increasing trend towards a maximum in the humid tropical biome, where physical factors like heat and moisture put very little constraints on processes of reproduction and diversification. It is estimated (Tisdell, 1999) that 35% of known species live in the tropics, 59% in mid-latitudes and 6% at high latitudes; yet tropical species would probably constitute at least 86% of them. In contrast, biodiversity reaches a minimum in the cold deserts and marine environments of the poles.

II. Biodiversity through Time

Fossil records are useful to understand biodiversity through time. However, there are many drawbacks with fossil records. For instance, of the less than 1% up to a few percent of species leaving a fossil record, only a fraction has actually been recovered. The portion of the fossil record is biased towards the more abundant and more widespread species (95% of all fossil species are marine animals, but 85% of today's recorded plants and animals are terrestrial) (Pullin, 2002). Nevertheless, the fossil record is still an invaluable "**pictorial history"** of life on Earth.

History of life on earth (by geological eras and periods):

Precambrian ("Origin of life")	First multicellular organisms existed. (5 billion years ago)
Paleozoic ("Ancient Life")	The origin of plants, most invertebrate types, the first vertebrates (back-boned animals, including fishes, amphibians, and reptiles). (600-230 million years ago)
Mesozoic ("Age of Reptiles")	The origin of flowering plants, dinosaurs, birds, and mammals. (245-65 million years ago)
Cenozoic ("Age of Mammals")	The diversification of flowering plants, insects, birds and mammals, and the appearance of humans. (1.8-65 million years ago)

(extracted from Abe, Levin and Higashi, 1997)

There has been an **increase in biodiversity**, from the appearance of the first organism up to the present day, despite the fact that more than 90% of all species that have existed have become extinct. This increase has not been continuous but is composed of radiations and stabilizations, paused by mass extinctions, of different groups at different times.

III. Importance of Biodiversity

Human beings could not survive without biodiversity. All our food and many industrial materials and medicines are provided by plants, animals and micro-organisms. Animals supply meat, leather and insulin. Plants give us rubber, timber and cotton, and are used to manufacture glue, soaps, photographic film and plastics. Antibiotics like penicillin and tetracycline come from micro-organisms.

In the developing world, biodiversity provides sufficient food, countless raw materials such as fibre for clothing, materials for shelter, fertilizer, fuel and medicines, as well as a source of work energy in the form of animal traction. The poor depend upon biological resources for almost all of their needs. In the modernized world, access to diverse biological resources is crucial to support a vast array of industrializations. In the continuing effort to develop efficient and sustainable agriculture for many different conditions, these resources provide raw materials for plant and animal breeding, as well as the new biotechnologies. In addition, biodiversity keeps the ecological balance necessary for planetary and human survival.

Energy Pyramid

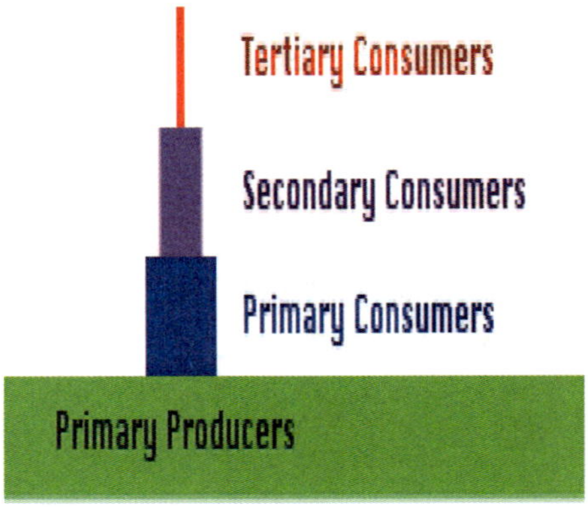

(modified from Jarvis, 2000, p.45)

Different values of biodiversity (Pullin, 2002; Jeffries, 1997):

Direct use values of biodiversity are concerned with the consumption or production of marketable commodities. They include food, medicine, and raw materials for biological control, industries, recreational harvesting and ecotourism.

Indirect use values of biodiversity are more difficult to quantify, because they are not related to direct trading in the marketplace. Being crucial for humans and wildlife, they can be provided by ecosystems such as photosynthesis, nutrient cycling and hydrological regulation.

Option value is the economic value of future benefits that a natural area could provide. For instance, the number of wild species that we utilize is currently increasingly rapidly, because of the potential for new drugs from the chemicals naturally occurring in plants. Pharmaceutical companies are competing to survey plant species for their medicinal properties and exploit the traditional use of plant chemicals by native people. Another example is the hunt for new food sources.

Existence value refers to the economic value of a habitat or species based on the amount people would be willing to pay to avoid its destruction. For example, how much would you pay to save a tiger from becoming extinct?

In addition, Cox (1997) described the four instrumental values for biodiversity:

Animals as diet
© Frank Lee

1. **Goods**
   Biodiversity in species and biomes contains potential nutrition from plants and animals to be explored. It is possible that many of these could be grown on a horticultural or agricultural scale, as well as harvested in the wild, adding variety at least to the human diet, and possibly even saving us from starvation, apart from the production of conventional crops. Finally, it has medical uses too.

Plant bug
© Samuel Sit

2. **Services**
   Examples of service of nature:
   - Green plants put new supplies into the atmosphere with oxygen and remove carbon dioxide.
   - Bats, insects and birds pollinate flowering plants and agricultural species.
   - Fungi and microbial life forms in the soil decompose dead organic materials and play a key role in recycling plant nutrients.

   Maintaining the diversity of the biomes is essential for the sustainable functioning of these species.

Nature to be treasured
© Frank Lee

3. **Information**
   Nature's knowledge is contained in the DNA within the living cells. The variety of genetic information determines the direction of the immune system, adaptation and evolution. Thus, genetic information is a potential economic good. Diversity also contains the aggregated wisdom of nature and the key to its future. However, only about 5–20% of the planet's species are known to science so far.

4. **Psycho-spiritual**
   This refers to the aesthetic and religious value of Nature. To many people, natural variety is something that is preferred and treasured.

IV. Threats to Biodiversity

No species is everlasting, and extinction is a natural phenomenon of life on Earth. Most species exist from 1 to 10 million years before being made extinct. For example, the loss of a terrestrial mammal takes place every 400 years on average, whereas a bird species every 200 years. Mass extinctions of species are sharp drops in biodiversity when many taxa disappear (Abe, Levin and Higashi, 1997). They occur from time to time, and are often followed by periods of steady re-population and re-colonization.

Another aspect of biodiversity loss is **genetic erosion**, or the gradual loss of genetic resources. Through agricultural practice, humanity has drastically reduced the complexity of ecosystems. Further reductions occur along the increasing tendency of single crops selection and the use of very few strains or varieties. The risks in this are huge: not only may the varieties become vulnerable to catastrophic pathogens, but there may be fewer opportunities to develop more varieties to meet new demands and conditions.

Single crop production © Frank Lee

Causes of biodiversity loss

Nowadays, the principle cause of biodiversity loss is **human activity**. Cox (1997) mentioned that land is being transformed at an exponential rate towards the simplification and standardization of ecosystems. As our population sizes or living standards continue to grow, so does the rate of land transformation. Pollution and the industrialization of agriculture and forestry are harmful to species, while over-harvesting can even have devastating effects on marine fisheries, wild animals and wild plants. The accidental or deliberate introduction of exotic species has often led to a reduction in the diversity of indigenous organisms, which are out-competed by the newcomers. Most importantly, the anthropogenic impact on global climate will produce changes in the natural landscapes that are difficult to estimate with respect to their negative impact on species.

The biggest threat to biological diversity posed by human activities is loss of habitat, which reduces other species' reserves of food, shelter and the size of their breeding areas. By definition (Reiss and Chapman, 2000), **ecological extinction** refers to the small population with the inability to fill an ecological role. **Threatened species** could become endangered while the endangered ones could become extinct.

Parrot – threatened by trade
© Frank Lee

The tendency of humans to trade exotic species has caused the deaths of some of the more prominent reptiles, amphibians, pachyderms, birds and flowering plants. Though such species are now protected by the **Convention on International Trade in Endangered Species of Flora and Fauna**, activities like poaching, smuggling and other forms of evasion still occur (Cox, 1997). For example, about 40 species of parrot have been threatened by international trade as pets. Trade in ivory and furs have also depleted stocks of many rare animals.

In addition to degrading natural habitats, humans have modified the environment by:

- industrial-scale agriculture and forestry;
- strengthening pest species;
- eliminating predators;
- introducing new species; and
- over-harvesting renewable resources.

Wasteful consumption – modern society, particularly in the rich developed countries, waste large quantities of natural resources. Usually people just throw biodiversity away – paper tissues, for example, may be made from wood pulp from forests which once hosted a healthy variety of animals and plants.

🡐 Waste paper © Frank Lee

V. Conserving Biodiversity

Early philosophies and religions provide thoughts on conservation of the human race above the rest of the living world (see "Anthropocentric Ethics" mentioned in Chapter 5). Conservation ethics were gradually developed after realizing that species and habitats, especially fertile land, were being lost through repeated exploitation.

Conservation Biology studies the human impacts on biodiversity and develops practical approaches to maintaining biodiversity (or preventing extinction) (Cox, 1997). Since the 20th century, the science of conservation biology has become an applied branch of ecology, but has only been recognized in its own right during the last few decades.

The 3 principles of conservation biology are:

1. Human action should not reduce biodiversity and ecological integrity.
2. Humans should not hasten premature extinction of wildlife populations or disrupt ecological processes.
3. Intact ecosystems must be protected.

Remember:

- We are totally dependent on the sun and earth.
- Every human intrusion has side effects.
- Everything is connected to everything else.
- We must reduce and minimize our damage.
- We should use care, restraint, humility and cooperation with nature as we alter the ecosphere to meet our needs and wants.

Some of the reasons for saving wild species include:

- Important scientific, ecological, economic, medical, aesthetic and recreational value
- Provides knowledge of evolution
- Provides ecological services
- Wild organisms provide beauty
- Ecotourism benefits travel industry
- Ethically important
- Pollination
- Economic benefits from products
- Wild strains needed for domesticated strains
- Plant-based medicines in developing areas and in developed countries

Aesthetic value of nature © Frank Lee

Biodiversity can be maintained by identifying the endangered species and protecting their populations and habitats. It is a complicated process requiring the taking back of land from the rapidly growing human population. In fact, there have been numerous initiatives to give legal protection to individual species, and to conserve them in seed banks and zoos. After all, the problem of maintaining overall biodiversity relies on putting efforts in saving as much species as possible, not just a single one.

Nature reserve: a way to preserve habitats

The protection of many groups of species is best served by preserving integrated habitats. We can start with the large parks and reserves, which cover a wide variety of environments, but they may still be insufficient for some wide-ranging migratory species. Governments have shown commitment to the conservation of habitats. We have seen successful cases in managing both the preserved lands and the competing interests that threaten them. (see the application of environmental impact assessment (EIA) in the case of Long Valley in Hong Kong in Chapter 14)

Convention on Biodiversity in 1992

One formal attempt to provide for future genetic and species diversity is the ***Convention on Biological Diversity***, which was signed by 158 governments at the **United Nations Conference on Environment and Development**, at Rio de Janeiro, Brazil in June 1992 (Cox, 1997).

The Convention recognized the need for (Melchias, 2001):

- a global data bank on biodiversity;
- an international fund; and
- ways of returning the benefits of genetic resource development to countries that possess and conserve the resources.

The United Nations Convention on Biodiversity noted, "so far, about 1.75 million species have been identified, mostly small creatures such as insects. Scientists reckon that there are actually about 13 million species, though estimates range from 3 million to 100 million … yet another aspect of biodiversity is the variety of ecosystems such as those that occur in deserts, forests, wetlands, mountains, lakes, rivers, and agricultural landscapes." It also stated, "species have been disappearing at 50 to 100 times the natural rate, and this is predicted to rise dramatically. Based on current trends, an estimated 34,000 plant and 5,200 animal species, including one in eight of the world's bird species, face extinction" (United Nations Environmental Programme, 2005)

The main objectives of the Convention include the conservation of biological diversity, the sustainable use of its components, and the fair and equitable sharing of the benefits arising from the utilization of genetic resources (Reiss and Chapman, 2000; Jeffries, 1997).

One of the major concerns of the treaty deals with the protection of genetic diversity. With tremendous genetic resources in hand, the developing countries are becoming reluctant to cede genetic resources to developed nations, while the latter also refuse to provide their expertise and investment in biotechnology to the former.

In general the Convention offers the chance to coordinate efforts to recognize the scope of the problem of species and genetic loss and start to tackle it. In addition, governments must enforce the laws and confront the problem at local levels, which is where the losses are occurring, in order to achieve success.

Ethical Issues

Many ethical issues are concerned with biodiversity and species extinction, and the subject has been one of considerable interest to environmental ethicists. A number of central, interlocking questions are raised:

- Is maintaining biodiversity (or aiming to increase it) a good thing?
- Do we have an ethical duty to preserve species?
- Is the extinction of species - a process that has been occurring since life evolved on earth - necessarily a bad thing?

The most common answer to these questions is that **biodiversity is ethically significant**. Clearly it is useful for human beings and should be protected as a valuable resource. By protecting biodiversity, we can preserve a big stock of genetic material for possible use in the future - materials that might, for example, prove useful in genetically engineering new crop strains (Soule, and Orians, 2001). Many species may also have a vast range of medical and industrial uses that are to be found out. Biodiversity is essential for the maintenance of ecosystems and the "services" they provide human beings such as the recycling of water, air, and nutrients. Furthermore, the loss of biodiversity might result in the collapse of ecosystems and cause problems for future generations of human beings. The consequences are by far beyond our imaginations.

Conclusion

Without biodiversity, the future is grim. People living in rural areas in the developing world are most at risk, because they are more likely to depend on what they grow and find in the wild. Less diversity means that they have to travel further to find supplies. As life becomes more difficult in the countryside, people are forced to move to towns (urbanization).

Economic values have been placed on biodiversity to enable us to compare benefits of conservation versus development, but the spiritual value of biodiversity remains fundamentally important to many people.

Despite the gloomy forecasts of biodiversity loss, efforts are being made to save the situation. Many of them aim to identify and list species, to analyze their habitats, and to suggest remedies for biodiversity losses. Surveys have been conducted to draw up lists of threatened species. Above all, it may require both a more generous attitude and global perspective towards conservation, despite the relentless pressure of population and economic growth.

第二章　生物多樣性及保育

經過幾億萬年，地球上的生物多樣性增加至我們今天看見。至今，大約一百七十萬個物種已被確認，但大多數仍然是未知。「生物多樣性」是指自然界物種豐富和多樣化。

I. 生物多樣性是什麼？

生物多樣性是指物種在生境中的數量、品種和種群大小。它是地球上各種各樣的生命，包括個體的多樣性，基因多樣性，種類多樣性 (Pietra, 2002; Melchias, 2001)：

- 所有生物的構造取決於父母遺傳下來的基因。例如：米包含數以萬計的基因變化而獵豹則非常少。基因多樣性是指物種擁有不同的基因和染色體，經繁殖和重組而傳下去。農民利用基因多樣性來促進農作物的生產量和抗蟲抗病的能力。
- 物種是一群在基因上相似的生物，它們可以交配及繁殖可生育的下一代，例如：馬或狗。物種多樣性是指在地球和某一生境的不同物種。物種越多，可用的自然資源越多。
- 生態系包括植物、動物、微生物和無生命元素如土壤、水和礦物。例如，昆蟲如：蜜蜂和蝴蝶，從花朵獲得食物，同時傳播花粉，令植物繁殖。**生態系(群落)多樣性**是指在生態系之內的不同食性層次裏，物種的相對數量。

在第一章曾提及生物可分為**五個界別**：
- 原核生物界，如細菌
- 原生生物界，如藻類，原生動物
- 真菌界，如菇類
- 植物界，如開花植物，以及
- 動物界，如雀鳥，魚，哺乳動物

物種的總數量仍是未知，只有約1.7-1.8百萬的物種被確認。不過，這個數目可推算為總物種數目的14%，其餘大多數為昆蟲和微生物，如病毒、原生動物和細菌 (Cox, 1997)。

在潮濕熱帶的生物群落，其物理因素如溫度和濕度，令繁殖和多樣化的過程順利，所以生物多樣性有一種增長趨勢。現估計 (Tisdell, 1999) 35% 的已知物種居住在熱帶，59% 在山腰和 6% 在高緯度；熱帶的物種已佔至少86%。相反，在寒冷的沙漠和海洋環境，生物多樣性最少。

II. 生物多樣性的演進

化石紀錄可用作瞭解以往的生物多樣性。不過，化石紀錄有許多缺點。例如：留下化石紀錄的物種只有少於1%至數個百份比，當中只有少部份可以還原。化石紀錄中，百分之九十五是海洋動物的化石, 但陸地植物和動物的化石較少 (Pullin, 2002) 。然而，化石紀錄仍然是地球生命的珍貴「圖像歷史」。

地球的生命史 (地質時代)：

前寒武紀（「生命起源」）	出現第一個多細胞生物 (五十億年前)
古生代（「古代生命」）	植物，大部分的無脊動物，首次出現的脊椎動物，如：魚，兩棲類及爬蟲類 (二億三千萬至六億年前)
中生代（「爬蟲類時代」）	有花植物，恐龍，雀鳥，哺乳動物 (六千五百萬至二億四千五百萬年前)
新生代（「哺乳類時代」）	有花植物，昆蟲，雀鳥，哺乳動物的多樣性；人類首次出現 (一百八十萬至六千五百萬年前)

(錄自 Abe, Levin and Higashi, 1997)

從首個生物的出現至現今，生物多樣性不斷增加。雖然超過90% 物種已經絕種。生物多樣性的增加不是連續的，在不同的時間有不同的升幅。

III. 生物多樣性的重要性

沒有生物多樣性，人不能生存。我們的食物及很多工業材料和醫學皆由植物、動物和微生物提供。動物供應肉、皮革和胰島素。植物提供橡膠、木材和棉花，以及製造膠漿、肥皂、膠片和塑料。抗生素如青黴素和四圜素均來自微生物。

在發展中國家，生物多樣性提供充足的食物、原材料如衣物纖維、房屋材料、肥料、燃料和醫藥，以及動物的勞動力。窮人依賴生物資源為作為他們的需要。在現今的世界，生物資源支持工業化的社會，發展有效及可持續的農業，並且為動植物的繁殖、生物科技提供原材料。此外，生物多樣性維護生態的平衡，令地球和人類得以生存。

生物多樣性的不同價值 (Pullin, 2002; Jeffries, 1997)：

生物多樣性的**直接使用價值**與商品的消耗或生產有關，包括食物、醫藥、生物原材料、工業、娛樂和生態旅遊。

生物多樣性的**間接使用價值**更難量化，原因是它們不涉及市場交易。生態系提供的光合作用，營養循環和水文調控，對於人類和野生生物來說，是極為重要的。

選項值是指提供未來的經濟價值。例如，由於從植物中的化合物發掘新藥物的潛力，我們使用野生物種的數量越來越多。製藥公司爭相研究藥用植物及開發土人對植物的傳統用途。其它例子是尋找食物的新來源。

存在價值是指人類會願意支付多少，以避免生境或物種的破壞。例如，你會支付多少來保存老虎以免絕種？

此外，Cox (1997) 描述四種生物多樣性的工具價值：

<table>
<tr><td>1. 商品
在物種和生物群落的生物多樣性中，植物和動物的潛在營養，有待發掘。大部分物種可以園藝或農業方式培植，在野外收成，增加食物的種類，甚至可以解決糧荒，除常規莊稼的生產之外。終於，它有藥用價值。</td></tr>
<tr><td>2. 貢獻
自然界貢獻的例子：
• 綠色植物提供氧氣和吸收二氧化碳
• 蝙蝠、昆蟲和雀鳥為開花植物和農作物傳播花粉。
• 真菌和土壤中的微生物分解已死的有機物，並且在植物營養的循環中，扮演重要的角色。
為了令維護這些物種持續發揮功能，維持生物群落的多樣性是必需的。</td></tr>
<tr><td>3. 資訊
在細胞之內的DNA包含自然的知識。不同的遺傳訊息決定免疫系統、適應和進化的方向。因此，遺傳訊息是潛在的經濟財產。多樣性聚集了大自然的智慧，也是未來的關鍵。到目前為止，只有約5-20% 的物種是為人所知。</td></tr>
<tr><td>4. 心靈
這是指自然的審美和宗教價值。很多人是喜歡和珍惜自然的多樣性。</td></tr>
</table>

IV. 生物多樣性受到的威脅

没有物種是永恆的。在地球上，絕種是一種自然現象。大部分物種能夠存在一百至一千萬年。例如：每400 年一種哺乳動物會消失，每 200 年一種鳥類會消失。大量絕種是指生物多樣性急劇減少 (Abe, Levin and Higashi, 1997)，每隔一段時間發生，隨後就穩步增長及擴張。

生物多樣性的消失是指基因資源逐漸減少。農業急劇減少生態系的複雜性。單一農作物的選擇和少用其他品種進一步減少生物多樣性。這存在巨大的風險：物種對病菌變得脆弱，而且減少物種的發展，以符合新的需求和環境。

生物多樣性減少的原因

現今，**人類活動**是生物多樣性減少的主要原因。Cox (1997) 提及，土地急速變換，令生態系簡單化和標準化。當人口增加或生活水準不斷提高，土地變換越來越快。農業和林業的污染和工業化對物種是有害的，過度捕獲可能摧毀海洋漁類、野生動物和野生植物。意外或故意引入外來物種會減少當地物種的多樣性。更重要的是，人類影響全球性氣候，改變自然地貌，對於物種的負面影響是難以估計。

人類活動對生物多樣性的最大的威脅是令生境消失，減少其他物種的食物儲備、庇護和繁殖區域。定義上 (Reiss and Chapman, 2000)，**生態絕種**是指群落太小，難以擔當生態角色。受威脅物種可能成為瀕危物種，而瀕危物種最後可能絕種。

珍貴物種的貿易導致一些爬蟲動物、兩棲動物、雀鳥和開花植物死亡。雖然這些物種受到瀕危野生動植物種國際貿易公約保護，不過，偷獵，走私和其他形式仍然發生 (Cox, 1997)。例如：由於國際貿易的關係，約四十個作為寵物的鸚鵡品種受到威脅。象牙和皮草貿令很多稀有動物絕跡。

除破壞自然生境外，還有人為改變環境，包括工業化的農業和林業、加強害蟲品種、消滅捕食性動物、引入新物種；以及過份採用可再新資源等。

浪費 — 現代社會，特別在發達國家，浪費很多自然資源。通常人們隨意拋棄，如紙張。紙張來自森林的木材，而森林住著不同的動物和植物品種。

V. 生物多樣性的保育

早期的哲學和宗教認為保護人類種族是較其他物種重要 (參見第五章中「**以人為本的倫理**」)。當意識到不斷開發後，物種和生境（尤其是肥沃的土地）消失，保育的倫理逐漸發展起來。

保育生物學是研究人類對生物多樣性的影響，並且運用實際的方法去維持生物多樣性，以避免絕種 (Cox, 1997)。保育生物學在二十世紀已成為生態學的一個應用分支，不過直到過去幾十年期間，才獲承認。

保育生物的三項原則是：

1. 人類活動不應該減少生物多樣性和生態的完整性
2. 人類不應該使野生生物過早絕種或擾亂生態過程
3. 生態系必須受到完整的保護

緊記：

- 我們十分依靠太陽和地球
- 人類的入侵有副作用?
- 萬物互有連繫
- 我們必須減少破壞
- 當取之及用之於自然時，我們必須小心，克制，合作及恭敬。

挽救野生物種的原因：

- 重要的科學，生態學，經濟，醫療，審美和消遣價值
- 提供進化的資訊
- 具生態學的作用
- 野生生物的美感
- 生態旅遊有助於旅行業
- 道德
- 傳播花粉
- 經濟效益
- 用於發展中及已發展國家的藥用植物

確認瀕危物種及保護族群和生境，可以維持生物多樣性。這是一個複雜的過程，須要從迅速增長的人口中取回土地。實際上，要主動保護個別物種，以及把它們保存在種子銀行和動物園。要維持整體生物的多樣性，不只是挽救其中一個品種，而是更多的品種。

保存生境是保護不同物種的最好方式。我們建立大型公園和保護區，覆蓋各種不同的環境。不過，對於一些廣泛遷移物種來說，仍然是不足的。政府承諾保育生境。已有成功的例子，把具爭議的土地變為保育區。(參見第十四章「**環境影響評估的應用**」— 香港塱原的個案)

生物多樣性條約

在一九九二年六月，在巴西里約熱內盧的聯合國環境和發展會議，一百五十八個政府簽署生物多樣性條約，以確保未來基因和種類的多樣性 (Cox, 1997)。

大會認可了需要為 (Melchias, 2001)：
- 一個全球性生物多樣性的資料庫；
- 國際基金；並且
- 回饋基因資源發展成果給提供資源的國家。

聯合國在生物多樣性條約提及：「到目前為止，確認了大約一百七十五萬個物種，大部份是細小的生物，如昆蟲。雖然估計範圍從三百萬個至一億個物種，不過科學家實際上總計只有約一千三百萬個物種……另一方面，生物多樣性是生態系的多樣性，如沙漠、森林、沼澤、山、湖泊、河流和農地。」並且陳述「物種消失的速度是自然消失速度的五十至一百倍，而且會繼續急升。根據現時的趨勢，約三萬四千種的植物和五千二百個的動物物種，包括八分之一的雀鳥種類，面臨絕種」(United Nations Environmental Programme，2005)。

條約的主要目的是保護和可持續運用生物多樣性，以及基因資源帶來的好處 (Reiss and Chapman, 2000; Jeffries, 1997)。

條約的主要關心保護基因的多樣性。當已發展的國家拒絕提供發展中國家生物科技的專門技術和投資，後者就不願提供龐大基因的資源給前者。

總括而言，會議提供機會去確認物種和基因損失的範疇，並且開始去解決問題。此外，政府必須執行法律，面對地方問題，以達到成效。

道德問題

許多道德問題與生物多樣性和絕種有關，環境道德者關注這個主題，並提出一系列的問題：
- 維護生物多樣性(或打算增加) 一件好事嗎？
- 我們有道德義務去保存物種嗎？
- 絕種 — 自生命進化的過程中，已經發生是一件壞事嗎？

對這些問題的答案，普遍認為在道德層面，生物多樣性是重要的。它明顯對人類有用，應視為一種可貴資源，受到保護。透過保護生物多樣性，我們保存大量的基因，作為未來的用途，例如：發展新農作物品種的基因工程 (Soule and Orians, 2001)。很多物種也許有廣泛的醫療和工業用途。生物多樣性是維護生態系的根本並為人類提供如：水、空氣和養份循環。此外，生物多樣性的損失或許導致生態系的崩潰，對下一代造成的問題，是無法想像的。

結論

沒有生物多樣性，未來是可怕的。在發展中國家鄉郊的居民最受威脅，因為他們依靠種植及捕獲。多樣化減少意味著他們必須到更遠的地方尋覓所需。當鄉郊生活變得困難時，人們被迫搬去城市 (即都市化)。

給予生物多樣性經濟價值，使我們可以比較保育和發展的效益，但生物多樣性的精神價值對許多人來說是十分重要。

儘管生物多樣性損失陰沉的展望， 努力被做保存情況。大多數打算辨認和名單種類，分析他們的棲所，和建議對生物多樣性損失的補救。勘測進行勾畫被威脅的種類名單。除此之外，儘管人口和經濟增長的壓力，不過保育是需要慷慨的態度和全球性的願景。

Questions for Further Thinking 思考問題

1. For what or for whom do we wish to conserve biodiversity?
   保育生物多樣性是為了什麼和為誰做呢?
2. How do we weight economic versus ethical reasons for conservation?
   我們怎樣衡量保育的經濟和道德價值?
3. Should we argue for the spiritual need for wildness as a reason for conservation?
   對大自然精神上的需要是保育的一個理由嗎?
4. Why are insects as animals so successful in this planet?
   為什麼昆蟲是地球上最成功的動物?

References 參考資料

Abe, T., Levin, S. and Higashi, M. (1997), *Biodiversity: An Ecological Perspective*, Springer, New York, p.59–60.

Cox, G. (1997), *Conservation Biology: Concepts and Applications*, Wm. C. Brown, Dubuque, IA, p. 2, 6, 33, 44–46, 308, 312.

Jarvis, P. (2000), *Ecological Principles and Environmental Issues*, Pearson Education Ltd., Harlow, England, p. 45.

Jeffries, M. (1997), *Biodiversity and Conservation*, Routledge, London, p. 139, 150.

Melchias, G. (2001), *Biodiversity and Conservation*, Science Publishers, Inc., Enfield, NH, USA, p. 3–5, 84–86.

Pietra, F. (2002), *Biodiversity and Natural Product Diversity*, Pergamon, Amsterdam, p. 3–4.

Pullin, A.S. (2002), *Conservation Biology*, Cambridge University Press, NY, USA, p.15–17.

Reiss, M. and Chapman, J. (2000), *Environmental Biology*, Cambridge University Press, Cambridge, p. 7, 73.

Soule, M.E. and Orians, G.H. (2001), *Conservation Biology: Research Priorities for the Next Decade*, Island Press, Washington D.C., USA.

Tisdell, C. (1999), *Biodiversity, Conservation and Sustainable Development: Principles and Practices with Asian Examples*, Edward Elgar Publishing Ltd, Cheltenham, UK.

United Nations Environmental Programme (2005), *Sustaining life on earth.* Retrieved January 20, 2005, from the Convention on Biological Diversity website: http://www.biodiv.org/doc/publications/guide.asp

Chapter 3. Human Impact on the Environment

As mentioned in the previous chapter, our natural environment is rapidly disappearing due to the influence of human activity. Now we explore more into various human impacts.

I. The Ever-growing Human Population

Homo sapiens have evolved from its ancestors around 3 million years ago, first appearing in East Africa. They survived as hunter-gatherers, expanding through the Middle East, Europe and Asia, down to Australasia and finally into the North and South American continents.

According to Reiss (2000), a population is "a group of individuals within a species that have the opportunity to breed with one another because they live in the same area at the same time". Basically, the population change equals to the sum of births and immigration minus the sum of deaths and emigration.

Today human population has reached to 6 billion. There has been an exponential rise for the last 100 years, and the reasons are due to the reduction in the death rate through advances in public health and hygiene, supported by the development in agriculture resulting from the industrial revolution. Notice the difference in populations between the less developed countries and the more developed countries on the right.

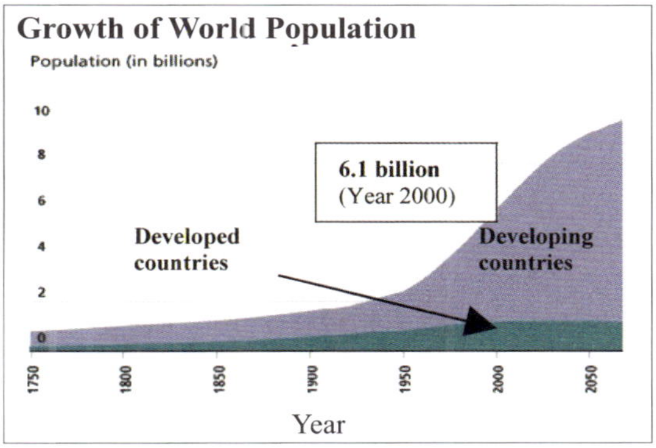

(modified from United Nations, 2005)

Due to urbanization, the number of individuals living in urban areas has tripled since 1950. Urban growth is the rate at which urban populations grow. The percentage of population of urban areas has increased from 14% to 42% between 1900 and 1993.

Population © Frank Lee

For a typical city or urban area, the natural increase of population occurs when there are **more births than deaths**, while immigration comes mainly from the rural areas. Urbanization, birth control methods as well as cost of raising and educating children are the factors affecting the **birth rate**. As for **death rate**, the factors include better nutrition, fewer infant deaths, longer life expectancy and improvements in medical and public health technology.

II. Our Relationships with the Environment

At the beginning, humans were gatherer-hunters. They used primitive tools to hunt animals and dig for vegetables. Tribes often had a home range and would move through it, perhaps on a seasonal basis, exploiting natural food sources as they became available. Their environmental impact was not significant as compared with other large animals of the time. Nevertheless, the ecosystem had more impact on the humans than they had on the ecosystem itself.

Until 5,000 years ago, deforestation in the lowlands began. The dominant forest cover was both burnt and exploited for construction materials, tools and firewood. Some of the cleared land was grazed by domestic animals such as goats, sheep and cattle; while some was cultivated for wheat and barley (Pu lin, 2002).

Farmland © Frank Lee

Agriculture, or cultivation of land for crop production, has been practiced for thousands of years. The natural ecosystem is manipulated to produce plant and livestock products. Gradually, human population density rose rapidly in areas where agriculture developed.

Farming techniques © Frank Lee

As farming techniques and skills were improved, individual plots of land were used on a more sustainable basis. Meanwhile, agricultural communities were based around marketplaces that developed for selling and exchange of produce.

In short, human populations survive with a change from exploiting the natural resources to exploiting the land itself to grow the products of their choice. Yet at the same time, the practices caused many environmental problems (to be discussed in next section).

Post-industrial Accelerated Change

Impact from industrial revolution © Frank Lee

The **industrial revolution** in the 18th century in Western Europe brought a whole new range of impacts, affecting human activity in many ways. Since then, humans would have a global rather than local impact on the environment. Unfortunately, the dominant view of the society believed that nature was for humans to control and the resources were to be exploited as wish (also known as the "*frontier ethic*" philosophy).

The *frontier ethic* includes the following principles:

1. There is always more and its all for us – the earth is an unlimited supplier of resources for human use.
2. Humans are apart from nature and immune to natural laws – humans view themselves as so special that they are somehow unrelated to the rest of the living world.
3. Human success derives from the control of nature – humans must dominate and control nature to be successful.

These principles have been greatly supported by the modernized world. The result is the rapid growth of industrial economies and increase in the people's standard of living in the developed world. While the population grows, resource per capita has also increased. As a result, the standard of living in many less developed countries has decreased and the earth's ecosystems have been seriously deteriorated.

Energy Consumption

Growth in world energy production and consumption in the next 30 years is projected expand by almost 60% between 2002 and 2030, reaching 16.5 billion tonnes of oil equivalent (OECD, 2004). Almost all growth in energy production will come from non-OECD countries. In fact, more than 70% of energy demand growth over the next 30 years will come from outside the OECD.

III. Various Human Impacts

Surviving through tough habitats
© Zoe Lee

1. Habitat destruction

The most direct threat to biodiversity comes from destruction of the habitat on which it depends. Throughout human history, habitat destruction has been probably at their greatest in Europe, where population density is among the highest and industrial activity has long been established. Virtually no natural habitats remain in most of these countries. Thanks to high awareness and proper management, many semi-natural habitats have been sustained and maintained their high biodiversity value.

Since 1950, intensification of agriculture in many regions has been driven by advanced technology such as the use of inorganic fertilizers. It assisted in fertilizing and ploughing of soils, as well as drainage of wetlands. Unfortunately, they also have negative impact and have contributed to the disappearance of these precious natural habitats. (Pullin, 2002).

Wetland © Frank Lee

In developing countries, habitat destruction has been an ongoing process and its impact is even more serious than that in the developed world. Clearing of tropical forest for subsistence agriculture has been practiced for many years. As populations have risen, forests have disappeared and fertility of the soil has declined rapidly.

Apart from the demand for modernization, another reason for habitat destruction in developing countries is because they are heavily in debt and being forced to sell off their natural assets, particularly tropical forest timber. Desperate for cash, many timber companies have exploited the valuable forests as wish. Road construction for the removal of timber has forced the peasant farmers to move away from the area and allowed even further forest destruction.

2. Use of fire for forest clearance

Humans learned to make and use fire about half a million years ago – very useful in vegetation. Trees are cut down to form a clearing and the sun dries the leaf litter and allows the area to be set alight thus clearing it for cultivation. The practice of burning and grazing allowed vegetation in many areas of the world developing into grasslands.

After forest fire © Frank Lee ➔

Since a typical standing forest is resistant to fire, when the ground litter is heavily shaded and the humidity is high, the fire does not normally spread beyond the boundaries of the clearing. However, when severe droughts occur, forests with the reduced canopy will become drier and be susceptible to fire. This causes the irreversible destruction of large areas of tropical forest. For example, many local fires spread out of control and devastated large areas in Southeast Asia in 1997. According to Pullin (2002), the health of over 70 million local people was affected with an economic loss of US$4.4 billion.

3. Desertification

Today, about 6 million ha of land worldwide are turning into desert each year. While both local and global climate changes exert their influence on deserts, humans remain as an important environmental factor. One controversial issue pertaining human impact on the environment is to what extent the recent spread of deserts has been due to human activity. Increasing human population pressure in these land areas definitely leads to reduced tree coverage as well as overgrazing through increasing livestock densities. These factors accelerate soil erosion and also desert formation.

Desert © Frank Lee

4. Pollution

E-waste problem © Anna Leung

Landfill: solution for waste pollution? © Frank Lee

Pollutants are emitted from agricultural and industrial processes around the globe. Fertilizers, pesticides and animal waste become pollutants when they enter the environment other than crop fields and cause the death of organisms in the natural ecosystem.

Large-scale industrial processes involving burning coal and extracting metals often produce pollutants such as sulphur dioxide and nitrogen oxides. They become local pollution or are transported by prevailing winds to pollute nearby regions (i.e. acid rain). Other toxic waste products including heavy metals and synthetic organic compounds were most conveniently and cheaply dumped into rivers, causing severe pollution in many industrialized nations.

Also see these chapters for more details:
Chapter 9 – Persistent Toxic Substances
Chapter 11 – Solid Waste Management
Chapter 12 – Sewage Treatment
Chapter 13 – E-waste

5. Introduced species and diseases

Throughout the years, humans have traveled around and colonized most of the terrestrial surface of the planet. At the same time, they have deliberately introduced species from other places to serve as crops or something to be aesthetically pleasing (e.g. garden plants and songbirds).

Undoubtedly, these introduced species have a negative impact on the native flora and fauna. For example, the introduction of goats to areas where plants are not resistant to heavy grazing, particularly in tropical and subtropical islands, would result in the reduction of some unique plant communities.

6. Overexploitation of resources

The increasing population and per capita demand for energy have resulted in direct overexploitation of our natural resources. Non-renewable resources, such as fossil fuels, are being used up too quickly and will be exhausted soon in this century. Likewise, the use of renewable resources such as water, fish stocks and timber require to be managed wisely and sustainably.

The ultimate goal is to prevent overexploitation, allowing natural regeneration and sufficient supply of food, water and materials for the present and future generations.

Water resources © Frank Lee

7. Extinction of species

As mentioned in previous chapter, humans have had a significant effect on biodiversity for many years. Hunting large animals for food has led to the extinction of species such as wild mammoths and tigers (Reiss, 2000). Besides, clearing natural vegetation for farmland and dwelling space and pollution of land, marine and atmosphere have all reduced the natural habitats and nurturing grounds of wildlife.

Panda: China's treasure to be extinct
© Frank Lee

Case Study – the Exxon Valdez Oil Spill Incident (Jackson, 1996)

On 24 March 1989, a supertanker called *Exxon Valdez* ran aground on reefs in Prince William Sound, Alaska. A total of 38,000 tonnes of crude oil (about 22% of the cargo) spewed out into the waters, covering an area of 2,500 km^2 and contaminating about 1,700 km of shoreline. This "Alaskan oil spill" has become the biggest environmental disaster of its kind in the world.

The amount of oil lost from the *Exxon Valdez* was not large in comparison with other oil spills, but its impact was undoubtedly among the most severe in history. Since the shores and waters of Prince William Sound had been rich in wildlife diversity, the incident caused intolerable devastation to them. The results included the death of numerous marine mammals such as sea otters, Pacific halibuts, herrings, pink salmons and clams. Over 360,000 seabirds were killed. The fish populations were severely affected by the oil spill, threatening the livelihood of many local fishermen. In terms of long-term effect, oil is persistent in some mussel beds and subsequently biomagnifies up the food chain and causes serious damage on reproduction and the health of predators.

Since then, the disastrous *Exxon Valdez* oil spill has received enormous attention and debate worldwide for its ecological and economic impacts as well as its recovery measures. One of the outcomes was the birth of Valdez (later known as the CERES) principles, as environmental responsible guidelines for corporations and organizations to follow.

For more details on the incident and the ethics applied, please refer to Chapter 5 ("Environmental Ethics").

Conclusion

Humans have had a significant effect on this planet for over tens of thousands of years. They first evolved as hunter-gatherers, similar to many large mammals in the ecosystem. The use of tools and fire allowed them to dominate over the environment, and eventually changing it. The industrial revolution and technological advances increased use of resource per capita, enabled explosive population growth and brought tremendous environmental impact. Statistics have shown that humans have had a negative impact on biodiversity by means of habitat destruction, habitat disturbance and overexploitation of natural resources. The *Exxon Valdez* oil spill occurred in Alaska in 1989 is one recent example to illustrate the devastation of natural environment caused by human.

第三章 人類對環境的影響

前章提及，由於人類活動的影響，我們的自然環境迅速消失。現在，我們進一步探索各種人類的影響。

I. 人口持續增長

大約三百萬年前，人類的祖先首先在東非出現。他們以狩獵為生，由中東擴展至歐洲和亞洲，南下澳洲，以及南北美洲大陸。

根據 Reiss (2000)，人口是指「在同一時間內，一個地區的一群同種的個體可以進行繁殖」。基本上，人口轉變的數量是出生和移居的總和減去死亡和遷移的總和。

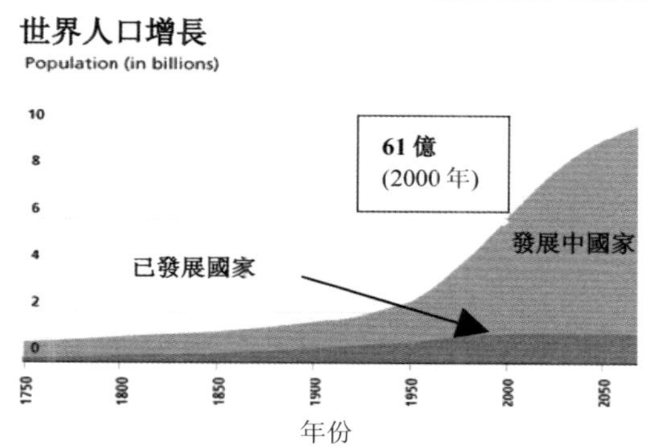

(modified from United Nations, 2005)

今天，人的人口已達至六十億。在過去100年，人口不斷上升，原因是工業革命及農業發展，促進公眾健康和衛生，令死亡率減少。請留意右圖所示，較少發展的國家和已發展的國家之間的人口差異。

由於城市化，自從1950年，住在市區的人數增加三倍。城市的增長即是城市人口增長率。市區的人口百分率從1900年的14% 增加了到1993年的42% 。

在一個典型的城市或市區，人口的自然增長指出生人數比死亡人數多，加上從鄉郊移居到市區。城市化，節育，以及教育小孩，是影響出生率的因素。至於死亡率，因素包括營養充足，嬰兒夭折率低，平均壽命增加，以及健康檢查和公共衛生技術的進步。

II. 我們與環境的關係

起初，人類是以狩獵為生。他們使用原始的工具追獵動物，挖掘蔬菜。部族按季節遷移，以開發天然食物資源。他們對環境的影響較其他動物少。因此，生態系對人類的影響較人類對生態系的影響為大。

直至5,000年前，砍伐森林在低地開始。主要的森林被燃燒及開發，以獲取建築材料，工具和木柴。一些土地被山羊，羊和牛的家畜放牧；一些土地為了種植小麥和大麥（Pullin, 2002）。

農業已有一段很長的歷史。自然的生態系被用作種植植物和家畜產品。逐漸地，人口密度在農業發展地區迅速上升。

當耕種技術改良，土地被持續使用。另一方面，農業社群圍繞著市場，以作買賣。

簡言之，人類由開發自然資源到開發土地種植。與此同時，產生很多環境問題（在下一節詳述）。

後工業加速轉變
十八世紀西歐的工業革命為人類活動帶來一系列的影響。自那時候起，人類會對環境的影響不只是本地，而是全球性。不幸地，社會主要的見解是自然資源是為人類所控制及開發（也稱為「拓荒者倫理」）。

拓荒者倫理包括以下原則：
1. 總是有更多資源，並且全是我們的 – 地球的資源是無限地為人類使用。
2. 人類遠離自然和不受自然法影響 – 人類認為自己是特別的，與自然界無關。
3. 人類的成功起源於對自然的控制 – 人類必須支配和控制自然，獲取成功。

這些原則被現代化的世界認同。結果是在已發展國家，工業經濟和生活水平迅速增長。當人口增長的時候，人均資源也增加。結果，很多發展較少的國家的生活水平減少，以及生態系受到嚴重的破壞。

能源消耗
在未來30年，2002年至2030年，世界能源產量和消耗增加至60%，達到一百六十五億公噸（經濟合作發展組織，2004）。能源的生產幾乎來自非經濟合作發展組織的國家。事實上，在未來30年，來自經濟合作發展組織國家的能源需求增長超過70%。

III. 人類各種的影響

1. 生境破壞
破壞生境直接威脅生物多樣性。在整個人類歷史，歐洲的生境破壞是最嚴重，原因是人口密度高，和工業活動發展悠久。實際上，大部分這些國家沒有自然的生境。幸好，還有合適的管理，很多半自然的生境得到保存。

自從1950年，在很多地區的農業因技術進步已急速發展，例如使用無機的肥料，以協助施肥及翻土。不幸地，這產生負面的影響和導致寶貴自然生境的消失(Pullin, 2002)。

在發展中國家，生境破壞持續，而且較已發展國家嚴重。清除熱帶雨林，用作耕種，已實踐多年。當人口上升時，森林急速消失，土壤變得貧瘠。

除了對現代化的需求外，發展中國家的生境受到破壞的另一個理由是負債嚴重，被迫廉價出售他們的自然資產，尤其是熱帶雨林的木材。很多木材公司開發珍貴的森林。為了運輸木材，建造道路，迫使農民遷離，以開發更多的森林。

2. 燃燒森林

人類約五十萬年前懂得使用火來作耕種。樹木被砍倒，曬乾樹葉，利用燃燒和放牧，以作種植。

森林是可以抗火的，原因是地面被樹葉遮蔽，濕度高，火很少向外擴散。不過，在極乾旱時，森林變得乾燥，容易發生火災，導致熱帶森林不可逆轉的破壞。例如，在 1997年，東南亞森林大火。根據 Pullin (2002)，超過七千萬當地人的健康受影響，經濟損失達四十四億美元。

3. 沙漠化

現時，每年大約六百萬公頃的土地變成沙漠。當本土和全球氣候變化對沙漠造成影響的時候，人類仍然是一個重要的環境因素。人類活動對沙漠化影響的程度仍然是一個爭議性的問題。人口增長減少樹木的覆蓋，加上過度放牧，令土壤腐蝕，形成沙漠。

4. 污染

全球的農業和工業的過程中產生污染物。肥料，殺蟲劑和動物廢物成為污染物，引致生態系中的生物死亡。

大規模的工業過程，涉及燃燒煤炭，抽取金屬，經常產生污染物，如二氧化硫和氮氧化物。它們成為本地污染，或被風吹送附近的區域（即酸雨）。其他有毒的廢物，包括重金屬和合成的有機化合物，傾倒到河流，導致嚴重的污染。

詳情請參考以下章節：
第 9 章 – 持續性有毒物質
第 11 章 – 固體廢物管理
第 12 章 – 污水處理
第 13 章 – 電子廢物

5. 外來物種和疾病

人類四處旅行，並且開拓殖民地。同時，他們故意引入外來物種作為種植或點綴（例如園藝和鳴鳥）。

無疑地，這些外來物種對本地植物和動物造成負面影響。例如，引入山羊到抵抗放牧較低的地區，尤其是熱帶和亞熱帶的島嶼，減少一些獨有植物品種。

6. 過度開發資源

人口和人均能源需求增加導致了我們的天然資源的過度開發。不可更新的資源，如化石燃料，迅速被耗盡。同樣地，例如水，魚類和木材等資源需要得到明智及可持續的管理。

最終目標是防止過分開發，讓自然再生，以及有充足的食物，水源和物質供應至後代。

7. 絶種

前章提及，人類對生物多樣性的影響已持續很多年了。為食糧而追獵大型動物，導致野生的長毛象和老虎絶種 (Reiss, 2000)。此外，為了農地而清除樹林，加上陸地，海洋和空氣污染，減少了自然生境和野生動物。

個案研究 — 艾克森凡德茲(Exxon Valdez) 石油洩漏事件 (Jackson, 1996)

在1989年3月24日，一艘超級油輪叫艾克森凡德茲在阿拉斯加王子威廉海灣擱淺。38,000 公噸的原油（貨物重量的 22%）泛蓋 2,500 km2 的面積，以及污染大約 1,700 公里長的海岸線。這「阿拉斯加石油洩漏」成為其世界上最大型的環境災難。

相對其他石油洩漏，艾克森凡德茲洩漏石油的數量不是最大，但是造成的影響無疑是最嚴重。因為王子威廉海灣的水域有很多野生動物，對它們造成破壞。例如大量的水獺，太平洋大比目魚，鯡，粉紅的鮭魚和蛤相繼死亡。超過360,000隻海鳥被殺死。魚獲嚴重地受石油洩漏影響，威脅很多當地漁夫的生計。以長期影響而言，石油積聚在一些貽貝類，並且在食物鏈累積，嚴重影響捕食者的生殖功能和健康。

自此，艾克森凡德茲石油洩漏令全世界關注其生態和經濟的影響。其中一個結果是創立凡德茲法則（稍後稱為 CERES 法則），是公司和機構對環境責任的指引。

事件的詳請及對環境倫理學的應用，請參考第五章「環境倫理」。

結論

人類在過往數萬年，對地球有很大的影響。他們以狩獵為生，與生態系中的哺乳類動物相似。工具和火種的使用令他們支配環境。工業革命和技術進步增加人均資源的使用，人口急劇增長，帶來極大的環境影響。統計數字說明人類破壞生境，過分開發天然資源，對生物多樣性造成負面影響。在1989年，在阿拉斯加發生的艾克森凡德茲石油洩漏，說明人類嚴重破壞自然環境的一個近期例子。

Questions for Further Thinking 思考問題

1. Which is more globally sustainable: a world in which the population is concentrated in towns, or one in which cities become less important and most people live in suburbs or scattered in rural communities?
   哪一個是可持續：人口集中在城市，或城市較為次要，人口分佈於郊區呢?
2. Construct a food web for a local park. Identify the effects of human activities in the web.
   試畫一幅本地公園的食物網，說明人類活動對食物網的影響。
3. Note the food you eat during a day. Construct a diagram showing food chain linkages you have made.
   留意你每日所進食的食物，試畫圖顯示其食物鏈的關係。

References 參考資料

Jackson A. and J. (1996), *Environmental Science: The Natural Environment and Human Impact*, Longman, Harlow, England, p. 342–345.

OECD (2004), *SourceOECD: factbook*. Retrieved February 10, 2005, from OECD Factbook 2005 website: http://haly.sourceoecd.org/vl=6701579/cl=16/nw=1/rpsv/factbook/11-01-12.htm

Pullin, A.S. (2002), *Conservation Biology*, Cambridge University Press, NY, USA, p.15–17, 54–74.

Reiss, M. and Chapman, J. (2000), *Environmental Biology*. Cambridge University Press, Cambridge, p.3–8.

United Nations (2005), *Population Growth*. Retrieved February 10, 2005, from Population Reference Bureau website: http://www.prb.org/Content/NavigationMenu/PRB/Educators/Human_Population/Population_Growth/Population_Growth.htm

Chapter 4. Hong Kong Environment and Precious Species

Victoria Harbor of Hong Kong © Frank Lee

Hong Kong's 1,103 km^2 of land contains 6.8 million people and has become one of the largest trading economies in the world. Steep mountains and efficient urban planning have allowed most of the population to reside in the 215 km^2 of urban area, while over 400 km^2 has been designated as "protected areas", including country parks, special areas and conservation zones. The heavy concentration of population and economic activities, within such a small area, leads to tremendous pressures on the environment. The ecological impact incurred by these conditions, particularly upon air quality, is compounded by the development across the Pearl River Delta region of Mainland China.

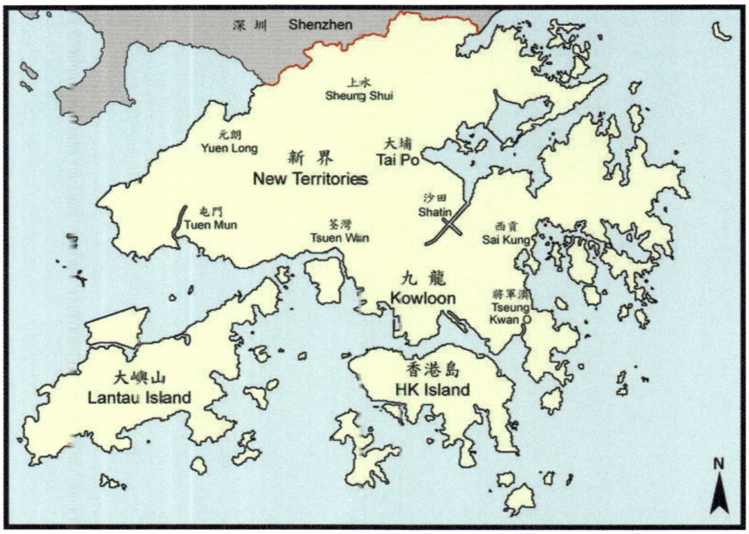

Map of Hong Kong Special Administrative Region © Kitty Siu

I. Environment-related Government Structure

In Hong Kong, there are a number of government authorities in charge of issues concerning the environment. Primarily, the **Environment, Transport and Works Bureau** is responsible for overall policy. It receives professional support from several government departments, and advice from the Advisory Council on the environment.

Under the Bureau, the **Environmental Protection Department** (EPD) executes environmental policies and programmes, vets environmental planning and assessment findings, enforces and reviews environmental laws, plans and develops facilities for liquid and solid waste disposal, and promotes environmental management, audit and reporting. EPD also promotes environmental awareness in the community. The **Environmental Campaign Committee** (ECC) organizes events to promote environmental awareness and manages the Environment and Conservation Fund.

The **Agriculture, Fisheries and Conservation Department** (AFCD) is the main authority for nature and wildlife conservation. It manages country parks, special areas, marine parks and a marine reserve, which are designated for nature conservation, recreation and education purposes. The department enforces the law to protect wildlife and plants, and examines ecological aspects of environmental impact assessment reports and planning studies. Finally, it promotes public awareness of nature conservation.

Lion Nature Education Centre of AFCD © Frank Lee

Other government authorities related to environmental management include:

- Planning Department - plans future land use and controls developments.
- Drainage Services Department - designs, builds, operates and maintains sewerage and sewage treatment facilities.
- Civil Engineering and Development Department - provides outlets for the re-use of inert construction and demolition material.
- Electrical and Mechanical Services Department - promotes energy conservation.
- Marine Department - enforces the law on marine pollution.
- Others - Government Laboratory and Hong Kong Observatory

In 2003-04, the Government spent $6.13 billion on the environment, or about 2.14% of total public expenditure (Information Services Department, 2003).

II. Laws and Regulations

Hong Kong's urban areas are besieged with numerous environmental problems such as air, water and noise pollution, and high accumulation of waste. To tackle these problems, the Government has undertaken various environmental protection measures ranging from legislation and regulatory control to implementation of numerous programmes and services. The ultimate responsibility lies with the EPD, which implements pollution prevention and control measures, and helps to formulate environment-related policies, including the planning of Hong Kong's sewage and waste management programmes.

Since 1998, all designated projects must follow the statutory **environmental impact assessment** (EIA) process and carry out mitigation measures or other actions recommended. Appropriate legislation is essential in governing air, water, waste and noise pollution.

Hong Kong's environmental ordinance and its functions (EPD, 2005):

Environment-related Ordinance	Functions
Air Pollution Ordinance	Controls air pollution; extended in 1998 to strengthen emission standards for diesel cars and tight duty vehicles other than taxis.
Ozone Layer Protection Ordinance	Bans the import and manufacture of substances that deplete the ozone layer and allows Hong Kong to meet its obligations under the 1987 Montreal Protocol on the issue.
Water Pollution Control Ordinance	Enacted in 1980; controls water pollution; manages discharges through a licensing system.
Dumping at Sea Ordinance	Prevents dumping of waste into the marine environment.
Waste Disposal Ordinance	Controls the pollution caused by livestock waste; covers the import and export of waste; bans shipments of hazardous waste from developed countries to Hong Kong since 1998.
Noise Control Ordinance	Amended in 1997; controls percussive piling noise as well as vehicle burglar alarms.
Environmental Impact Assessment (EIA) Ordinance	It provides the legal framework for conducting EIA on designated projects and implementation of agreed environmental measures through Environmental Permits. Since 1998, over 500 development projects were ensured by EIA recommendations.

Enforcement of the above pollution control ordinances, is undertaken by four regional offices and the territorial control office of the EPD, through investigation of pollution complaints, inspection and licensing of pollution sources, issuing pollution abatement notices and prosecution of offenders.

In 2003, EPD conducted more than 68,000 inspections to enforce control on air, noise, waste and water pollution. They included regular checks on environmental compliance and investigations of pollution complaints from the community. The enforcement work resulted in more than 500 prosecutions and $5 million in fines (Information Services Department, 2003).

Besides legislation, a number of programmes and services have been introduced, to fight against pollution problems, with special emphasis on waste. Three strategic landfills were built, a Chemical Waste Treatment Centre on Tsing Yi Island began its operation, and a **Waste Reduction Framework Plan**, which aims to reduce waste through incineration, recycling and other measures, was unveiled. The **Harbour Area Treatment Scheme (HATS)** aims to collect and treat the sewage discharged at Victoria Harbour via an outfall off East Lamma Island by 2008 (see Chapters 11 and 12 for details).

Hong Kong Environmental Protection Festival 2002 © Frank Lee

Community awareness and participation is encouraged, through a series of campaigns, organized by the Environmental Campaign Committee, under which, a Waste Recycling Competition is undertaken. Many secondary and primary schools are provided with education packages and environmental programmes are extended to universities. At the same time, the city celebrates the annual **World Environment Day** and **Environmental Protection Festival**, during which, most of the talks and workshops are organized for various sectors of the community.

Regional and International Co-operation

Pollution does not respect boundaries and the pollutants from Hong Kong and Guangdong often mix, making the problem even more serious. To tackle regional environmental issues, Hong Kong has been cooperating with its Mainland partners since 1990 through the Joint Working Group on Sustainable Development and Environmental Protection between the HKSAR and Guangdong, focusing on addressing pollution problems in Deep Bay and conserving Mirs Bay.

Deep Bay and Mirs Bay © Frank Lee

The series of actions including strategic and regulatory planning, supported by appropriate legislations, promotion of cleaner technology and community awareness and participation have enabled Hong Kong and its neighbors, not only to clean up the environment, but also to safeguard it from further deterioration (EPD, 2005).

Pearl River Delta region (modified from Lands Department, HKSAR map)

III. Wildlife in Hong Kong

As mentioned in Chapter 2, biodiversity is an indispensable part of our capital. Hong Kong is home to a spectacular ecosystem. The biological diversity of this tiny city, relative to its size, is disproportionately large. About 740 square meters of land is countryside and home to 1,900 species of flowering plants (including 120 orchids), and over 2,800 species of vascular plants. Hong Kong has 175 different types of ferns and 210 seaweed species. The Mai Po marsh is one of the world's most important wetlands.

Mai Po marsh © Frank Lee

Hong Kong's climate and physical environment provides a wide range of habitats and supports a rich and diversified fauna. According to AFCD surveys, the 24 identified habitats consist of rocky shore, grassland, fishpond, forest, freshwater, etc. Those habitats of high ecological value include Fung Shui woods, mangroves, seagrass beds and wetland plant communities.

Overall, Hong Kong hosts 465 species of birds (1/3 of the total bird species in China), 21 species of bats, 2,000 varieties of moths, 238 species of butterflies (1/6 of the total butterfly species in China), 140 kinds of freshwater fish, 80 species of reptiles and 57 types of mammals.

Number of wild species in Hong Kong (as of December 2004):

Species (number)	Representative species of Hong Kong
Mammals: 57	- Chinese White Dolphin, Rhesus Monkey
Freshwater fish 140	- Green Grouper, Red Snapper
Birds: 465	- Black-faced Spoonbill, Crested Bulbul
Butterflies: 238	- Common Birdwing, White Dragontail
Dragonflies: 11[illegible]	- Chinese Greenwing, Hong Kong Flatwing
Reptiles: 80	- Chinese Soft-shelled Turtle, Burmese Python (snake)
Amphibians: 2[illegible]	- Romer's Tree Frog, Hong Kong Newt
Corals 84	- Stony corals (e.g. *Platygyra sinensis*, *Favia speciosa*)

(extracted from AFCD, 2005)

Country parks and beaches © Frank Lee

There are 23 country parks and 16 special areas (five of which are located outside country parks) with a total area of 41,582 ha (about 38% of the total land area of Hong Kong). They cover scenic hills, woodlands, reservoirs, islands, indented coastlines, marshes and uplands. All are carefully protected for the purposes of nature conservation, countryside education and compatible scientific studies. The four marine parks and one marine reserve covering 2,430 consist of scenic coastal areas, seascapes and biologically important habitats.

Ordinances related to nature and wildlife protection (Information Services Department, 2003):

Forests and Countryside Ordinance	Provides for the general protection of vegetation, gives special protection to certain rare plants such as magnolia and orchid.
Country Parks Ordinance	Designs, controls and manages the countryside as country parks and special areas for nature conservation, education and scientific research purposes.
Marine Parks Ordinance	Designs, controls and manages the ecologically important marine environment as marine parks and marine reserves for nature conservation, education and scientific research purposes.
Wild Animal Protection Ordinance	Under the Ordinance, it is prohibited to: - Hunt protected wild animals or use illegal hunting appliance to hunt wild animals; and - Buy, sell, export or offer for sale or export any protected wild animal killed or taken in Hong Kong.
Animals and Plants (Protection of Endangered Species) Ordinance	Over 30,000 wildlife species are protected under the Ordinance, which has been implemented since 1997, giving effect to the Convention on International Trade in Endangered Species of Wild Fauna and Flora (CITES) and regulating the import, export or possession of endangered species.
Fisheries Protection Ordinance	Regulates fishing practices and prevents destructive fishing activities such as using explosive or toxic substances for the purpose of fishing.

Let us take a look at some of the treasures of Hong Kong:

Black-faced Spoonbill (modified from Information Services Department, 2003)

- Scientific name: ***Platalea minor***
- With a black bill flattened with a spatulate end and extensive black facial skin as icon image, the bird is a large wading species with white plumage and black legs.
- Every winter, over 150 black-faced spoonbills, which are about 25% of the world's population, spend the winter in Mai Po Ramsar site.

HK's star animal - Chinese White Dolphin © Frank Lee

- Scientific name: ***Sousa chinensis***
- Mainly pink, especially darker at dorsal and pectoral fins; gray pigments on the forehead; eat shrimps, crabs and fish.
- Number of individuals found: 45 to 152
- Active areas: about 40% in west of Lantau waters; others in Sha Chau and Lung Kwu Chau Reserve Areas

Hong Kong Newt © Frank Lee

- Scientific name: ***Paramesotriton hongkongensis***
- Characterized by an elongated body, a long tail and an apparent neck.
- Can be found in pristine mountain streams at all elevations

Green Turtle © Frank Lee

- Scientific name: ***Chelonia mydas***
- The only known species of sea turtle breeding locally (i.e. Sham Wan of Lamma and in many remote beaches).
- Results from AFCD's satellite-tracking on the green turtles nesting in Hong Kong, have shown that they migrated to feeding grounds on Hainan Island.

Common Birdwing butterfly
© Frank Lee

- Scientific name: ***Troides helena***
- The black and yellow colouring on the wings is a warning pattern to predators that it is unpalatable.
- It attains a wingspan of 13-15cm for the males and up to 18cm for the females.
- This species and the Golden Birdwing are the only **<u>insects</u>** protected by Animals and Plants (Protection of Endangered Species) Ordinance of Hong Kong.

Other precious species include:

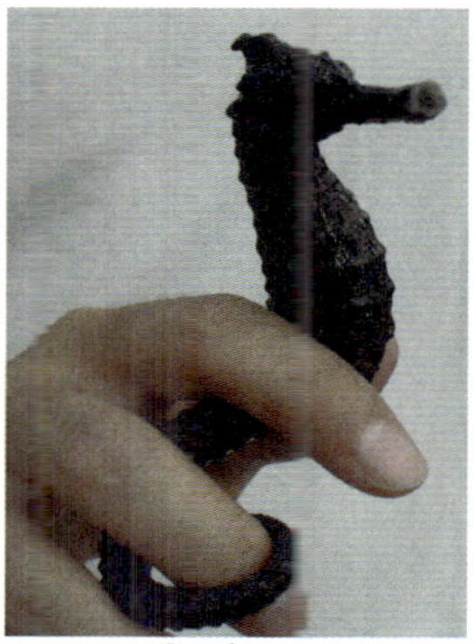

Seahorse
(Hippocampus spp.)
© Frank Lee

Lantern Bug
(*Pyrops candelaria*)
© Frank Lee

Corals and coral fishes
© Frank Lee

The following pictures best describe the real image of Hong Kong – an energetic urbanized city with very unique rural coverage:

Birds-eye view of Victoria Harbour from the Peak © Frank Lee

Aberdeen, HK Island © Frank Lee

Questions for Further Thinking

1. What is the current government structure and political framework for environmental affairs?
2. Comment on these bodies in handling environmental issues, with examples.
3. Which department formulates and carries out consultations of environmental policy and decision? Do we have any conservation policy or law to follow?
4. Suggest major items to be included in a conservation policy suitable for Hong Kong's rural area.
5. Which environmental issue should we, in Hong Kong, set as our first priority? Why? How can we tackle the environmental issue?

Urban development: another typical image of Hong Kong?
© Frank Lee

第四章　香港環境及珍貴物種

香港的土地面積為1,103平方公里，人口共680萬，是世界最大的貿易經濟體系之一。由於地勢陡峭及有效的城市規劃，大部分人口均居住於佔地215平方公里的市區內，其餘超過400平方公里的土地則已列為「受保護地區」，如郊野公園、特別地區及保育區。由於稠密的人口及頻繁的經濟活動都集中於一小片的土地，自然對環境構成沉重壓力。本港情況對生態所帶來的影響，近年因中國大陸珠江三角洲發展蓬勃而使問題越趨惡化，其中尤以空氣質素的問題最為明顯，令本來已棘手的環境問題更趨嚴峻。

I. 與環境有關之政府機構

香港的環境事宜由數個政府部門負責，其中以**環境運輸及工務局**為首，負責整體政策，其轄下有數個政府部門提供專業支持，並有環境諮詢委員會提供意見。

環境運輸及工務局轄下的**環境保護署**負責推行環保政策及計劃，調查環境計劃及評估結果、執行及檢討環境法例、規劃發展污水及固體廢物處理設施、推廣環境管理、審計及報告。環境保護署亦會提高公眾環保意識。**環境保護運動委員會**舉辦各類活動以提高公眾環保意識及管理自然保育基金。

漁農自然護理署負責存護本港的自然生境及野生動物，管理郊野公園、特別地區、海岸公園及海岸保護區，以達致自然存護、康樂及教育等目的。該部門亦負責執行保護野生動物及植物的法例、就環境影響評估提供有關自然存護方面的意見，亦提高市民對自然存護的認識。

其他與環境管理有關的政府部門包括：
- 規劃署 — 規劃土地用途及控制發展；
- 渠務署 — 負責有關污水及污水處理設施的設計、建築、運作及維修等工作；
- 土木工程拓展署 — 提供惰性拆建物料循環再用的機會；
- 機電工程署 — 推廣節約能源；
- 海事署 — 執行有關污染本港水域的法例；
- 其他 — 政府化驗所及香港天文台。

政府於2000–2004年度用作環境的支出共61.3億，約佔整體公共開支百分之2.14% (政府新聞處，2003)。

II. 法律與規例

現時香港市區受到空氣污染、水質污染、噪音污染及廢物囤積等多種環境問題困擾。為了解決這些問題，政府實施了各項不同的環境保護措施，包括立例管制以及推行多項活動和服務。環境保護署擔當統籌的角色，負責推行減少污染及管制措施，並幫助制定與環境相關的政策，如規劃香港的污水及廢物管理計劃。

自1998年起，所有指定的項目必須進行法定的環境影響評估及採取消減污染的措施或其他建議方法。合適的法例對規管空氣、水質、廢物及噪音污染至為關鍵。

香港有關環境的法例及其功用：(EPD, 2005)

有關環境的法例	功用
空氣污染管制條例	管制空氣污染；條例於1998年擴展，以加強管制柴油車輪排放廢氣的標準。
保護臭氧層條例	本條例管制含損害臭氧物質產品的生產及入口，讓香港履行1987年《蒙特利爾議定書》訂明的國際義務。
水污染管制條例	本條例於1980年制定，用以管制水質污染；以排照制度管理污水排放。
海上傾倒物料條例	防止廢物傾倒於海上。
廢物處置條例	管制由禽畜廢物引致的污染；管制廢物的進出口；自1998年起禁止有害廢物從已發展國家運抵本港。
噪音管制條例	於1997年修訂，管制撞擊式打樁噪音及車輛防盜警鐘。
環境影響評估條例	本條例為指定工程提供法定架構，以進行環境影響評估，及實施環境許可證內訂明的紓緩措施。自條例於1998年實施至今已完成了500多宗環評項目。

上述的污染管制條例由環境保護署轄下的四個區域辦事處及總區辦事處負責執行，其職責包括調查有關污染的投訴、檢查、發牌規管污染源頭、發出「消減污染通知書」及檢控違例人士。

環境保護署於2003年進行了超過次68,000巡查，有關空氣、噪音、廢物及水質污染執法，執法行動包括例行巡查和調查市民投訴污染的個案。提出超過500次檢控，罰款總額超逾五百萬元。(政府新聞處，2003)

除法例管制外，環保署亦推行多項計劃和服務，以對抗污染，並將重點放於減廢方面。興建三個策略性堆填區、啟用青衣的化學廢物處理中心、推行**減少廢物綱要計劃**，希望透過焚化、回收及其他措施來減少廢物。維港現正進行「**淨化海港計劃**」，旨在利用南收集維港兩岸的污水，並經和處理。(詳情請參與第11及12章)。

環境保護運動委員會舉辦各式各樣活動，以提高公眾的環保意識及推動社區參與，例如廢物回收比賽、向多間中小學提供環保教材套及替大學舉辦環保活動。同時，香港每年都慶祝**世界環境日**及**環保節**，期間舉行一連串的講座及工作坊讓公眾參與。

亞洲及國際合作

污染無分疆界，香港的空氣污染物混和了廣東省的污染物，使問題越趨嚴重。自1999年起，香港特區政府與廣東省政府組成「粵港持續發展與環保合作小組」，以加強粵港兩地合作，致力解決地區環保問題，並將解決后海灣的污染問題及保育大鵬灣列為重點。

香港及鄰近的城市透過實施多項措施，包括定期制定策略性計劃、設立合適的法律、推廣環保科技及提高公眾的環保意識及參與，以清潔環境及防範問題進一步惡化 (環境保護署，2005)

III. 香港的野生動物

正如第二章所指出，生物多樣化是香港不可或缺的資產。香港擁有多樣化的生態系統，相對於其面積來說，這裡的生物多樣性是超乎比例地高。香港約有740平方米的土地屬郊區範圍，孕育了超過1900種顯花植物 (包括120種蘭花)、超過2800種維管束植物、175種蕨類植物及210種海藻，其中，米埔沼澤更是世界最重要的濕地之一。

香港的氣候及地理環境形成了不同的生境，從而孕育了豐富及多樣化的動物種類。根據漁護署所作的調查，生境可劃分為24個類別，包括岩岸、灌木草地、魚塘、樹林、淡水等，而風水林、紅樹林、海草床和濕地植物群更是具有高生境價值。

香港記錄到的鳥類品種數目達465種之多，相當於全中國野鳥數目的三分之一，此外，亦記錄到21種蝙蝠、2000種蛾、238種蝴蝶(相當於全中國蝴蝶品種的六分之一)、140種淡水魚類、80種爬行類及57種哺乳類。

香港野生物種數目(截至2004年12月數據)：

物種數目	本港知名物種
哺乳類：57 淡水魚類：140 鳥類：465 蝴蝶：238 蜻蜓：110 爬行類：80 兩棲類：23 珊瑚：84	- 中華白海豚、恆河猴 - 青斑魚 - 黑臉琵鷺、紅耳鵯 - 裳鳳蝶、燕鳳蝶 - 華艷色蟌、黃紋棘腹蟌 - 水魚、蟒蛇 - 盧氏小樹蛙、香港瘰螈 - 石珊瑚(如中華扁腦珊瑚、標準蜂巢珊瑚)

(資料來源：漁農自然護理署，2005)

本港41,582公頃土地，約佔本港總面績38%，已劃作23個郊野公園和16個特別地區(其中5個位於郊野公園範圍之外)，其範圍包括風景怡人的山嶺、叢林、水塘、島嶼。這些地方都得悉心的護理，作保育、教育及科學研究用途。此外，香港亦有四個海岸公園及一個海岸保護區，總面積為2,430公頃，其中包括風景優美的海岸、海景和重要的海洋動植物的棲息地。

有關保護自然及野生動物的法例 (政府新聞署，2003)：

林區及郊區條例	所有植物均受條例保護，而一些稀有品種如木兰花及蘭花更受特別保護。
郊野公園條例	規劃、管制及管理郊區為郊野公園和特別地區，作保育、教育及科學研究用途。
海岸公園條例	規劃、管制及管理具生態價值的海洋如海岸公園及海岸保護區，作保育、教育及科學研究用途。
野生動物保護條例	條例列明任何人禁止： - 狩獵任何受保護野生動物或使用非法狩獵器具狩獵野生動物；及 - 購買、售賣、輸出、要約出售或要約輸出在香港被殺死或取得的受保護野生動物。
動植物(瀕危物種保護)條例	動植物(瀕危物種保護)條例自1997年起推行，列出超過30,000種受保護的動植物，目的是要實施瀕危野生動植物種國際貿易公約(CITES)及管制進口、出口及管有瀕危物種。
漁業保護條例	規管捕魚方法及防止具破壞性的漁業活動如使用炸藥或有毒物質去捕魚。

香港的一些珍寶如下：

1) 黑臉琵鷺

- 學名：*Platalea minor*
- 大型涉禽，全身白色，腳全黑，其黑色的咀呈匙狀以及大片黑色裸露面部皮膚已成為其標記
- 每年均有超過150隻黑臉琵鷺（約佔全世界黑臉琵鷺數目的25%）於米埔拉姆薩爾濕地渡冬

2) 中華白海豚

- 學名：*Sousa chinensis*
- 身體呈粉紅色，背部及背鰭更深色，前額灰色，以蝦、蟹和魚為食料
- 現時記錄得的數目：45至152
- 分佈地點：約40%出沒於大嶼山以西水域，其餘則在沙洲及龍鼓洲海岸公園

3) 綠海龜

- 學名：*Paramesotriton hongkongensis*
- 身體修長、有尾及明顯頸部
- 這品種可於山上清澈的溪流找到
- 學名：*Chelonia myda*
- 綠海龜是唯一在本港繁殖的海龜品種(即南丫島深灣和多個偏遠沙灘)
- 漁農自然護理署以衛星追蹤在本港繁殖的綠海龜，結果顯示牠們遷徙至海南島的覓食地

4) 裳鳳蝶

- 學名：*Troides helena*
- 其翅膀上的黃色與黑色是警戒色，讓其捕獵者知道其味道不好
- 雄蝶展翅可達13–15厘米，而雌蝶則可達18厘米
- 裳鳳蝶和金裳鳳蝶是本港唯一受動植物(瀕危物種保護)條例保護的昆蟲

思考問題

1. 現時負責處理環境事務的政府組織和政治架構是如何的?
2. 你如何評價上述組織處理環境問題，請附以事例說明。
3. 哪個部門負責制定及執行環保政策的諮詢和決定? 現時香港有保育政策或法例可遁嗎?
4. 試建議於制定香港的郊區的保育政策時應包括什麼重要事項。
5. 香港應優先處理的哪種環境問題? 試加以解釋? 我們應如何解決這個問題?

References 參考資料

AFCD, HKSAR (2002), *New Viewpoints – Country Parks in Focus* (2nd ed.), Cosmos Books Ltd, Hong Kong, p. 6–7.

AFCD, HKSAR (2005), *Hong Kong Biodiversity Online*. Retrieved February 10, 2005, from Hong Kong Biodiversity Online website: http://www.hkbiodiversity.net/html/en/home.html

EPD, HKSAR (2005), *Environment Hong Kong 2004*. Retrieved February 5, 2005, from Environment Hong Kong website: http://www.epd.gov.hk/epd/misc/ehk04/english/main/contents.html

Information Services Department, HKSAR (2003), *Hong Kong 2003*. Retrieved January 25, 2005, from Hong Kong Yearbook 2003 website: http://www.info.gov.hk/yearbook/2003/eindex.html

Lands Department, HKSAR (2003), *Hong Kong and Its Neighboring Region map* (8th ed).

Chapter 5. Environmental Ethics

Ethics are concerned with what is right and wrong; it is a matter of what one ought to do or not to do. Different applications of ethics include: Business Ethics, Social Ethics, Medical Ethics, Ethics of Science, Legal Ethics, Bioethics, as well as Environmental Ethics.

Environmental ethics is the study of how humans ought to interact with the environment. It consists of a systematic account of the moral relations between human beings and their natural environment. Aspects to be studied in environmental ethics include:

- Humanity's relationship to the environment;
- Humanity's understanding of and responsibility to nature; and
- Humanity's obligations to future generations through the use of the natural environment.

Our relationship with the environment?
© Janet Chan

I. History of Environmental Ethics

Throughout human history, the worldwide view that has prevailed is that human rights should take precedence over nature. This has been reflected by behaviour such as mass production of non-durable, non-recyclable goods, over-use of private motor vehicles, mass tourism, etc. However, some opposite views have emerged since the 19th century:

- In 1864, George Marsh published "*Man and Nature*" which was thought to be the first comprehensive description in English of human destruction of the natural environment.
- 1908 marked John Muir's unsuccessful attempt to protect an area of wilderness in California, USA.
- The death of the last passenger pigeon — an extinction that gained a high public profile in 1914, focused attention on Man's destructive influence on Nature.

An intellectual climate developed during the last few years of the 1960s, in large part due to the publication of two papers in ***Science***:

- Lynn White's "*The Historical Roots of our Ecologic Crisis*" (published in March 1967)
- Garett Hardin's "*The Tragedy of the Commons*" (published in December 1968)

Silent Spring — the Milestone of Modern Environmental Movement

A reawakening of environmental issues took place when a marine biologist named Rachel Carson published her book *Silent Spring* in 1962. The book focused on the widespread use of agricultural chemicals and pesticides, which had occurred since World War II, and warned of their potential threat to human and other animal life. The ideas that pollution in one area can affect the environment in another, and that humans cannot live without the goods and services provided by ecosystems, are now commonly understood facts.

Silent Spring was first released in the *New Yorker* magazine and then as a book which sold half a million copies in 16 countries within a year. It subsequently inspired other authors and activists, and engendered an almost immediate grass-roots sentiment for improvement of environmental quality and passage of related federal legislation.

The book that brought new thoughts to mankind ➔

(modified from Carson, 1994)

Carson's writings made the world community aware of the interrelationships of humans and ecosystems. Undoubtedly, the publication of *Silent Spring* is often regarded as the event, perhaps more than any other, which initiated the modern era of environmental concern and activism.

© Frank Lee

"Tragedy of the Commons"

In 1968, Garrett Hardin addressed the philosophy that a self-interested person would keep common resources (like air and water) in his own possession and accumulate them. Ultimately this reflects why wealth and poverty exist in this planet.

Hardin applied the theory to population growth and its effects on the earth's scarce resources. His essay appeared in *Science* and has been included since then in works on many subjects, including ecology, environmentalism, health care, philosophy, political science, ethics, sociology, law, psychology and economics. His thesis illustrated the need to combine political and social theory with biological data. Hardin stated, "a finite world can support only a finite population; therefore, population growth must eventually equal zero" (Hardin, 1968).

We can see today that Hardin's solution has limited application, and the pollution problem continues to spiral out of control. The tragedy is being played out in the air we breathe and the water we drink.

II. Varieties of Environmental Ethics

Many philosophers argue that our responsibilities to the natural environment are only indirect, that the responsibility to preserve resources, for example, is best understood in terms of the responsibilities that we owe to other humans. ***Anthropocentric*** (human-centered) ***ethics*** deems that only human beings have moral value. Thus, although we may be said to have responsibilities regarding the natural world, we do not have direct responsibilities to the natural world (Bowen, 1994).

Many controversies in the early decades of the environmental movement, such as air and water pollution, toxic wastes, and the abuse of pesticides, arose because of a perspective of anthropocentric ethics. Pesticide-contaminated food and polluted drinking water pose direct threats to man. Thus, anthropocentric environmental ethics involves simply applying standard ethical principles to new social problems.

As a form of environmental ethics, anthropocentrism argues that Man's responsibility to the natural environment is only indirect. This responsibility towards nature is a trigger-off from the responsibility we owe to other humans. In fact, future generations are morally relevant under this notion.

Other philosophers have argued that we do have direct responsibilities to natural objects other than human beings. ***Non-anthropocentric ethics*** grants moral standing to such natural objects as animals and plants. Typically, this approach requires further extensions and revisions of standard ethical principles. Controversies surrounding the ethical treatment of animals and the threatened extinction of many plant and animal species are among the best-known issues of non-anthropocentric ethics (Bowen, 1994).

Animals to be granted moral standing
© Frank Lee

A further development of environmental ethics occurs by shifting from a focus on individual living things - for example, spotted owls or redwood trees - to a focus on collections or "wholes" such as species, populations, or ecosystems. ***Holistic ethics*** believes that we have moral responsibilities to collections of (or relationships between) individuals rather than (or in addition to) those individuals who constitute the whole. For example, a holistic environmental ethics might allow selective hunting of individual animals so long as the population of that species is not endangered. Greatly influenced by ecology, holism raises more serious philosophical challenges than do more individualistic ethics (Bowen, 1994).

Advocated by Aldo Leopold, ***Land ethics*** is a form of holistic environmental ethics (Leopold, 1949). It argues that a thing is right when it tends to preserve the integrity, stability and beauty of the biotic community. It is wrong when it tends otherwise. Land ethics should be seen as an ethic extended to include land and does not replace human ethics (Bell *et. al*, 1996).

The *Gaia hypothesis* ("*Gaia*" is the Greek name for the Earth) by Lovelock suggested that the Earth behaves like a single living organism and have self-regulating ability. Thus, everything on earth will act together to modify the physical conditions to produce the best conditions for life. After all, it implies that the ultimate purpose to protect the ecosystem is to preserve the domination of human beings on Earth (Lovelock, 1979).

Summary of Varieties of Environmental Ethics:

Varieties	*Explanations*	*Examples*
1. Anthropocentric (human-centered) Ethics	Evaluates environmental actions solely in terms of their effects on humans.	Some urge the adoption of soft energy options because he or she is fearful of the effects on climate of burning fossil fuels.
2. Non-anthropocentric (animal-centered) Ethics	Counts all animals as morally considerable.	If cyanide pollution in the river cause nonhumans to suffer, then this is a moral minus that must be taken into account independently of how things will be for humans.
3. The Holistic ("Everything") Ethics	Considers each action by considering consequences for individuals and adding them.	Mining involves smashing up rocks, disturbing geological structures, spoiling fossils and the like.

III. Philosophies in Environmental Ethics

Different types of ethics about the environment exist, and at present there is no consensus regarding which is most appropriate. Generally speaking, environmental ethics fall into one of the two paradigms, *shallow ecology* or *deep ecology*. Other new concepts include *social ecology* and *ecofeminism*.

Shallow Ecology

Shallow ecology considers the values of nature to be instrumental to humans, and is thus said to be strongly anthropocentric. Although members of some nonhuman species can be said to have certain rights, in general their interests are said to be secondary to the interests of humans. Shallow ecology emphasizes the relationships between individuals (Bell *et. al*, 1995).

There are some branches under the concept of shallow ecology:

- **Utilitarianism**'s central goal is to achieve the greatest good for the greatest number of people. Consequently, actions are said to be morally correct if they produce the greatest net balance of good over evil consequences.
- **Concepts of justice** assume the fundamental equality of individuals, and therefore focus on questions of fairness in the distribution of costs and benefits when decisions are made about the environment.
- **Concepts of freedom** are used to maximize freedom from coercion, the presence of opportunities for choice, and civil liberties and democratic forms of governance.
- **Theism** focuses on the role of religious traditions and beliefs in guiding our environmental actions.

Deep Ecology

Deep ecology is a diverse philosophical movement originally associated with the work of the Norwegian philosopher **Arne Naess** in 1975. At first deep ecology was characterized by the conviction that humans should more fully recognize their relationships with nature and should agree that all living organisms are of equal intrinsic value. Deep ecology is receiving increased attention from environmental philosophers and scientists who contribute to the development of environmental ethics (Naess, 1975; Belshaw, 2001).

Deep ecology: never harm Nature
© Frank Lee

Deep ecologists consider everything to be morally relevant in a direct sense. What matters ultimately is everything, not just what is personal, human, conscious, or alive.

Deep ecologists also believe that humans should not interfere with nature; instead, they should cooperate with it and maintain its beauty, integrity and stability. This ethic also holds that everything is morally relevant, but gives greater emphasis to undisturbed nature. Not only are individual organisms said to have moral relevancy in their own right, natural ecosystem functions are to be maintained.

One form of deep ecology is predicated by a **biocentric viewpoint**, which maintains that all species have an intrinsic right to exist in the natural environment. Another form is predicated by the **ecosystem concept**, which emphasizes the interdependence of members of the biotic community, the importance of species diversity for ecological stability, the limits of populations and natural resources, and concern for long-term spatial and temporal effects. The ethical behavior is defined in terms of consequences to the whole ecosystem rather than aggregate functions derived from the total benefits to individual members. Accordingly, deep ecology is said to be strongly holistic (Belshaw, 2001).

Shallow ecology vs. Deep ecology

Shallow ecology	Deep Ecology
• Dominance over Nature • Natural environment as resource for humans • Materials or economic growth for growing human population • Believe in ample resource reserves • High technological progress and solutions • Consumerism • National/centralized community	• Harmony with Nature • All nature has intrinsic worth/species equality • Elegantly simple material needs (material goals serving the larger goal of self-realization) • Belief that Earth's "supplies" are limited • Appropriate technology; non-dominating science • Economical use and recycling • Minority tradition/bioregion

Social Ecology

This refers to a philosophical and political movement associated with the work of Murray Bookchin in 1980. Social ecologists argue that environmental problems are the result of oppressive hierarchical relationships in human society. Changes in human society, social ecologists argue, will also lead to changes in the way the environment is understood and treated.

In theory, the philosophy suggests that many human societies are organized into hierarchical layers of dominance and oppression, where some classes of people and some kinds of human qualities or abilities are regarded as superior to others - for instance, men are thought superior to women, intellectual skills superior to physical skills. The idea that the natural world is inferior to human beings and to be exploited and abused stems from, and is an expression of, the hierarchical nature of human relationships with one another.

Ecofeminism

The term was first used in 1974 by Francoise D'Eaubonne. Since then ecofeminism has become a large but diverse movement, encompassing a wide range of perspectives from within the feminist movement and the environmental movement. Like social ecologists, ecofeminists do not argue that human society must be altered before changes in attitudes toward the natural world can occur. Rather, they argue that oppression of women and the natural world are "twin oppressions" that must be addressed together. It promotes an ethics of relationship and care between humans and non-humans and the particular environment in which they are located (Bowen, 1994).

Ecofeminists of environmental ethicists have argued for much more local, context-specific approaches to environmental ethics based on specific human relationships to particular environments. It is an ethic of relationship and care, the manifestation of which varies from place to place and relationship to relationship.

IV. Putting Environmental Ethics into Practice

Environmental ethics is interconnected with political and social structures concerning the use of natural resources, so the field also touches the areas of social and political philosophy. In the struggle to conserve the environment, ethicists also use the knowledge and theories of science, for example, in issues such as those dealing with global warming and pollution.

In 1970, the Environmental Protection Agency of the USA was established to oversee the various federal environmental laws that had been enacted. One of its major functions is to review the environmental impacts of highway projects, large-scale commercial and reconstruction, power plants, and other large undertakings in the federal government.

The *Exxon Valdez* Case

In 1989, the disastrous *Exxon Valdez* oil spill in Alaska helped fuel demands for environmental controls, since cases like this clearly show the damage that can be caused to fish, birds, and the natural environment. The *Exxon Valdez* oil spill was also disastrous to Alaskan fishermen, who blame the accident for degrading ability to fish and make a living. What is always being weighed legally and ethically is how much damage afflicted on the environment and its inhabitants can be judged as reasonable and how much cannot.

Formerly known as the Valdez principles developed after the tanker *Exxon Valdez* spilled its cargo of oil in Alaska in 1989, the **CERES principles** have gained its popularity by corporations as an explicit public commitment to sound environmental practices (*Exxon Valdez* Oil Spill Trustee Council, 2005). Basically the principles include protection of the biosphere, sustainable use of natural resources, reduction and disposal of wastes, energy conservation, risk reduction, etc.

These principles emphasize the importance of corporate responsibility for the environment. The fundamental philosophical position they reflect is environmental stewardship, which seems to be defined as looking after the environment on behalf of future generations. In fact, the principles suggest that the reason for environmental protection is indeed the benefit of future generations.

The CERES principles commit the signing organization to ethical behavior toward the environment in a variety of contexts; the protection of wilderness, biodiversity, human health, and scarce resources are all explicitly mentioned. In addition, the principles urge companies to open their records to the public about their environmental affairs and any environmental hazards they may be creating. Thus, the principles link an environmental ethic with the idea of freedom of information.

Codes of Practice

In recent years, codes of practice, professional codes of conduct, and ethical codes have been adopted by a growing number of organizations, industries, corporations, professions, and governments. Such codes can have a variety of functions, but they are all public commitments to particular standards of individual and/or organizational behavior. They act as an internal benchmark against which individuals or organizations may be measured to see whether their performance reaches the standards they have set for themselves.

Conclusion

In its broadest sense, environmental ethics is the study of our moral relationship with everything that is a part of the environment, within which questions of right and wrong can arise. It includes other human beings and the things human beings make, such as machines and cities.

There are many different ways of ethically analyzing environmental issues, namely ***Shallow Ecology***, ***Deep Ecology***, ***Social Ecology*** and ***Ecofeminism***. Of course, these approaches lead to disagreement and controversy, both in regard to the general principles that should be followed (if any) in environmental ethics and in the response to particular environmental issues and problems.

© Frank Lee

第五章　環境倫理

倫理學與對錯有關；是應該做或不應該做什麼。倫理學可分為：商業倫理學，社會倫理學，醫學倫理學，科學倫理學，法律倫理學，生物倫理學，以及環境倫理學。

環境倫理是研究人類和環境的相互作用。它解釋了人和自然環境之間的道德關係。環境倫理包括以下幾方面：

- 人類與環境的關係；
- 人類對自然的理解和責任；以及
- 人類對後代及自然環境的義務。

I. 環境倫理的歷史

人類歷史當中，普遍認為人類的權力是超過大自然的，例如：大規模生產非耐用，不可循環的貨品，過度使用私家車輛，大眾旅遊等。不過，自從十九世紀，一些不同的想法不斷出現：

- 1864年 — George Marsh 出版「人和自然」，描寫人類對自然環境的破壞。
- 1908年 — John Muir 未能成功保護位於美國加州的自然環境。
- 1914年 — 最後一隻的信鴿死亡 ，喚醒人類對大自然破壞的關注。

1960年代，在*科學*發表的兩篇文章引起學術界的迴響：

- Lynn White's "*The Historical Roots of our Ecologic Crisis*" (發表於1967年3月)
- Garett Hardin's "*公共的悲劇*" (發表於1968年12月)

寂靜的春天 — 現代環保運動的里程碑

1962年，海洋生物學家Rachel Carson出版*寂靜的春天*，再次喚醒人類對環境問題的高度警覺。書中強調自從第二次世界大戰，農業化學物和殺蟲劑的廣泛的使用，警告對人類和其他動物的潛在威脅。一個地區的污染能夠影響另一區的環境，而且沒有生態系提供的產品與服務，人類不能生存。

寂靜的春天在*紐約人*雜誌中首次發表後，一年內在十六個國家賣出五十萬本。隨後啟發其他作者和活躍分子，以及引發民眾的情緒，改善環境質素，通過相關的聯邦法規。

Carson的著作令世界意識到人類和生態系之間的關係。無疑地，*寂靜的春天*的出版通常被認為是開啟現代環境學研究先河的重要事件。

「公共的悲劇」

在1968年，Garrett Hardin 提出利己的人會把共同的資源（如：空氣和水）據為己有。最終，反思地球為甚麼存在富有和貧窮。

Hardin把理論應用於人口增長及對地球稀有資源的影響。他的文章在*科學*發表，而且包括了很多主題：生態學，環境論，保健，哲學，政治科學，倫理學，社會學，法律，心理學和經濟。他的論題把政治，社會學和生物的資訊的理論結合起來。

Hardin陳述，「一個有限的世界只能夠支持有限的人口；因此，人口增長最終會等於零」(Hardin, 1968)。現在，我們可以知道Hardin的解決方案是有限的，污染問題繼續失控。悲劇結束於我們呼吸的空氣和飲用的食水。

II. 環境倫理的類別

很多哲學家爭論我們對自然環境的責任是只是間接的，保護資源的責任是為了其他人類。**以人類為中心**的倫理認為只有人類有道德價值。因此，儘管我們對自然界應負責任，不過不是直接的責任 (Bowen, 1994)。

環境運動的初期，很多爭議是關於，空氣和食水的污染，有毒廢物的排放，殺蟲劑的濫用問題是因為以人類為中心的倫理學。受殺蟲劑污染的食物和受污染的食水直接威脅人類。因此，以人類為中心的環境倫理只涉及是把標準倫理的原則應用於新的社會問題。

這一種環境倫理，以人類為中心的倫理認為人類對自然環境的責任只是間接的。對自然負責任是基於對其他人類負責任。在道德方面，這概念與下一代有關的。

其他哲學家則主張了除了人類外，我們對自然有直接的責任。非以人類為中心的倫理認為自然體，如：動物和植物，皆有道德的地位。這種方法需要擴大和修改標準倫理的原則。非以人類為中心的倫理的爭論圍繞著合乎道德對待動物和瀕臨絕種的物種(Bowen, 1994)。

環境倫理學的發展從集中在個別生物，如：斑梟或紅杉樹，轉移到集中在「整體」，如：種類，族群，或生態系。**整全倫理**相信我們不單對個別物種，而是對整體有道德責任。例如，整全環境倫理學可以容許狩獵個別動物，但不危害整個族群。受生態學的影響，整全主義較個體倫理引起更多的哲學挑戰（Bowen, 1994）。

Aldo Leopold提倡，土地倫理是一種整全環境倫理學（Leopold, 1949）。當一件事情關乎保護生物族群的完整，穩定和美麗時，就是正確，否則是錯的。土地倫理應該被視為延伸至土地的倫理，並且不取代人類的倫理（Bell *et. al.*, 1996）。

Gaia論說（「Gaia」是希臘文，意思是地球）由Lovelock提出，地球像一個具生命的有機體和有自動調節的能力。因此，地球上的一切活動，互相作用，產生最適合生命的環境。畢竟，保護生態系的最終目的乃是保存人類在地球的優勢 (Lovelock, 1979)。

不同類別倫理的撮要:

類別	解釋	例子
1. 以人為中心的倫理(以人類為中心)	單以對人類的影響作為評估環境的行為	鼓勵採用軟性能源，原因是他 / 她擔心燃燒化石燃料時，影響氣候
2. 非以人為中心的倫理(以動物為中心)	所有動物也包括在道德的考慮	氰化物污染河流，令生物和環境受苦，是不合乎道德
3. 整全倫理(整體)	作出行動時，考慮對整體的結果	採礦對岩石造成猛烈的衝擊，破壞地質結構及化石

III. 環境倫理學的哲學

目前，有不同種類的環境倫理學，對於哪個是最合適，則沒有共識。一般而言，環境倫理學分為兩類，*膚淺的生態學*或*深刻的生態學*。其他新的概念包括*社交的生態學*和*環境女權運動*。

膚淺的生態學

膚淺的生態學認為自然對人類的價值是工具性的，因此是強烈以人類為中心。儘管某些非人類的物種無疑有一定的權利，不過它們的權利對於人類是次一等的。膚淺的生態學強調個人之間的關係 (Bell *et al*, 1996)。

膚淺的生態學的一些分支：

- **功利主義**的目標是令大多數人得到最大的好處。因此，當行為帶來的好處多過壞處時，就是道德上正確。
- **公平概念**假設個人平等，因此當作出關於環境的決定，著重資金及利益的分配。
- **自由概念**用來從強制使自由最大化，對選擇，和公民自由和民主政體種類的統治的機會的存在。
- **有神論**著重於宗教傳說和信念指導我們在環境方面的行為。

深刻的生態學

深刻的生態學是一個多元的哲學運動，是挪威哲學家Arne Naess於1975年所創。首先，深刻的生態學認為人類應該更全面認識他們與自然的關係，以及所有生命體都有同等的內在價值。深刻的生態學得到環境哲學家和從事環境倫理學發展的科學家的關注 (Naess, 1975; Belshaw, 2001)。

深刻的生態學者認為一切都與道德有關。一切都是要緊的，不只是人類，有意識的，或活著的。他們也相信人類不應該干擾自然；反之，他們應該與它合作，保持美麗，完整和穩定。這倫理也認為一切與道德有關，特別強調不干擾自然。不只是個體生物在道德上有它們的權利，也要保持自然的生態系的運作。

深刻的生態學的其中一個觀點是以生物為中心，聲稱所有物種在自然環境有其本身的權利。另一個觀點則是以生態系為中心，強調族群成員之間的相互依賴，物種多樣性對生態穩定的重要，人口和天然資源的限制，並且關注長遠及短期的效果。倫理行為是根據整個的生態系的結果，而不是個別成員的累積得益而定。因此，深刻的生態學是強調整體性 (Belshaw, 2001)。

膚淺的生態學與深刻的生態學

膚淺的生態學	深刻的生態學
• 凌駕自然之上 • 自然環境是作為人類的資源 • 物質或經濟的增長是為了人口增長 • 相信地球有充裕的資源 • 高技術的發展和解決方案 • 消費主義	• 與自然和諧共處 • 自然有本身的價值/物種同等 • 物質需求簡單 • 相信地球的「供應」是有限的 • 合適的技術；非支配性的科學 • 節約使用和再循環

社會生態學

這是Murry Bookchin於1980年提出的一個哲學和政治行動。社會生態學者主張環境的問題是專制階級社會的結果。社會生態學者認為社會的變化改變了人類理解和對待環境。

理論上，這哲學認為社會是由統治和受壓迫的階級組成，某些階層的人因其出眾的素質和能力處於高人一等的地位。例如，一般被認為男人比女人優越，智能技巧比體能優越。這理論認為人類可以淩駕自然之上，所以過度地開採和揮霍自然資源，並且一個人與其他人之間的關係是決定其社會地位的根本。

生態女權運動

這名詞是Francoise D'Eaubonne 於1974年所創。自此，環境女權運動成為一個龐大而多元的運動，從女權主義運動到環境運動。像社會生態學者一樣，環境女權主義者認為首先改變對自然界的態度，才可以改變人類社會。反而，女性和自然界受壓迫是"孿生"的，具有很強的共性，必須加以綜合考慮。它提出應該加強人類和非人類與環境之間的關係（Bowen, 1994）。環境女權主義者認為環境倫理應該是地區性的，著重人與人的關係和關懷。

IV. 實踐環境倫理學

環境倫理學認為天然資源的使用與政治和社會結構有關，所以領域也涉及社會和政治哲學。在保護環境的過程中，倫理學家也使用科學的知識和理論，例如：全球暖化和污染。在 1970 年，美國成立環保局，監督各種有關環境的法律。它的主要的功能之一是評審高速公路計劃，大規模商業和重建，發電廠，以及其他工程對環境的影響。

艾克森凡德茲(Exxon Valdez)案例

1989 年，Exxon Valdez在阿拉斯加洩漏石油事件引致魚類，鳥類和自然環境的損害。Exxon Valdez洩漏石油事件令到阿拉斯加的漁獲減少和影響漁夫生計。在法律和倫理學上，怎樣去判斷環境和原居民受到多少程度的傷害呢。

在1989年阿拉斯加ExxonValdez洩漏石油事件後，發表CERES原則 (之前稱Valdez原則) 反映公司對環境實踐作出一個明確的公眾承諾（Exxon Valdez Oill Spill Council, 2005）。基本上，原則包括保護生物圈，可持續使用天然資源，減少棄置廢物，保存能源，減低風險等等。這些原則強調公司對環境責任的重要性。它們的基本哲學立場是環境的義務，為下一代保護環境。事實上，原則表明保護環境的理由確實是為了下一代著想。

CERES 原則指定簽署組織對環境應有的倫理行為，包括保護荒地，生物多樣性，人類健康，和稀有資源。此外，原則鼓勵公司向大眾公開關於環境事務和損害環境的記錄。因此，原則把資訊自由的概念加入到環境倫理。

在近幾年，越來越多的機構，工業，公司，行業，和政府採用了行業約章，專業操守約章，和倫理約章。這樣的約章可以有各種功能，特別是個體和/或組織對公眾承諾。這些約章作為內部的基準，以測量個體或機構是否達到他們自己設有的標準。

結論

廣義來說，環境倫理學是研究我們與有關環境一切的道德的關係，涉及對錯的問題。而且，包括人類和人類所創造的事物，如：機器和城市。在倫理學上，分析環境問題有很多不同的方式，如：膚淺的生態學，深刻的生態學，社會的生態學和生態女權主義。當然，從一般性的原則到對個別環境問題的回應，這些方法都引起爭論和辯論。

Questions for Further Thinking 思考問題

1. Modern transportation systems, especially private motor vehicles, provide huge benefits to some human beings, yet the environmental costs in terms of pollution, loss of habitat, environmental harm, and wildlife deaths from road accidents are very high. How do we make ethical judgments about such costs and benefits?
現代運輸系統，如私家車輛，對人類有很大效益。不過，它們帶來的污染，生境破壞，環境損害，交通意外引致野生動物死亡等環境代價亦非常高。倫理上，我們應該怎樣作出的判斷呢?

2. Regarding issues such as tourism, energy and waste generation, deforestation, resource consumption, pollution, and climate change, what are the environmental ethical values of human interactions in these?
關於旅遊，能源，廢物，砍伐森林，資源損耗，污染和氣候轉變等問題，它們的環境倫理價值是什麼呢?

References 參考資料

Bell, P., Greene, T., *et al*. (1996), *Environmental Psychology* (4th ed.), Harcourt Brace College Publishers, Fort Worth, p. 190, 508.

Belshaw, C. (2001), *Environmental Philosophy: Reason, Nature and Human Concern*, McGill-Queen's University Press, Montreal, Canada, p. 202–203.

Bookchin, M. (1980) *Toward an Ecological Society*, Black Rose, Montreal.

Bowen, J. (1994), *Environment Education: Imperatives for the 21st Century*, James Nicholas Publishers, Albert Park, Australia, p. 79–103.

Carson, R. (1994), *Silent Spring*, Houghton Mifflin, Boston, USA (book cover).

Exxon Valdez Oil Spill Trustee Council (2005), *Exxon Valdez Oil Spill Trustee Council-Home Page*. Retrieved February 20, 2005, from the website: www.evostc.state.ak.us

Hardin, G. (1968), "*The Tragedy of the Commons*" Science, 162 December: 1243–1248.

Leopold A. (1949), *A Sand County Almanac*, Oxford University Press, New York.

Lovelock, J. (1979), *Gaia: A New Look at Life on Earth*, Oxford University Press, Oxford.

Naess, A. (1975), "*The Shallow and the Deep, Long-Range Ecology Movements*" Inquiry 16 (1975) 95–100.

White, L. (1967), "*The Historical Roots of Our Ecological Crisis*" Science, 155 March (10): 1203–1207.

Chapter 6. Ecological Footprint

Imagine there is a physical barrier covering our land, making us isolated from the outside world. Then we must live on our own resources and products only.

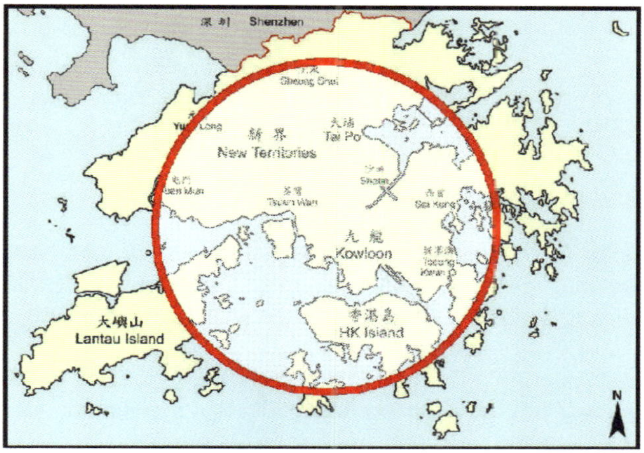

© Frank Lee

Fortunately this is not true! In reality, we import natural resources and discharge the unwanted waste. In today's world, Nature is being used up by many different countries. The relationship between human consumption and nature exploitation affects the **HUMAN IMPACT** (i.e. population, consumption and technology) on the natural environment.

The fact is, **we all rely on Nature!**

© Frank Lee

Now here comes the dilemma:

- Is the relationship between human beings and nature breaking up?
- Will our next generations be threatened by the deteriorating environment?
- Can Hong Kong be sustainably developed? Can the world be sustainably developed?

We must learn the concept of Ecological Footprint.

I. What is Ecological Footprint (EF)?

Many environmentalists and scientists have noticed the problem, particularly after the industrial revolution. Among them, William Rees *et al.* developed the concept of Ecological Footprint in 1960-1970s. Rees published the book ***Our Ecological Footprint - Reducing Human Impact on the Earth*** in 1996, and since then the concept has been firmly established in the foundation of sustainable development, ecological economics and urban studies.

(extracted from Rees *et al.*, 1996)

As a tool to evaluate the actual reality of nature exploitation, the **Ecological Footprint** is an indicator to measure the use of natural resources needed to sustain a person or a society's daily consumption at a particular place.

Imagine for the case of Hong Kong, the theoretical ecological footprints for its people look somewhat like the chart on the right. ➔

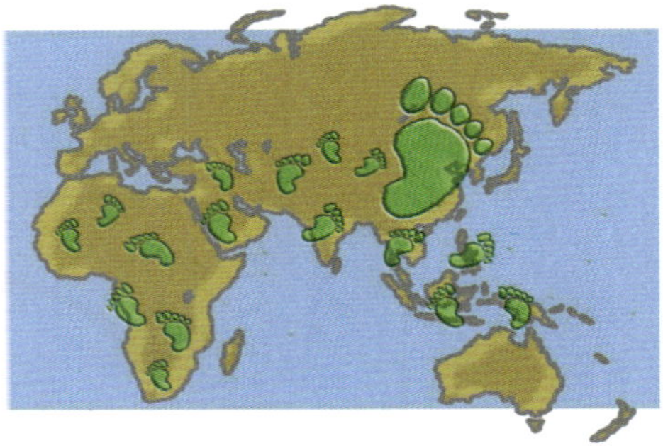

© Elly Leung

The ecological footprint is a simple environmental accounting method that can provide a measure of the material intensity of our economy (Footprint, 2003). The target size for the footprint is known as the **Fair Share**, i.e. the productive areas divided by the world population. Basically, the ecological footprint of a population is estimated by calculating how much land and water area is required on a continuous basis to produce all the goods consumed, and to absorb all the wastes generated by that population. **In other words, the more resources one country consumes, the larger its footprint.**

II. Calculation of EF

In order to calculate the ecological footprint of energy consumption, you estimate how much land is needed to produce the same energy from the sun. For example, plants convert energy from the sun into plant material (biomass) and people turn biomass into fuel such as ethanol and methanol. You calculate the amount of energy used from a certain piece of land (joules/hectare/year).

The **resources needed** (i.e. energy, food, clothing, transportation, etc.) and **biodegradation of wastes** together are converted to biologically productive land (unit in hectare).

70% of the world is covered by sea while 30% is land. After deducting iceberg and lifeless sea, the biologically productive land for sea is actually **3 billion ha** (3% of total). Similarly, after deducting desert and plateau, the biologically productive land consists of about **83 billion ha**, which is 50% of total.

1 hectare (ha) = 100m x 100m

© Frank Lee

Resources entitled for each global citizen (in terms of biologically productive land)	=	**global biologically productive land** / **total world population** (6.8 billion)

Five types of biologically productive land:

Sea

Forest

Pasture

Arable land

Built area

(all images © Frank Lee)

Based on previous formula, we find out the resources entitled for each citizen to be calculated as follows:

Sea	Arable land	Pasture	Forest	Built area	Total
0.5	0.25	0.6	0.6	0.03	1.98 ha

According to the ***Brutland Commission*** (World Commission on Economic Development, 1987), 12% of land should be designated as conservation area, in which no human disturbance is allowed. As a consequence, the resource one can use is **1.7 ha** of biologically productive area (about one and a half football field).

III. Ecological Footprint of Hong Kong

Suppose the population of Hong Kong is 6.8 million, we then use the same formula and estimate that the average Hong Kong resident has a footprint of **6 ha per capita**. This EF is much larger than some developed countries like Germany, United Kingdom and Japan. Hong Kong people actually rank the world's 13th biggest users of natural resources out of 151 countries and regions. In fact, Hong Kong needs at least 444 times its land size to sustain itself (Footprint, 2003).

Regarding the five components of Hong Kong's EF, we have:

Sea	Arable land	Pasture	Forest	Built area	Total
1.6	0.5	0.4	0.2	3.3	6 ha

Hong Kong's EF per citizen is 3 times larger than global share. © Elley Leung

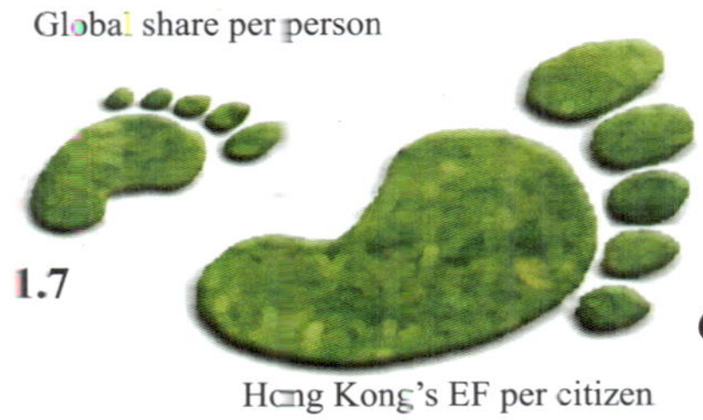

That means we need 3 Earths to sustain human beings if all the people in the world have the lifestyle as the Hong Kong citizens.

Why is Hong Kong's Footprint so large?

Hong Kong appropriates and exploits resources from elsewhere, particularly from south China, in order to sustain its activities. Natural resources include agricultural land in China (where most of Hong Kong's food is grown) and water (i.e. Hong Kong buys 80% of its water from Guangdong Province). With much of its manufacturing industry stationed in the Mainland over the last twenty years, Hong Kong's significance as a polluter has extended across the border.

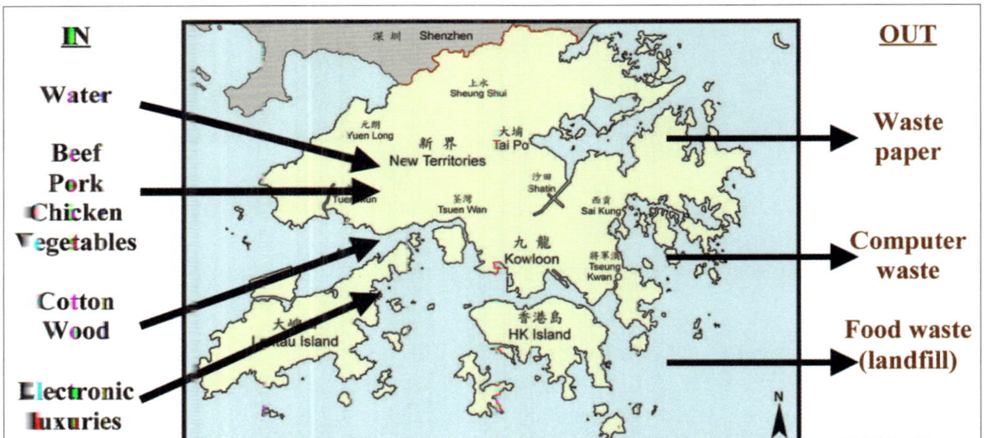

© Kitty Siu

Energy: our footprint is 8 times larger than the global share.

© Frank Lee

Lack of renewable energy facilities (i.e. wind power)

© Roy Tam

Water footprint – supply from Tung Jiang river

© Frank Lee

Calculate your own EF on the next page!

How Big is Your Ecological Footprint (EF)?

ENERGY USE

1. What kind of transportation do you use most often?
I walk / use bicycles..............................0 marks
I take public transportation..............25 marks
I carpool / take taxi with friends....50 marks
I drive / take taxi myself..................80 marks

2. How much time do you spend on computers/ playing TV games every day?
No use at all..0 marks
Less than 1 hour..................................20 marks
1 to 3 hours..50 marks
more than 3 hours...............................80 marks

3. How much time did you spend on flying on an airplane last year?
0 hours..0 marks
1 to 10 hours...80 marks
11 to 30 hours...................................150 marks
more than 30 hours.........................300 marks

FOOD CONSUMPTION

4. How much meat is in your meal?
I am a vegetarian.................................0 marks
One out of three meals has meat....20 marks
Two out of three meals have meat..30 marks
All meals have meat in them............40 marks

5. How much of your food is imported / packaged?
More than 75%...................................25 marks
75%...15 marks
25% to 75%..5 marks
Less than 25%.......................................0 marks

MATERIAL USE

6. Which of the following do you have?
PC / mobile phone / TV set / air conditioner / heater / digital camera / DV

Add 5 points for each.......total _____ points

7. How often do you buy clothes/shoes/bags based on fashion trend?
Never.........0 marks Seldom............30 marks
Quite frequently...........50 marks

8. How many books (including textbooks and comics) have you bought in the last 3 years?
1 to 50.........20 marks 51 to 100.........40 marks
More than 100.........60 marks

9. How do you use paper?
I write on both sides of paper.............0 marks
I photocopy on both sides of paper....0 marks
None of the above................................10 marks

WATER USE

10. How much water do you use when bathing?
I shower for less than 10 mins.............0 marks
I shower for 11 to 20 mins...................15 marks
I shower for more than 20 mins........35 marks
I use the bath tub.................................40 marks

URBAN USE

11. What is the size of your flat/house?
0 to 200 square feet..............................0 marks
more than 200 square feet.................10 marks

WASTE MANAGEMENT

12. I do NOT recycle the following:
Waste paper / aluminium cans / plastic bottles / old clothing (5 marks for each)...... Total _______

13. Do you practice composting food/fruits leftovers?
Yes ....0 marks Never................20 marks

FORMULA: Your EF = Total marks / 100

My EF is ______________ **hectares**

That means it takes (your EF / 1.7)

__________ **Earths in order to sustain your current standard of living!**

(source: Roy Tam)

IV. Ecological Footprints of Other Countries

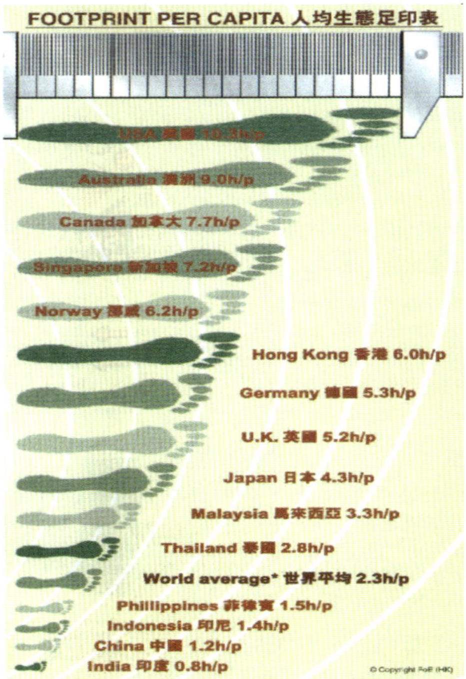

© Friends of the Earth (HK)

As you can see from the chart on the right, the EF's of many countries are above 1.7 – e.g. USA has the biggest EF (10.3), followed by Australia (9.0), Canada (7.7), Singapore (7.2), etc. – meaning that many of these regions cannot provide enough natural resources for their people to use to sustain the society's daily consumption.

For example, London's ecological footprint is 120 times the land area of the city. This is the area needed to supply London's total consumption needs. Also, the Netherlands has an EF fifteen times larger than its actual territory.

Here comes the question: **with these unreasonable facts, how is the world able to sustain itself at the moment?**

The answer is that while many of us are enjoying our over-exploiting lifestyle, two parties are suffering:

1. Poor people

Notice the ecological footprints of developing countries – China (1.2), India (0.8) and some African countries (0.3–0.9) – indicating that many in those part of the world are living with fewer resources less than the global fair share, in order to compensate our greedy over-consumption.

2. Our environment

At the same time the earth's environment is being sacrificed, because there are too many people who consciously over-consume resources that Nature can never regenerate.

Furthermore, a theory called **Sustainable Development** has emerged recently. It states that economic gains are made only if the potential for future generations is not sacrificed. For details, see Chapter 7 of this book.

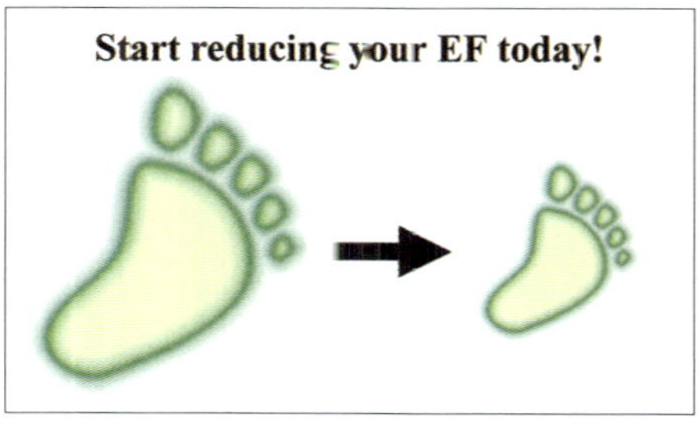

第六章　生態足印

幸好，這不是真的! 實際上，我們輸入自然資源和排放不需要的廢物。現今，自然資源將被耗盡。**人類對自然環境的衝擊**就是人類消耗和自然開發之間的關係，如：人口、消耗和技術。事實上，**我們完全依賴自然界！**

現在，困境來了：
- 人和自然之間的關係已經破壞嗎?
- 我們的下一代會受到惡化的環境威脅嗎?
- 香港可以持續發展嗎?世界可以持續發展嗎?

我們必須學習生態腳印的概念。

I. 什麼是生態足印?

在工業革命以後，很多環境學家和科學家注意到問題。在1960–1970年代，William Rees等發展了生態足印的概念。Rees於1996出版了「我們的生態腳印 – 減少人類對地球的衝擊」，自此之後，這個概念成為可持續發展、生態學經濟和都市研究的基礎。

作為評估自然開發的工具，**生態足印**是量度一個地方對自然資源的運用，以維持個人或社會的消耗的一個指標。

生態腳印是一個簡單的環境計算方法，量度我們經濟的物質程度(Footprint, 2003)。足印的大小為具生產力的土地面積除以人口數量。生態足印計算維持人口的消耗及清除所產生的廢物所需的土地。**換句話說，一個國家消耗資源越多，足印就越大。**

II. 生態足印的計算方法

爲了計算能源消耗的生態足印，你估計要多少土地才能產生同樣的太陽能。例如，植物太陽能轉化成生物量，並且人們把生物量轉化成燃料如：乙醇和甲醇。你計算一塊土地所用的能量(焦耳/公頃/年) 。

轉換所需的資源 (即能量、食物、衣物、交通等。) 及廢物為具生物生產力土地(以公頃為單位) 。海洋佔70%，而陸地佔30% 。扣除冰山和無生命的海洋以後，具生物生產力的海洋面積實際上是三十億公頃 (佔全部的3%) 。

同樣，扣除的沙漠和高原以後，具生物生產力的土地面積約八十三億公頃，佔全部的50%。

每人賦予的資源 (以具生物生產力的土地量度單位)	=	全球具生物生產力的土地 --- **總人口(六十八億)**

五種具生物生產力的土地為：**海洋**、**森林**、**牧場**、**耕地**及**建築面積**。

根據以上的方程式，計算出每人賦予的資源如下：

海洋	耕地	牧場	森林	建築面積	總數
0.5	0.25	0.6	0.6	0.03	1.98 公頃

根據布倫特蘭委員會 (世界經濟發展委員會，1987)，土地的12%應該被選為保護區，不被干擾。因此，你可使用的資源是1.7公頃的具生物有生產力的土地 (大約一個半足球場的面積)。

III. 香港的生態足印

假設香港的人口是680萬，我們然後使用方程式去計算，香港居民平均有6公頃的腳印。這個數字比一些已發達國家為大，如：德國、英國和日本。在自然資源的運用方面，151 個國家當中，香港人排行第13名。實際上，香港需要至少444 倍的土地面積來維持自己的需要(Footprint, 2003)。

關於香港生態足印：

海洋	耕地	牧場	森林	建築面積	總數
1.6	0.5	0.4	0.2	3.3	6 公頃

香港的人均生態足印是全球人均的生態足印三倍。如果全世界都以香港的生活模式，我們需要三個地球。

為什麼香港的生態足印這樣大呢?

香港從南中國開發及利用資源。自然資源包括在中國的農田(種植大多數香港的食物)和食水(香港從廣東省購買80%的食水)。在過去二十年的製造業，香港已成為嚴重的跨越區域污染者。

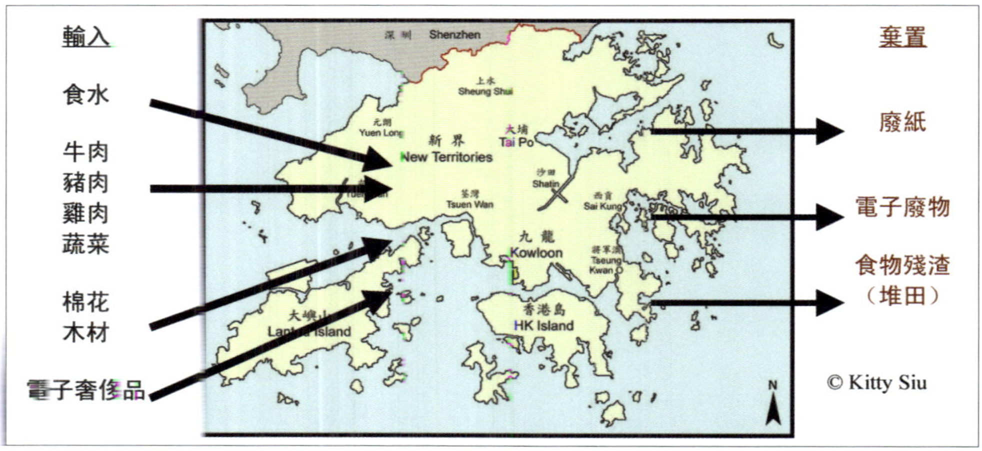

能量：我們生態足印是全球的八倍

缺乏可再新能源設施，如：風力發電

水的足印－來自東江水

在下一頁，計算自己的生態足！

看看你的生態足印有多大?

能源

1.平日外出，你多用哪種交通工具?

我走路 / 踏單車……………………………0分
我乘交通公具……………………………25分
我和朋友一起乘的士……………………50分
我乘私家車 / 的士………………………80分

2. 你每日花多少時間用電腦，玩遊戲機?

没有用……………………………………0分
少於一小時………………………………20分
一至三小時………………………………50分
多過三小時………………………………80分

3. 在過去的一年，你共搭了多少過小時飛機?

0小時……………………………………0分
1至10小時………………………………80分
11至30小時……………………………150分
多過30小時……………………………300分

食物

4. 肉類佔你的飯餐有多少?

我食素……………………………………0分
每天三餐中有一餐肉……………………20分
每天三餐中有兩餐肉……………………30分
每餐都有肉………………………………40分

5. 你的食物中，有多少是進口／包裝?

多過75%…………………………………25分
75%………………………………………15分
25%至75%…………………………………5分
少過25%…………………………………0分

物質生活

6. 你擁有以下哪一樣物品?

個人電腦 / 手提電話 / 電視 / 冷氣機 / 暖爐 / 數碼相機 / 數碼播錄機

每項加5分…………………………共 _____ 分

7. 你會趕潮流而買衣服／鞋／袋嗎?

不會………… 0分　　間中會………30分
會……………50分

8. 你過去三年共買多少書本（包括教科書和漫畫書）?

1至50…………20分　　51至100………40分
多過100…………60分

9. 你用紙的習慣是怎樣?

我會用紙的兩面…………………………0分
我會用雙面影印…………………………0分
没有以上的習慣…………………………10分

用水

10. 你每天洗澡平均用多少水?

我洗澡時間少於10分鐘…………………0分
我洗澡時間是11至20分鐘………………15分
我洗澡時間多過20分鐘…………………35分
我會用浴缸洗澡…………………………40分

建成地

11. 你個人的居住面積是多少?

0至200平方呎……………………………0分
多過 200 平方呎…………………………10分

廢物處理

12. 我不會回收以下哪項物品:

廢紙/鋁罐/膠樽/舊衣(每個5分)………共______

13. 你會把吃剩的食物／果皮堆肥嗎?

會 ……………… 0分　　不會……………20分

方程式: 你的生態足印= 總分 / 100

我的生態足印是 ____________ **公頃**

你需要（你的生態足印/ 1.7）

________ 地球以維持你現時的生活方式!

(source: Roy Tam)

IV. 其它國家的生態足印

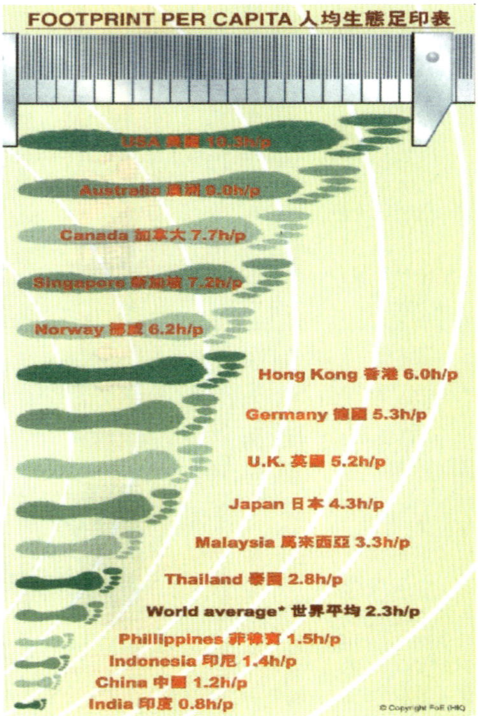

© 香港地球之友

右圖所示，許多國家的生態足印是1.7以上，如：美國有最大的生態足印(10.3)，其次是澳洲(9.0)，加拿大(7.7)，新加坡(7.2)等。這些地區無法為他們的人民提供足夠的自然資源，以承受社會的消耗。

例如：倫敦的生態足印是該城市的120倍。這是提供倫敦的總消耗的土地面積。再者，荷蘭有比它的面積大15倍的生態足印。

問題是: 從這些不合理的事實，世界怎能夠承擔呢？

答案是當大部份人享受過份開發的生活方式時，有兩類人受苦：

1. 窮人
發展中國家的生態足印較低，如：中國(1.2)，印度(0.8)，一些非洲國家 (0.3–0.9)，表示這些地區得到較少的資源，以補償我們過量的消耗。

2. 我們的環境
因為有太多人故意消耗資源，地球的環境被犧牲，不能再生。

此外，理論可持續發展最近出現。它指出我們可以得到經濟上的利益，但不可以犧牲下一代的資源。詳見本書第七章。

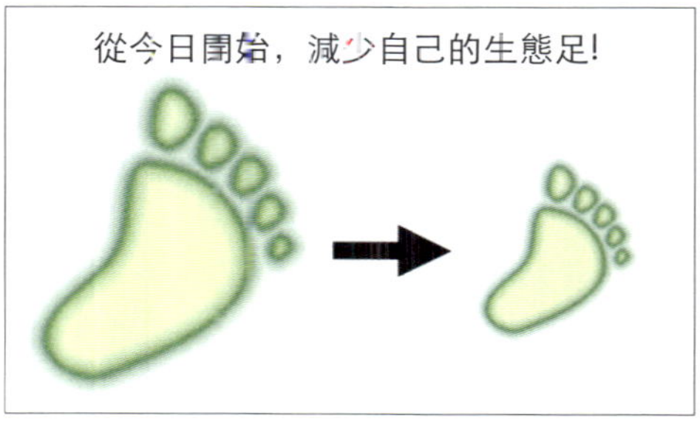

Questions for Further Thinking 思考問題

1. What is your own Ecological Footprint? Which component (energy consumption, land use, urban use, food consumption, etc.) contributes the biggest in your EF? How can you reduce the size of your EF?
   你的生態足印是多少?佔最大部份的是什麼？（能源消耗，土地使用，都市使用，食物消耗等）你怎樣減少你的生態足印呢?

2. Does globalization promote or prevent the growth of EF of a country? Name one practical example of globalization that affects the size of EF in Hong Kong.
   全球化是促進或防礙一個國家的生態足印的增長?試舉一個全球化的例子，影響香港生態足印的大小。

References 参考資料

Footprint (2003), *香港人 — 地球母親的敗家子*，Footprint, vol. 2, p. 12–13.

Rees, W. *et al*. (1996), *Our Ecological Footprint — Reducing Human Impact on the Earth*, New Society Publishers, Gabriola Island, BC.

World Commission on Economic Development (1987), *Our Common Future (the Brundtland Report)*, Oxford University Press.

Acknowledgement

Special thanks to **Ms. Elly Leung** of *Footprint* and **Friends of the Earth (香港地球之友)** for providing some of the images and materials for this chapter.

Chapter 7. Sustainable Development

Take a look at these facts about our environment:

- Over the past 30 years, we've consumed one third of our planet's resources.
- Our ozone layer is depleting at an annual rate of 3%, resulting in more people dying of skin cancer. According to research, a 10% loss in global ozone levels will lead to more people getting skin cancer and will also cause about 1,600,000 new cases of blindness.
- The world's population will double in 49 years at the current growth rate.
- The rapid increase in the world's population puts great pressure on our ecological systems and contributes to poverty worldwide.
- Every day we throw away an average of 2,800 tons of paper and paperboard in Hong Kong. Every ton of paper means the destruction of 17 trees, so that means we are cutting down 48,000 trees per day.
- Experts estimate that within 40 years we will have no more rainforest if we continue clearing forests at the current rate. We are already losing more than 50,000 species of plant, insect, bird and animal every year.
- In Hong Kong, it is estimated that all our landfills will be full in 10–15 years if we continue to create and dispose of waste in the current manner.

(source: Sustainable Development Unit, 2002)

The point is if we continue with our current living style, we will leave little natural resources to our future generations. Therefore, if we are to provide a social, economic and natural environment that will give them a chance to enjoy a better quality of life, we need to understand what **sustainable development** really means.

I. What is Sustainable Development?

The concept of "sustainability" is gaining popularity throughout the world, and has become one of the terms most frequently referred to by the media. This concept involves the demand at present and in the future, sovereignty over countries, equity among nations, natural resources, ecosystem capacity, and the integration of environmental protection and development.

Sustainability: enough harvest for the future?

© Frank Lee

Sustainability originates from ecology. First appeared in fishery and forestry, the term refers to a resources management strategy. The strategy allows the harvesting a reasonable portion of resources, as long as the availability of the resources is not threatened and the resources can be replenished well enough to compensate for the harvesting. For example, the fishery in a region is said to be sustainable production when the capture rate of fishes is reasonably lower than their breeding rate.

"Sustainable Development" has become a popular term in environmental education since 1990s. Countries like the United Kingdom, China and Australia have used it as a main concept, from ecology studies to a combination of social, economics and environmental studies. Most people regard lifelong learning and lifestyle change necessary to achieve it.

The definition of sustainable development laid down by the **World Commission on Environment and Development** is:

> "Development that meets the needs of the present without compromising the ability of future generations to meet their own needs."
> (World Commission on Economic Development, 1987)

In simple terms, social development is considered sustainable when resources, economy, society and environment are in harmony.

The primary concept of sustainable development is promoting the advance and development of the society and simultaneously upholding environmental protection principles. Sustainable development highlights protection and improvement to the ecology when we are promoting production. Resources should be sustainable, to the extent that they are available for human use for a longer period, or even forever, to avoid exhaustion, which would undoubtedly undermine the life and production of human beings.

Features of Sustainable Development

It achieves equity both within generations (intra-generational equity) and across generations (inter-generational equity). Also, it consists of three aspects:

1. Social sustainability: reducing poverty, maintaining a stable level of population and enabling equal opportunity to the deprived group.
2. Economic sustainability: full employment and optimization of human welfare.
3. Environmental sustainability: sustainability in resource use, conservation and that all physical and economic activity should be made compatible with the surrounding.

(source: Council for Sustainable Development, 2004)

II. The World in Action

International agencies, commissions and organizations including the United Nations (UN) have actively been researching and producing publications statements, principles, reports and recommendations, and have held many meetings on creating a common framework for a more sustainable world over the last three decades. In 1980, the UN appealed to all nations to work out the fundamental relationship between the nature, society, ecosystem, economy and usage of natural resources, so as to maintain sustainable development.

In November 1983, the UN established the World Commission on Environment and Development (UNCED). The prime minister of Norway, G.H. Brundtland, presided over the commission. The United Nations appealed for the commission to formulate the "Global Agenda for Change" on the basis of sustainable development.

In 1987, the commission submitted the report titled "Our Common Future" to the General Assembly of the United Nations. A model of "Sustainable Development" was thus put forward. In 1992, the United Nations Conference on Environment and Development (UNCED), held in Rio de Janerio, Brazil, carried out the *Agenda 21*.

Regarded as sublimation of the modern knowledge of sustainable development, *Agenda 21* is a comprehensive plan of action to be taken globally, nationally and locally in every area in which humanity impacts on the environment. It was adopted by more than 178 Governments at the UNCED and advocated 8 Guiding Principles:

- Economy
- Natural Resources
- Environmental Quality
- Biodiversity
- Health & Hygiene
- Mobility
- Leisure & Cultural Vibrancy
- Society & Social Infrastructure

(extracted from ESCAP, 1998)

World Summit on Sustainable Development

The latest UN's World Summit on Sustainable Development was held in Johannesburg, South Africa in August 2002. During the summit, every country agreed to join hands to protect the environment because the development of one country will affect other countries' economics, society and environment. As resources are limited, even if the superpowers seize the resources from other countries, they cannot be sustained in the long term. All countries in this global village should gather together and shoulder the responsibility for a sustainable future.

At the same time, the World Summit reminded us of the situation with some shocking figures: 1. billion people without access to safe freshwater, 1.7 billion living in water scarce areas, and 1.3 billion living in extreme poverty (Johannesburg Summit, 2003).

"Development" of Sustainable Development:

1962	***Silent Spring*** by Rachel Carson published
1970	First **Earth Day** established in the US
1971	International Institute for Environment and Development established in UK
1972	Conference on Human Environment in Stockholm, Sweden
1983	UN established The World Commission on Environment and Development
1987	UN published ***The Brundtland Report*** *(Our Common Future)*
1990	International Institute for Sustainable Development established in Canada
1992	**Conference on Environment and Development** in Rio de Janerio, Brazil
1994	Conference on Population and Development in Cairo, Egypt
1997	**Earth Summit II** in New York and **Kyoto Conference/Protocol**
2002	**World Summit on Sustainable Development** in Johannesburg, South Africa

III. Significance of Sustainable Development

The concept of "sustainable development" was endorsed globally shortly after it was put forward. As science advances, the deeper knowledge of the earth makes people wonder if the earth has any limit to sustaining life on the planet. In short, the issue of concern is the "carrying capacity of the environment", which is the population and economy that a certain set of resources can support at a given time while maintaining stability of the environment.

Development is considered sustainable if the total reserve of natural resources remains unchanged or even increases. Sustainability is related to the interaction of environmental, economic, social, cultural and political factors. From an ecological view, the main concern of sustainability is the stability of the biological system. From a global view, the crux would be maintaining cultural diversity. The cultural perspective of sustainability would focus on maintaining the stability of each society and its cultural regime, maintaining cultural diversity, promoting domestic and international fairness.

Ultimately, sustainable development calls for unselfish behavior. It urges us to consider the consequences of our actions, for the natural environment as well as for ourselves and future generations. The above-mentioned *Brundtland Report* argued that to balance economic development and environmental protection, society would need to implement a new set of objectives, namely (World Commission on Economic Development, 1987):

Economic development
© Frank Lee

- Changing the quality of growth
- Meeting essential needs for jobs, food, energy, water and sanitation
- Ensuring a sustainable level of population
- Conserving and enhancing natural resources
- Re-oriented technology to serve real needs
- Merging economic and the environment in decision-making
- Preserving essential ecosystem functions
- Providing for the well being of future generations
- Integrating economic and environmental policies
- Improve the overall quality of life, not just income growth
- Ending poverty, and distributing resources more equally

All these ideas result in the new "Earth Citizen" concept with the following characteristics:

Traditional concept	New "Earth Citizen" concept
• Human-oriented • Man is owner of the Earth • Man dominates over everything • Environmental problems only exist in certain areas • Earth's resources are unlimited • Organisms can live alone • Economic comes first in social development	• Earth-oriented • Man is just the guardian of the Earth • Every organism including human is equal • Environmental problems are everybody's problems • There is an end to the Earth's resources • Organisms depend on each other • Social development deals with the balance of environment, economics and society

IV. Sustainable Development in Hong Kong

Sustainable Development first came to light in Hong Kong in 1997 when the Government launched the "Study on Sustainable Development for the 21st Century" (**SUSDEV 21**). Conducted by the Planning Department, the Study aimed to define what sustainability should mean for Hong Kong, how to measure it and what policy areas and institutional structures needed to be changed to facilitate better decision-making. The SUSDEV 21 developed a set of indicators for measuring the sustainability of the development of Hong Kong.

© Eric Yuen

The SUSDEV 21 Guiding Principles for Hong Kong:

Economy • Hong Kong should achieve a competitive and prosperous market-based economy which provides the resources to meet the needs and aspirations of the population, both now and in the future.
Health and Hygiene • Hong Kong should provide a living and working environment and pursue policies which promote and protect the physical and mental health and safety of the people.
Natural Resources • Hong Kong should promote the sustainable use of natural resources to minimize its ecological footprint through improving consumption efficiency, minimizing the use of non-renewable resources and re-using, recycling waste and recovering energy from wastes.
Society and Social Infrastructure • Hong Kong should foster a stable, equitable, ethical and progressive society and enable present and future individuals to contribute to and fulfill their potential by providing universal access to adequate and appropriate educational opportunity and social infrastructure.
Biodiversity • To maintain the biodiversity of Hong Kong and to minimize any threat which consumption in Hong Kong may have on biodiversity elsewhere
Leisure and Cultural Vibrancy • Protect and enhance the vibrancy of Hong Kong's recreational opportunities, leisure activities, cultural diversity, archaeological, historical and architectural assets.
Environmental Quality • Hong Kong should be pro-active in avoiding environmental problems for present and future generations, seek to find opportunities to enhance environmental quality, and minimize the unwanted side effects, locally, nationally and internationally, of development and inefficiencies such as air, noise and water pollution or land contamination.
Mobility • Hong Kong should provide safe, accessible, efficient and clean transport systems and pedestrian facilities along with an efficient transport network for the movement of goods and facilitation of services for the community.

(source: Planning Department, 2001)

After all, Hong Kong has its own definition of sustainable development as stated in SUSDEV 21 (**Planning Department, 2001**):

> *"Sustainable Development in Hong Kong balances social, economic and environmental needs, both for present and future generations, simultaneously achieving a vibrant economy, social progress and better environmental quality, locally, nationally and internationally, through the efforts of the community and the Government."*

In 1999, the Chief Executive made a clear statement of commitment towards sustainable development for Hong Kong in his Policy Address. In this policy address the Chief Executive focused on making Hong Kong a sustainable, world-class city by (**Mottershead, 2004**):

- Finding ways to increase prosperity and improve the quality of life while reducing overall pollution and waste;
- Meeting our own needs and aspirations without doing damage to the prospects of future generations; and
- Reducing the environmental burden we put on our neighbors and helping to preserve common resources.

The **Sustainable Development Unit** (SDU) has been set up under the Administration Wing of the Chief Secretary for Administration's Office since April 2001. The main tasks are to facilitate the integration of sustainable development into new Government initiatives and programmes and in the community generally, and to provide administrative and secretarial support to the Council for Sustainable Development.

The **Council for Sustainable Development** was established on 1 April 2003. The Council provides expert advice to the Government, help to build a better understanding of the concept of sustainable development and encourage the community to put it into practice. More specifically, it aims to play a pivotal role in formulating strategy, fostering community support and promoting public awareness of sustainable development in Hong Kong. Specifically, the outlines have been set up as follows (**Council for Sustainable Development**, 2004):

(source: **Council for Sustainable Development**, 2004)

- To advise the Government on the priority areas it should address in promoting sustainable development;
- To advise on the preparation of a sustainable development strategy for Hong Kong that will integrate economic, social and environmental perspectives;
- To promote public awareness and understanding of the principles of sustainable development; and
- **Sustainable Development Fund**: HK$100 millions granted over 10 years for developing a strong public awareness of the concept of sustainable development and to encourage sustainable practices in Hong Kong. In the first round of application, out of a total 127 applications, only 8 projects were approved (HK$7.75 million).

Members of the Council are appointed by the Chief Executive and include people with experience and expertise in the environmental, social and business sectors, as well as senior government officials. The Council is a focus and a forum for exchanging views on key issues related to Hong Kong's long-term sustainability. Some of the new policies and measures include:

- Protect the water resources;
- Protect and minimize the loss of good farmland;
- Protect the ecologically sensitive areas;
- Enhance efficiency and minimize loss of non-renewable resources through coordinated efforts of road, bridges, railway building, as well as major infrastructure such as ports, airports, and sewage treatment facilities;
- Control the population growth
- Attain the right sector policies for enhancing efficient use of facilities and human resources.

Wind power for HK?

The development of the new airport and the Tsing Ma Bridge are perfect examples of applying the principles of sustainable development in Hong Kong. For more details, please see **Chapter 4.3** of the **Field Trip Guide**.

V. How Can You Be Involved?

We can all contribute to sustainable development in our daily lives. Every small action plays a crucial part in the course of sustainable development. For examples:

At Home:

- Reduce, reuse and recycle.
- Use biodegradable products.
- Bring your own bag.
- Reduce greatly the amount of packaging we use for various goods, and ensure that we dispose of our waste in an appropriate manner.
- Reduce the use of air-conditioners or other energy inefficient appliances.
- Use the stairs rather than the lift to go up and down a few floors.
- Enhance our personal development by pursuing opportunities for continuous education.
- Take a positive attitude towards life and our relationships with friends and family.

Remember to use recycle bins!
© Frank Lee

At Work:

- Cultivate plants in the office to help improve the indoor air quality and create a visually soothing environment.
- Consult colleagues, rather than write numerous memos or e-mail messages.
- Don't waste paper. Use **both sides**.
- Properly maintain office equipment so as to save both energy and money.
- Don't switch on office machinery or appliances unless they are in regular use.

At School:

- Carry out projects that will help develop an understanding of the meaning and importance of sustainable development.
- Take part in co-curricular activities, such as sports and social welfare programmes.
- Establish and then improve sustainability indicators for schools in terms of energy, water and other resources used.

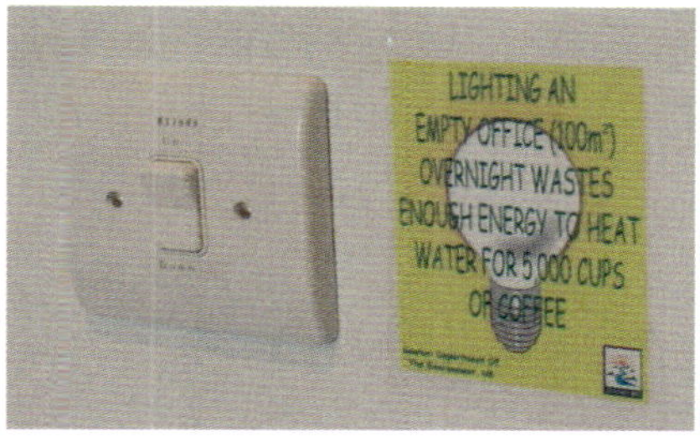

Show others how to conserve energy.
© Frank Lee

Within the Community:

- Spend more time getting to know your neighbours and take an active role in building your community. Be a helper in voluntary organizations.
- Walk or cycle rather than taking motorized transport for short trips.
- Stop car engines while waiting.
- Improve your skills and knowledge through continuous education and training.
- Enjoy and respect nature and countryside.

Spread the news and importance of sustainable development to others!!

Conclusion

To sum up, sustainable development balances social, economic, environmental and resource needs, both for present and future generations, simultaneously achieving a vibrant economy, social progress and a high quality environment, locally, nationally and internationally, through the efforts of the community and the Government.

© Frank Lee

The future development of Hong Kong should not depend on economic achievement alone at the expense of the environment. We must become more efficient and reduce our reliance on environmentally inappropriate forms of transport, power generation and consumption, wastewater discharge and solid waste disposal. Recycling industries and clean technology should be encouraged, and ecologically important sites be protected.

All in all, we should not observe the natural environment from the sideline, but instead we must live mutually with it. Our goal is to preserve and protect the Earth, so that we and our future generations can continue to enjoy permanent health and wealth.

第七章　可持續發展

讓我們先了解一些現況：

- 在過去三十年，我們消耗了地球資源的三分之一。
- 我們的臭氧層每年耗盡3%，很多人死於皮膚癌。根據研究，10%臭氧層的損失會導致更多人有皮膚癌，也導致1,600,000個失明的新例。
- 以現時的增長率，世界人口會在49年後會加倍。
- 世界上人口迅速增長對我們的生態系造成重大的壓力和導致全球貧窮。
- 在香港，每天我們平均扔掉2,800噸廢紙。每噸廢紙破壞17棵樹，所以我們每天砍倒48,000棵樹。
- 專家預計如果我們繼續以目前的速度砍伐森林，40年內我們便沒有雨林。我們每年失去超過50,000種的植物，昆蟲，鳥和動物。
- 在香港，如果我們繼續以目前的方式製造和處理廢物，所有的堆填區在10–15年內會被填滿。

（來源：Sustainable Development Unit, 2002）

重點是如果我們繼續以目前的生活方式，我們便沒有自然資源給下一代。因此，如果要供社會，經濟和自然環境給下一代，以享受更好質素的生活，我們就需要理解可持續發展的意思。

1. 什麼是可持續發展？

「可持續性」的概念在全世界越來越普及，成為很多媒體使用的名詞。這個概念涉及現在和未來，國家之間的主權，國家之間的公平，自然資源，生態系的承載量，環境保護和發展整合的需求。

可持續性源於生態學。這個名詞首先在漁業和森林學中出現，是指資源管理策略。策略容許運用資源的一部分，而不威脅資源的供應和資源亦能夠被補充。例如，在一個漁業區域，捕魚率比魚類的出生率低，就是可持續的生產。

自從二十世紀九十年代，「可持續的發展」成為環境教育的一個普遍使用的名詞。像美國，中國和澳洲那樣的國家，把它作為一個主要的概念，從生態學研究到社會，經濟和環境研究。大部分人認為持續學習和改變生活方式，以達致可持續發展。

在環境與發展世界委員會制訂的可持續發展的定義是：

> 「既能滿足我們現今的需求，又不損害子孫後代能滿足他們的需求的發展模式。」
>
> (World Commission on Economic Development, 1987)

簡單來說，當資源，經濟，社會和環境融合，就是可持續的社會發展。

可持續發展最初的概念是促進社會發展，同時要保護環境。在促進生產時，可持續發展著重保護和改善生態。資源應該是可持續的，為人類提供一段長時間，甚至永遠，避免耗盡，破壞生命和產量。

可持續發展的特徵

達到世代（世代以內的公平）和跨世代（世代之間的公平）的公平。它由三方面組成：

1. 社會的可持續性：減少貧窮，保持穩定的人口，給予被剝奪的團體平等的機會。
2. 經濟可持續性：充分就業和優化人類福利。
3. 環境的可持續性：在資源使用的可持續性，保育而協調經濟活動和環境。

II. 國際行動

國際委員會和機構，包括聯合國 (UN) 在過去三十年積極研究和出版聲明，原則，報告和建議，並召開會議討論關於創造可持續世界的共同框架。在 1980 年，聯合國向所有國家提出在自然，社會，生態系，經濟和運用自然資源之間的基本關係，以保持可持續的發展。

在1983年11月，聯合國成立環境與發展世界委員會（UNCED）。挪威的首相G.H. Brundtland主持委員會。聯合國要求委員會在「可持續的發展」的基礎上闡明「全球改變議程」。

在1987年，委員會向聯合國提交報告「我們共同的未來」。「可持續發展」的模型因此再踏出一步。在1992年，在巴西里約熱內盧，聯合國環境與發展會議（UNCED）實行廿一世紀議程。

可持續發展被視為現代知識的昇華，廿一世紀議程是一個完整的計劃，範圍包括全球，各國和受人類影響的環境。在聯合國環境與發展會議，超過178個政府採納議程，以及提倡8項原則：

- 經濟
- 自然資源
- 環境質素
- 生物多樣性
- 健康與衛生
- 流動性
- 康樂及文化活力
- 社會及基礎設施

© Frank Lee

世界可持續發展高峰會

在2002年8月，聯合國在南非約翰內斯堡舉行世界可持續發展高峰會。在高峰會議期間，各國同意聯合保護環境，因為一個國家的發展會影響其他國家的經濟，社會和環境。當資源有限，即使超級強國奪取其他國家的資源，它們也不能長期維持下去。在這個地球村，所有國家應該聚集一起，肩負一個可持續的未來的責任。

同時，世界高峰會提醒我們處境：十一億人沒有衛生的淡水使用，十七億人生活在缺水的地區和十三億人過著赤貧的生活（Johannesburg Summit, 2003）。

可持續發展的發展過程：

1962	Rachel Carson 發表*寂靜的春天*
1970	聯合國首次舉行**地球日**
1971	英國成立國際環境與發展組織
1972	在 瑞典的斯德哥爾摩舉行人類環境的會議
1983	聯合國成立環境與發展世界委員會
1987	聯合國出版布蘭特報告（我們共同的未來）
1990	加拿大成立國際可持續發展組織
1992	在巴西里約熱內盧舉行**環境與發展會議**
1994	在埃及開羅舉行人口與發展會議
1997	紐約的**地球高峰會II**及京都會議 / 協議
2002	南非約翰內斯堡舉行**世界可持續發展高峰會**

II. 可持續發展的重要性

「可持續發展」的概念提出不久後，得到全球的簽署。因為科學進步，對地球有更深刻的認識，使人類知道地球對維持生命的限制。簡單來說，關注的問題是「環境的承載量」，是資源能夠在一段時間內支持人口和經濟，並同時維持環境的穩定。

如果自然資源的總數儲備剩餘沒有變化或甚至增加，發展就算是可持續的。可持續性與環境，經濟，社會，文化和政治因素的相互關聯。從生態的角度，可持續性關注的主要是生物系統的穩定性。從全球的角度，癥結在於保持文化多元化。可持續性的文化把重點放在保持社會和文化制度，維持文化多元化，促進公平。

最後，可持續發展呼籲無私的行為。它鼓勵為我們和後代的自然環境，考慮我們行動的後果。布蘭特報告主張平衡經濟發展和環境保護，社會需要實施新一套的目標

- 改變增長的質素
- 滿足基本的工作，食物，能源，水和衛生需求
- 確保人口達到一個可持續的水平
- 保育和增加天然資源
- 調整新科技，以達到真正的需求
- 合併經濟和環境決策
- 保護必要的生態系功能
- 為後代謀求福利
- 整合經濟和環境政策
- 改善整體生活質素，不只是增加收入
- 終止貧窮，公平分發資源

新地球公民的概念有以下特性:

傳統概念	新「地球公民」概念
• 以人類為中心 • 人類是地球的主人 • 人類控制一切 • 環境問題只存在某些地區 • 地球的資源是無限的 • 生物能夠單獨生存 • 社會發展以經濟為先	• 以地球為中心 • 人類是只是地球的看護者 • 所有生物包括人類是同等的 • 環境問題是每個人的問題 • 地球的資源會耗盡的 • 生物是互相依賴的 • 社會發展要平衡環境，經濟和社會

IV. 香港的可持續發展

香港政府於1997年提出廿一世紀可持續發展（SUSDEV 21）的研究。由規劃署進行，研究目的是香港可持續性的定義是什麼，如何測量，以及政策和制度上的結構需要怎樣配合。廿一世紀可持續發展制訂一套指標，以測量香港發展的可持續性。

香港廿一世紀可持續發展指導原則：

經濟 • 香港應致力推動市場經濟的蓬勃發展，增強競爭力，從而提供充足的資源，以應付現今一代及子孫後代市民大眾的需求，並配合他們的意願。
健康與衛生 • 香港應提供適意的居住及工作環境，並制定政策以促進和保障市民的身心健康和安全。
自然資源 • 香港應鼓勵採用符合「可持續發展」原則的方式運用自然資源，透過改善資源消耗的效益、盡量減少使用不可再生的資源、善用可循環再造的廢物，以及從廢物中回收資源等方法，盡量減少對其他地區的生態影響。
社會及基礎設施 • 香港應致力建立一個穩定、公平、重視道德觀念及不斷進步的社會，並提供充足和合適的教育機會和社會基礎設施，令現今一代和子孫後代的個人潛力得以充分發揮，貢獻社會。
生物多樣化 • 維持香港的生物多樣化，盡量避免香港在耗用資源時會對其他地方的生物多樣化構成威脅。
文化活動 • 保障及提高市民參與康樂和消閒活動的機會，維持文化多元化，以及保存具考古、歷史及建築價值的文化財產，並促進其價值。
環境質素 • 香港應採取防患未然的措施，同時為現今一代和子孫後代設想，預防環境問題，並積極改善環境質素，盡量減低發展計劃對本港、國家及世界其他地方所產生的不必要副作用及影響發展效益的問題（例如空氣、噪音、水質或陸上污染等）。
交通運輸 • 香港應為市民提供安全、四通八達、高效率及低污染的運輸工具及行人設施，並提供一個高效率的運輸網，以促進貨運及客運服務。

(來源：Planning Department, 2001)

在廿一世紀可持續發展，香港對可持續發展的定義 (Planning Department, 2001)：

> 「在社會大眾和政府群策群力下，均衡滿足現今一代和子孫後代在社會、經濟、環境和資源方面的需要，從而令香港在本地、國家及國際層面上，同時達致經濟繁榮、社會進步及環境優美。」

行政長官在《一九九九年施政報告》中清楚闡述香港可持續發展的承諾，在施政報告，行政長官著重令香港成為可持續發展的世界級都會 (Mottershead, 2004)：

- 在追求經濟富裕、生活改善的同時，減少污染和浪費
- 在滿足我們自己各種需要與期望的同時，不損害子孫後代的福祉；以及
- 減少對鄰近區域造成環保負擔，協力保護共同擁有的資源。

自二零零一四月起，**持續發展組**（SDU）隸屬政務司司長辦公室轄下的行政署。持續發展組的主要工作是推動將可持續發展融入政府新的措施及計劃和社會，並在可持續發展委員會成立後向其提供支援。

可持續發展委員會在2003年4月1日成立。委員會向政府提供專業意見，建立更清晰的可持續發展概念，鼓勵社會實踐。它制訂策略，促進社會支持，加強可持續發展的公眾意識。要點如下：

- 就推動可持續展的優先處理項目，向政府提供意見；
- 就為香港籌劃一套融合經濟，社會及環境因素的可持續發展策略，提供意見；
- 增進大眾對可持續發展原則的認識和了解。

可持續的發展基金：為數一億元的可持續發展基金（基金）已經成立超過十年，以集中資助那些有助加深市民認識可持續發展概念及鼓勵市民在香港實踐可持續發展原則的措施。在第一輪的申請，127申請當中，只批准8個計劃書（775萬港元）。

委員會的成員由行政長官委任，成員包括環境，社會和商業領域中的專業人仕，以及高級政府官員。委員會的焦點是討論香港長期可持續性問題的一個論壇。幾個新的政策和措施包括 (Council for Sustainable Development, 2004)：

- 保護水資源；
- 保護和減少優質農地的損失；
- 保護生態敏感地區；
- 透過協調道路，橋梁和鐵路的建築，以及主要基礎設施，如：港口，機場和污水處理設施，以增加效率和減少不可更新資源的損失；
- 控制人口增長；
- 利用合適的政策，提高設施和人力資源的效率。

在香港，新機場和青馬大橋的發展是應用可持續發展原則的極佳例子。詳情請看**野外研習指引**第4.3章。

V. 你如何參與？

我們在日常生活可以為可持續發展出力。每個小小的行動在可持續發展的過程中扮演重要的角色。例如：

家居：

- 減少，再用和循環。
- 使用生物可降解的產品。
- 攜帶自己的購物袋。
- 減少貨物的包裝，並確保以合適的方式處理廢物。
- 減少使用空氣調節機或其他能源效率低的產品。
- 如果只是幾層數的話，上落可用樓梯。
- 持續進修，以加強我們的個人發展。
- 以積極態度面對生命及朋友和家庭的關係。

上班時：
- 在辦公室種植植物，幫助改善室內空氣質素，建立視覺上使人平靜的環境。
- 直接向同事查詢，避免使用備忘錄或電子郵件。
- 不浪費紙張，雙面使用。
- 適當保養辦公室設備，以節省能源和金錢。
- 不用時，關掉辦公室的機器或設備。

上課時：
- 撰寫報告書，幫助理解可持續的發展的意思和重要性。
- 參與課程活動，如運動和社會服務活動。
- 建立和改善能源，水質和資源運用的可持續性指標。

社區內：
- 花更多些時間認識你的鄰居和主動參與建設社區。在志願機構當義工。
- 短路程，可步行或騎腳踏車。
- 等侯時，關掉汽車引擎。
- 透過持續教育和訓練，改進你的技術和知識。
- 享受和尊重自然和郊野。

最重要的是要向其他人宣揚可持續發展的消息和重要性！

結論

總括以而，可持續發展為現今和後代平衡社會，經濟，環境和資源需要，同時透過社區和政府的努力，令香港在本地、國家及國際層面上，同時達致經濟繁榮、社會進步及環境優美。

香港的未來發展不應該只發展經濟，犧牲環境。我們必須提高效率，減少在運輸，發電，消費，污水排放和固體廢物的不適當處理方式。鼓勵再造工業及潔淨技術，保護生態上重要的地區。

總而言之，人類不應該從旁觀察自然環境，反之我們必須和諧共處。我們的目標是保護地球，那麼我們和後代都能夠繼續享受健康和財富。

Questions for Further Thinking 思考問題

1. Section IV has mentioned the eight SUSDEV 21 Guiding Principles for Hong Kong. Which one of them do you think is the most important? Justify your answer.
   第 IV 節提及八項香港廿一世紀可持續發展指導原則。你認為哪項最重要？試闡釋。

2. Apart from the recommendations for supporting sustainable development mentioned in Section V, do you have any other suggestions? How can you achieve?
   除了第 V 節提及支持可持續發展的建議外，你有其他建議嗎？你怎樣實行呢？

3. Globalization is a popular economics concept in modern society. Does globalization contribute to promote sustainable development? Or vice versa? Please explain.
   在現代化的社會，全球化是一個普遍的經濟概念。全球化是否在推動可持續發展？抑或反其道而行呢？試解釋。

References 參考資料

Council for Sustainable Development, HKSAR Government (2004), *Sustainable Development – What is Sustainable Development?* Retrieved January 10, 2005, from website: http://www.susdev.gov.hk/html/en/sd/index.htm

Economic and Social Commission for Asia and the Pacific (ESCAP) (1998), *Agenda 21 and Challenges for Asia and the Pacific*, New York: United Nations.

Economic and Social Commission for Asia and the Pacific (1999), *Energy, Environment and Sustainable Development III: Energy Infrastructure Policies and Issues*, New York: United Nations.

Johannesburg Summit (2003), *United Nations: World Summit on Sustainable Development.* Retrieved January 10, 2005, from website: http://www.johannesburgsummit.org/html/basic_info/basicinfo.html

Kirkby, J., O'Keefe, P. and Timberlake, L. (1995), *The Earthscan Reader in Sustainable Development*, London: Earthscan Publications.

Lo, F. and Marcotullio. P. (2001), *Globalization and the Sustainability of Cities in the Asia Pacific Region*, New York: United Nations University Press.

Mottershead, T. (2004), Sustainable Development in Hong Kong, Hong Kong: Hong Kong University Press, p.25–35.

Planning Department, HKSAR Government (2001), *Study on Sustainable Development in the 21st Century: Final Report*. Retrieved January 10, 2005, from website: http://www.info.gov.hk/planning/p_study/comp_s/susdev/final_hp/ch1.htm

Sustainable Development Unit, HKSAR Government (2002), *Sustainable Development*, Hong Kong: Government Printing.

Stokke, O. (1991), *Sustainable Development*, London: Frank Cass & Co. Ltd.

World Commission on Economic Development (1987), *Our Common Future (the Brundtland Report)*, Oxford University Press, p.43–46, 49.

Zhou, J.M. (2001), *Sustainable Development in Asia, America and Europe with Global Applications: A New Approach to Land Ownership*, Cheltenham, UK and Nothampton, MA: Edward Elgar.

Chapter 8. Ecotourism

A moment with nature
© Frank Lee

The travel and tourism industry is currently one of the largest industries in the world. According to the estimates of the World Tourism Organization (a United Nations-affiliated research and support group), the number of tourists worldwide was 650 million in the year of 2000 and will reach 1,000 million in 2010. The annual growth rate up to 2010 will be around 4%. Furthermore, the tourist industry employs more than 130 million people worldwide and makes annual capital investments in excess of HK$3 trillion (Holden, 2000).

Tourism is not only a powerful economic force but a major factor in the physical environment as well. The inter-relationships of all earth systems and human systems have become significant over the past few decades. Apart from the benefits, we must understand the negative effects of tourism, especially its environmental impact.

Today nearly 70% of international tourists travel for vacation, leisure, and recreation, as opposed to visiting friends and relatives or traveling for other purposes. But the tastes of tourists are changing. There is an increase of more flexible and independent travelers who pursue more personalized experiences like exploring culture and nature.

It has been recognized that tourism must preserve and protect the environment and natural attractions so that people will continue to travel and the sites can be sustainable. Gradually, many non-governmental organizations (NGOs) and interested parties have promoted activities called ***ecotourism***.

I. What is Ecotourism?

One of the earliest and most quoted definitions of ecotourism was put forward by Hector Ceballos-Lascurian, a Mexican architect who set up a travel agency called "Ecotours" in 1983, serving people interested in nature and Mexican culture. The term was defined by him four years later as:

> *"... tourism that consists in traveling to relatively undisturbed or uncontaminated natural areas with the specific object of studying, admiring and enjoying the scenery and its wild plants and animals, as well as any existing cultural manifestations (both past and present) found in these areas." (Ceballos-Lascurain, 1987)*

Ecotourism is responsible travel to natural areas promoting conservation awareness of natural environment, wildlife and cultural heritage, and improving the welfare of local people at the same time (Ecotourism Society, 1991). As one of the fastest growing industries worldwide, it also includes landscape appreciation, wildlife watching and outdoor adventures.

Furthermore, ecotourism involves more than conservation. It is a form of travel that responds to a region's ecological, social, and economic needs. It enriches the understanding of the site's culture and natural history, fosters the ecosystem's integrity, and produces economic opportunities and conservation gains.

Ecotours promote conservation
© Frank Lee

In terms of benefits and importance, ecotourism (Wearing and Neil, 1999):

1. Provides jobs and income for local people.
2. Makes possible funds for purchasing and improving protected or natural areas to attract more ecotourists in the future.
3. Provides environmental education for visitors.
4. Encourages heritage and environmental preservation and enhancement (creating new or enlarged country parks, forest preserves, biosphere preserves, recreation areas, beaches, marine and underwater trails, and attractions).

Outdoor classroom for everyone
© Frank Lee

The **International Ecotourism Society** estimates that this form of travel is growing by 20% annually (compared with 7% for tourism overall) and that it generated about US$154 billion in receipts in 2000. (TIES, 2000)

According to Holden (2000), the **key principles** of ecotourism include:

1. It should not degrade natural resources and should be developed in an environmentally sound manner.
2. It should provide long-term benefits to the resources, to the local community and industry (benefits may be conservation, scientific, social, cultural, or economic).
3. It should provide first-hand, interactive and enlightening experiences.
4. It should involve education among all parties - local communities, government, NGOs, industry and tourists (before, during and after the trip).
5. It should encourage recognition of the intrinsic values of the resources.
6. It should involve acceptance of the resource on its own terms, and in recognition of its limits, which involves supply-oriented management.
7. It should promote understanding and involve partnerships among the government, NGOs, industry, scientists and locals (both before and during operations).
8. It should promote moral and ethical responsibilities and behavior towards the natural and cultural environment by all participants.

Ecotourism vs. Traditional Tourism (Weaver, 2001):

	Ecotourism	Traditional Tourism
Purpose	Educational and environmentally responsible tour	Sightseeing, shopping, dining based vacation
Destination	Nature, history and culture based destinations	No specific destinations
Rules	Minimize disturbance to environment and the locals	No specific rules
Guide's Role	Educational interpretation of nature and local culture and promotion of environmental protection	Introduction of fun and interesting areas to the tourists, may also introduce culture and custom
Participants' Role	Be ready before the trip, ask questions and learn actively	Enjoy the journey passively
Outcomes	Benefit society, economy and the environment, achieve sustainable development	Economic and social gains but not always sustainable

Ecotourism involves a combination of environmental, social and economic factors. Broadly speaking, those who participate in ecotourism activities can be divided into four groups: the business sector, ecotourists, the government sector and non-government organizations.

Business sector
Within the travel and tourism industry, movements are taking place in response to the growing interest by consumers and travelers in protecting the environment. The "product" must be offered with a sense of mission. Ecotourism suppliers must take responsibility for standards of education and training because clients are looking for quality experiences.

Ecotourists are a new challenge to both tour operators and policy-makers. They demand experiences different from mass-tourists and have more interest in participation and self-improvement in natural areas. How to arrange programs which both satisfy ecotourists and conservation and how to educate them are matters that have to be tackled.

Capturing wonders of nature © Frank Lee

Research has revealed that ecotourists are mostly individuals (Fennel, 2003):

- with advanced education
- more interested in traveling to different cultural, historical and environmental sites
- amicable and broad-minded, intelligent, self-assured and sociable
- or members of conservation organizations.

Above all, they seem to be much more concerned with seeking both personal and interpersonal rewards and opportunities rather than with escaping from personal and interpersonal environments.

Government sector

The government is actively involved in the administration, legislation and the financing of national parks, nature reserves and so on (Wearing and Neil, 1999). Some governments have established codes in order to monitor ecotourism. Ecotourism codes have been developed in the United States, Canada and Australia, providing both general and specific guidelines for the ecotourism administration body, tour operators and ecotourists. A number of policies, codes and principles oriented towards natural or cultural resources have been introduced by government bodies, associations and industries. Tourism codes tend to focus on the biophysical or cultural resources and regulate the conduct of tourists. Codes of ethics based on values are mixed with codes of practice or specific guidelines.

Ecotourism codes published by AFCD ➔

Non-government organizations

With increasing concern about environmental conservation, many recreational, historic, professional, ethnic, health and religious organizations are becoming involved in resource management, ecotourism research, policy-making and safeguarding of the environment (Fennell, 2003). This leads to the interests of local communities and the long-term development of ecotourism, sometimes working jointly with policy-makers and communities in dealing with ecotourism problems arising from business operations or from visitors.

II Development of Ecotourism

The concept of ecotourism has only been established for just over two decades. A brief summary on its worldwide development can be seen as follows:

Year	*Event*
1983	First attempt to use the word "ecotourism" by Mexican Hector Ceballos-Lascurain, boosting the local rural economy, creating new jobs and preserving the ecology.
1992	*Agenda 21*: though tourism is not a specific agenda item at the Rio Earth Summit, the conference report briefly mentions it in chapters on mountain and coastal ecosystems.
1995	Charter on Sustainable Tourism: delegates at the World Conference on Sustainable Tourism in Canary Islands agree on 18 priority points for sustainable tourism development.
1996	*Agenda 21 for the Travel and Tourism Industry*: backed by the World Tourism Organization, WTTC, and Earth Council, the report outlines key measures that governments and the tourism industry can take toward sustainability.
1997	Malé Declaration on Sustainable Tourism: tourism and environment ministers from Asia and the Pacific agree on general guidelines for the sustainable development of tourism.
1999	UN Commission on Sustainable Development's Working Programme for Sustainable Tourism: the first action-oriented work program on sustainable tourism recognizes the need for sustainable tourism development that safeguards local culture and the environment.

2000	UNEP's Principles for Implementation of Sustainable Tourism: a set of guidelines to help governments, industry, and other stakeholders - including UNEP conventions that address tourism issues - to apply the general concept of sustainable tourism in practice and to minimize environmental impacts from tourism.
2002	United Nations **International Year of Ecotourism**: Governments, the tourism industry, and other stakeholders review and share their experiences with ecotourism at the World Ecotourism Summit in May.

The World Travel and Tourism Council designed an ***Agenda 21 for the Travel and Tourism Industry*** in 1996. This specific ***Agenda 21*** mentions a number of points relating to the adaptation of tourism to the environment. If a tourism company adopts and follows the points given below, it will be included in the World Travel and Tourism Council's environmental classification in accordance with the so-called "Green Globe". ***Agenda 21*** for the travel and tourism industry includes the following points:

1. Waste - minimize, re-use and recycle.
2. Energy - save and use effectively, reduce emissions.
3. Fresh water resources - safeguard quality, avoid waste.
4. Drainage water - purify and reduce effluent.
5. Replace environmentally dangerous substances.
6. Transport - avoid harmful pollution and other impacts on the environment.
7. Use of land - plan, look after, prevent the deterioration of the landscape.
8. Involves employees, customers and local receivers.
9. Design products and techniques so they are more efficient, less polluting, locally appropriate and globally available.
10. Cooperation - with the local communities, tourists, small enterprises, sectoral bodies, the local councils, the government.

The United Nations designated the year of 2002 as the ***International Year of Ecotourism*** (Page and Dowling, 2002). A series of preparation activities and regional conferences were carried out. The topics of the conferences varied from ecotourism management in different ecosystems (such as mountain areas) to ecotourism as a tool of sustainable development. The results of the preparatory conferences were presented at the World Ecotourism Summit in Quebec, Canada during 19-22 May 2002.

III. Impacts of Ecotourism

Ecotourism is closely related to the concept of sustainable development (see Chapter 7), which considers three fundamental and equal objectives, namely economics, environment and society. As for the development of ecotourism, there are different types of positive and negative impacts on the society, economy and environment. They include:

Excessive dolphin watching: wildlife appreciation or disturbance?
© Frank Lee

	Positive Impact	Negative Impact
Society	• Social mix • Mix of cultures • Evidence of cultures • Support of cultures, and crafts	• Visitors outnumber residents • Conflicts • Congestion • Irresponsible visitor behavior • Commercialization of cultures • Crime
Economy	• Money to local economy • Job and local wealth creation • Improved service levels • Additional income source	• Poor job spectrum • Exports of profits • Demand for infrastructure • Neglect of other economic opportunities
Environment	• Environmental improvement schemes • Renovation and rehabilitation • Income for conservation	• Congestion and visitor over-use; • Vandalism • Threat to flora and fauna habitats • Visual pollution • Uncertainty

(extracted from Fennel, 2003)

Environmental impacts of ecotourism
The "been-there-done-that" style of travel causes a greater negative impact on the environment than its economic gains. With better transportation and communication, there are virtually no longer remote or unreachable places. Large number of visitors brings large-scale construction of infrastructure, accommodation and entertainment facilities. Uncontrolled development will damage the natural environment through pollution, noise and garbage to the area.

Carrying capacity for visitors must be determined. It is defined as the maximum number of carry visitors that the area can receive without damaging its attractive features (Fennel, 2002). Enforcement of this limit, along with good management and maintenance, is essential.

Generally speaking, the impacts of ecotourism on the environment include:

- Flora
  - Physical damage through removing samples, careless use of fires, logging by fires, clearing for camping, dumping of waste, or vehicle use;
  - Loss of fragile species;
  - Disrupting the soil texture;
  - Reducing vegetation cover and species diversity
- Water quality
  - As a result of waste disposal, swimming, boating, and oil spills (from small boats);
  - Contamination of fish and reef life;
  - Eutrification.
- Air quality
  - Carbon monoxide emission from vehicles and motor boats;
  - Exacerbation of local air quality problems

Flower © Zoe Lee

Pollutants emitted from vehicles © Frank Lee

4. Wildlife fauna

- Determined by the resilience of the species, the intensity and frequency of the visitor attention;
- Disturbance of the habitat (animals may vacate the site since their feeding and breeding behaviors are interrupted);
- Depended on visitor-related resources;
- Disrupted migration paths.

Chinese White Dolphin: victim of ecotour? © Frank Lee

5. Natural Landscape

- Disturbed by collecting rock samples as souvenirs, vandalism, tracks and mountain-climbing;
- Sites affected by recontouring to accommodate particular tourism facilities or panoramas, enclaves created on sites as viewing areas, or creation of access to tourism lodges.

In a nutshell, ecotourism does not work when ecotours are so popular that they actually destroy the very environment they seek to protect!!

IV. Ecotourism in Hong Kong and China

Facts of Hong Kong (Hong Kong Facts Sheet, 2003):

- Area of Hong Kong is about 1,100 km^2. About 70% is countryside.
- The tourism industry is one of the most important foreign exchange earners in Hong Kong. In 2002, tourism brought in HK$62.2 billion for Hong Kong.
- In 2002, visitors from the mainland accounts for 41% of all visitors coming to Hong Kong or 20% increase to a record of HK$16.6 million.

High ratio of nature conservation areas and rich species diversity

Mai Po Nature Reserve
© Frank Lee

Over 70% of Hong Kong's total area is some kind of conservation land, and such a ratio is among the highest in world cities. 23 country parks, located on Hong Kong Island, Lantau Island, Kowloon and in the New Territories, comprise over 40% of land area. The four marine parks and one marine reserve contribute to marine conservation in Hong Kong waters. The 1,500 hectares of Mai Po Nature Reserve is an internationally recognized important wetland. In addition, many special areas with ecological or geological values have been designated as **Sites of Special Scientific Interest (SSSI)** since 1975.

Government's role and support

The government of Hong Kong has made effort on protecting ecological resources through legislation, including Forest and Countryside Ordinance, Wild Animals Protection Ordinance, Marine Parks Ordinance (1995) and Country Parks Ordinance (1975). Various recreational and educational facilities in the protected areas greatly facilitate visitors in conducting different nature-based activities. Recently, the government and many interest groups have published a number of wildlife field guides and hiking books providing useful information to the public.

Unique cultural heritage and historical background

Ingredients of cultural heritage such as different kinds of buildings, vibrant festivals and traditional culture, and food, handicrafts and accessories are all closely related to our history and crucial in developing ecotourism. This learning-based ecotourism will make the tour more meaningful than simply appreciating the beauty of the landscape.

In addition, Hong Kong's geographical location, the favorable climate with its surprising biodiversity and prosperous economic development all contribute to the huge potential in developing ecotourism in the city.

Examples of ecotourism programmes in Hong Kong:

- Sea – coral watching and dolphin watching at Marine Parks
- Land – guided natural walks at Mai Po Nature Reserve and the Wetland Park
- Air – bird watching at Kowloon Park, butterfly watching at Fung Yuen, Tai Po

A local ecotour in Tung Ping Chau ➔

© Frank Lee

On the other hand, there are still challenges and constraints facing ecotourism development in Hong Kong. Some of them include:

1. Loopholes in existing legislation for protecting our wildlife resources and their habitats;
2. Pollution from Hong Kong itself and South China;
3. Conflicts among various government department towards environmental protection and ecotourism development;
4. Lack of well-trained ecotourist guides;
5. Conflicting views among different stakeholders; and
6. Rapid development of infrastructure and urbanization.

Ecotourism is not a new concept in China. With extraordinary landscapes and species richness, China has put huge effort on developing ecotourism as one of its major income in recent decades. It is generally accepted that ecotourism implies nature/culture-based tourist activities. Although it is difficult to estimate the importance of ecotourism in China, it may be reasonable to guess that a higher proportion of ecotourism exists in the inland areas than coastal China because more nature and cultural based tourism takes place in inland China (Weaver, 2001).

Beautiful scenery in China © Frank Lee

In March 1994, China approved and promulgated *China Agenda 21: White Paper on China's Population, Environment and Development in the 21st Century*, a document that promotes "green tourism" and uses ecotourism as a means to make the conservation of its biodiversity financially more attractive. China has accepted the basic concept that ecotourism can preserve the integrity of ecosystems while providing economic opportunities which make conservation of natural resources financially beneficial to local communities.

The **National Tourism Administration of China** designated 1999 as the year of "Ecotour China". Abundant ecotourism resources are located mainly in the inland areas of China. Currently more than 200 nature reserves have been established, covering an area of 72 million hectares, or 7.2% of China's territory. Fourteen marine nature reserves with national status have been declared, along with 710 forest parks, 60 large botanical gardens and 512 historic and scenic sites, and further protected areas have been established more recently (China National Tourism Administration, 2003).

Conclusion

Ecotourism involves travel to relatively well-conserved natural areas. It takes particular care of the environment and promotes conservation, aiming to provide non-consumptive experience and to educate participants. Ecotourism undoubtedly has the power to bring about both favorable and unfavorable changes to its resources.

In short, ecotourism should:

- provide in-depth learning to participants;
- respect for local culture, traditions and environment;
- be sensitive to the environment; and
- provide direct financial support for the local people and environment in need.

Journey in the nature is fun!
© Frank Lee

Last but not least, here are some recommendations for sustaining ecotourism in Hong Kong:

1. Comprehensive ecotourism policy
2. Development of accreditation system
3. Marketing of ecotourism
4. Follow Environmental Impact Assessment
5. Cooperation with the Pearl River Delta
6. Balance development and conservation
7. Aware of the *TRAGEDY OF THE COMMONS*
8. THE EARLIER THE BETTER

第八章　生態旅遊

旅遊業目前是世界其中一個最大的工業。根據世界旅遊組織（一個附屬聯合國的研究和支持團體）的估計，在2000年，全世界的遊客數目是六億五千萬，2010年，將達到十億。每年增長大約4%。此外，旅遊業在世界各地雇用超過一億三千萬人，每年投資超過三萬億港元（Holden, 2000)。

旅遊不單是一種強大的經濟力量，而且是自然環境中的一個主要的因素。在過去幾十年，地球和人類之間的關係值得關注。除了得到利益外，我們必須理解旅遊的不良影響，尤其是對環境的影響。

今天將近70% 的遊客主要是度假，閒暇和消遣，反而拜訪親友或其他目的較少。不過，遊客的品味正在改變。以探索文化和自然作個人體驗的遊客不斷增加。

旅遊必須保護環境和自然，令遊客繼續旅遊。逐漸地，很多非政府機構和相關組織促進了生態旅遊。

1. 生態旅遊是甚麼？

生態旅遊的其中一個和經常被引用的定義是由Hector Ceballos-Lascurian提出，他是一名墨西哥建築師，於1983年設立一間名為「Ecotours」旅行社，令人對自然和墨西哥文化產生興趣。四年後，他定義如下：

> *「…遊覽相對不受干擾或不被污染的自然地區，研究，欣賞，享受景色和野生動植物，及文化（過去和現在）。」* (Ceballos-Lascurain, 1987)

生態旅遊遊覽自然環境，以促進自然環境，野生動物和文化遺產的保育意識，同時改善當地人的福祉 (Ecotourism Society, 1991)。作為世界成長最快的工業之一，它也包括欣賞風景，觀看野生動物和戶外冒險。

此外，生態旅遊不單是保育，它是一種回應生態，社會，經濟需要的旅遊方式。以豐富對景點文化和自然歷史的認識，促進生態系的完整性，提供經濟機會和保育利益。

以利益和重要性而言，生態旅遊

1. 提供當地人工作和收入。
2. 成立基金，購買和改善受保護或自然地區，以吸引更多生態旅遊者。
3. 為遊客提供環境教育。
4. 鼓勵保存遺跡和環境（建立新的或擴建國家公園，保護森林，保護生物圈，消遣地區，海灘，海洋和水底行跡，和吸引景點）。

國際生態旅遊會估計這種旅遊方式每年增長20%（佔整體旅遊業7%）。在 2000 年，它帶來大約一千五百四十億美元的收益。(TIES, 2000)

生態旅遊的8項原則 (Holden, 2000)：

1. 它不可以破壞天然資源，應該以一種符合環保的方式開發。
2. 它應該為資源，本地社群和工業提供長遠好處（得益可以是保育，科學，社會，文化或經濟）。
3. 它應該提供即時，互動和有啟發性的體驗。
4. 它應該教育相關的團體，包括本地社群，政府，非政府機構，工業和遊客（在旅行之前，期間和之後）。
5. 它應該鼓勵認識資源本質的價值。
6. 它應該包括認識資源和其限制，及供應導向的管理。
7. 它應該促進政府，非政府機構，工業，科學家和當地人中之間的夥伴關係（在運作前和期間）。
8. 它應該促進參加者對自然和文化環境的道德及倫理責任和行為。

生態旅遊與傳統旅遊 (Weaver, 2001)：

	生態旅遊	傳統旅遊
目的	具教育性及環境責任的旅遊	觀光，購物，飲食為主
地點	地點的自然，歷史及文化	沒有特定的地點
守則	減少對環境和當地的影響	沒有特定的守則
導遊的角色	介紹自然及當地文化，以及推廣環境保護	介紹有趣的景點，可能介紹一些文化和習俗
參加者的角色	旅遊前有充分準備，主動發問及學習	較被動去參與
結果	社會，經濟及環境獲益，達致可持續發展	經濟及社會獲益，但沒有可持續發展

生態旅遊是環境，社會和經濟因素的組合。廣義來說，那些參加生態旅遊可分為四類：商業，生態旅遊者，政府和非政府機構。

商業
在旅遊行業內，透過實際行動去回應消費者和旅客在保護環境的興趣。「產品」要提供一種使命感。生態旅遊主辦者必須對教育和訓練的標準作出承擔，因為客戶需要有質素的體驗。

生態旅遊者對於旅遊業者和政策制訂者是一個新的挑戰。他們要求的體驗有別於一般遊客，著重於對自然的參與和自我改善。如何安排計劃才可以滿足生態旅遊者和保育，以及如何教育他們。這些問題是要面對的。

研究顯示生態旅遊者大部分是 (Fennel, 2003)：
- 有高等教育
- 有興趣前往不同文化，歷史和環境的景點
- 友善和寬宏大量，聰明，有自信和善交際
- 或者是保育機構的成員

此外，他們尋求個人和社交的機會，而不是逃避個人和社會。

政府

政府積極參與國家公園和自然保護區的管理，法規和財務。一些政府建立一些法則去監管生態旅遊。在美國，加拿大和澳洲均發展了生態旅遊法則，提供一般和專門的指引給環生態旅遊管理機構，旅遊業者和生態旅遊者。政府機構，協會和工業提出了很多關於自然或文化的政策，法則和原則。旅遊法則著重於生物物理學或文化的資源，協調遊客的行為。以價值為依據的倫理法則加入行業的規則或指引。

非政府機構

由於環境保護日漸受到關注 (Fennell, 2003)，很多消遣性，歷史性，專業人士，種族，健康和宗教組織都涉及資源管理，生態旅遊研究，決策和捍衛環境，保障當地社群的利益和生態旅遊的長期的發展，有時透過政策制訂者和社會與商務運作一起共同處理生態旅遊的問題。

II. 生態旅遊的發展

生態旅遊的概念建立了僅僅超過二十年。全球發展的概要如下：

年份	事件
1983	Mexican Hector Ceballos-Lascurain 首先使用「生態旅遊」一詞，提高本地效區經濟，創造就業機會，保護生態。
1992	議程 *21*：雖然旅遊不是地球峰會的一個專門的議程項目，會議簡要報告關於山和海岸生態系統。
1995	在可持續旅遊的章程：在金絲雀島舉行的世界可持續旅遊會議達成十八項首要可持續的遊覽發展的協議。
1996	旅遊業的議程 *21*：得到世界旅遊組織 (WTTC) 和 Earth Council 支持，報告略述那些政府和旅遊行業邁向可持續性的措施。
1997	可持續旅遊的男性宣言：來自亞洲和太平洋的旅遊業和環境部長為旅遊的可持續發展的指引達成協議。
1999	為可持續旅遊而設的可持續發展工作計劃的聯合國委員會：在可持續旅遊的第一個行動導向的工作計劃確認需要可持續旅遊發展，以保護當地文化和環境。
2000	為履行可持續旅遊而設的 UNEP 原則：一系列的指引，幫助政府，工業和其他相關利益者包括 UNEP 會議關注的旅遊問題，實際應用可持續旅遊一般概念，將旅遊對環境影響降到最低。
2002	聯合國國際生態旅遊年：在五月的世界生態旅遊高峰會，政府，旅遊業，和其他相關利益者審核和分享他們的生態旅遊經驗。

世界旅遊業理事會於1996年為旅遊業設計一個議程21。這個議程21提及很多關於旅遊業怎樣適應環境。如果一家旅遊公司采納和跟隨以下幾點，它會被世界旅遊理事會分類為「綠色地球」。議程21用於旅遊行業的包括以下幾點：

1. 浪費 — 減至最小，再用和再循環。
2. 能源 — 節省和有效使用，減少排放。
3. 淡水資源 — 確保質量，避免浪費。
4. 排水 — 純淨和減少污水。
5. 取代對環境危險的物質。

6. 運送 — 避免有害的污染和對環境造成影響。
7. 使用土地 — 計劃，照料，防止地貌退化。
8. 涉及雇員，顧客和當地居民。
9. 設計產品和技術，因此提高效率，更少污染，適用於本地和全世界。
10. 合作 — 跟本地社群，遊客，小企業，行業團體，本地議會，政府。

聯合國指定 2002 年為國際生態旅遊年 (Page and Dowling, 2002)。一系列活動和地方區會議。會議的主題包括由不同的生態系中的生態旅遊管理（例如山區）至作為可持續的發展的一件工具的生態旅遊。預備會議的結果於2002年5月19–22日加拿大魁北克世界生態旅遊高峰會期間公佈。

III. 生態旅遊的影響

生態旅遊與可持續發展的概念是密切相關的（看第七章），都是考慮三基本和相同的目標，即是經濟，環境和社會。至於生態旅遊發展，在社會，經濟和環境上有不同的正面和負面影響：

	正面影響	負面影響
社會	• 社會融合 • 文化融合 • 文化的證據 • 支持文化	• 訪客數目超過原居民 • 衝突 • 擁塞 • 不負責任的訪客行為 • 文化的商業化及罪行
經濟	• 促進當地經濟 • 創造當地工作和財富 • 改善服務水平 • 額外收入來源	• 粗略的工作範圍 • 出口利潤 • 對基礎設施的需求 • 忽略其他經濟機會
環境	• 環境改善計劃 • 整修和復原 • 用於保育的收入	• 擁塞和訪客過度使用 • 破壞的行為 • 威脅植物和動物棲息地 • 視覺污染

生態旅遊的環境影響

傳統的旅遊風格對環境造成重大的負面影響多過其經濟收益。由於有更好的運輸和通訊，不會再有遙遠或不能達到的地方。大量訪客帶來大型的基礎設施，住宿和娛樂設施。不受控制的發展會帶來污染，噪音和垃圾，破壞自然環境。

決定訪客的容納量。它定義為地區能夠接待日常訪客的最大數目而不損壞其有吸引力的特徵 (Fennel, 2002)。利用良好的管理和維護，強制執行這個限制，是必要的。

一般而言，生態旅遊對環境的影響包括：

植物

- 透過採集樣本，不小心使用火種，用火伐木，作為露營地區，廢物排放，或使用車輛；
- 損失脆弱物種類；
- 擾亂土壤結構；
- 減少草木的覆蓋和物種的差異。

水質
- 廢物處理，游泳，划船，以及石油洩漏（從小的船）；
- 污染魚類和珊瑚礁。

空氣質素
- 車輛和馬達船排放的一氧化碳；
- 空氣質量問題惡化。

野生動物
- 取決於物種受訪客注意的程度；
- 棲息地受到干擾（動物會離開該處因為他們的飲食和繁殖行為被打斷）；
- 依靠與訪客相關的資源；
- 擾亂遷移路徑。

自然地貌
- 收集岩石作為紀念品，故意破壞，足跡和攀石都會干擾自然地貌；
- 興建遊覽設施，觀景區，或住宿地方，對景點造成影響。

總括而言，當生態遊客太多時，破壞需要保護的環境時，生態旅遊就不能運作！

IV. 香港和中國的生態旅遊

香港實況 (Hong Kong Facts Sheet, 2003)：
- 香港的面積大約1,100平方公里。大約70%是郊區。
- 旅遊業是香港的其中一個最重要的外匯收入。在2002年，旅遊業為香港帶來六百二十二億港元。
- 在2002年，來自中國的遊客佔訪港遊客41%，有20%的增長，帶來一千六百六十萬港元。

高比例的自然保護區的面積及物種多樣性

香港的總面積超過70%是受保護土地，這個比例在全球最高。位於香港島，大嶼山，九龍和新界的23個郊野公園，佔土地面積超過40%。四個海岸公園和一個海岸保護保存香港水域資源。1,500公頃的米埔自然保護區是國際認可重要的濕地。此外，自從1976年，有生態或地質價值的很多特別地區被劃分為具特殊科學價值地點(SSSI)。

政府的角色和支持

香港政府透過立法盡力保護生態資源，包括林區及郊區條例，野生動物保護條例，海岸公園條例（1995年）和郊野公園條例（1975年）。保護區的康樂和教育設施促進訪客進行以自然為主的活動。最近，政府和很多團體出版了野生動物戶外考察和遠足書籍，提供公眾有用資訊。

獨特的文化遺產和歷史背景

文化遺產，如不同的建築物，充滿活力的節日，傳統的文化，食物和手工藝品與我們的歷史有關，對發展生態旅遊是很重要的。相對只是欣賞美麗風景的旅遊，這個以學習為主的生態旅遊更有意義。

此外，香港的地理環境，合適的氣候，生物多樣性和繁榮經濟發展，對發展生態旅遊方面有很大的潛力。

香港生態旅遊的例子：
- 海 — 在海岸公園觀賞海豚和珊瑚
- 陸 — 在米埔自然保護區和濕地公園的導賞
- 空 — 在九龍公園觀鳥，在大埔鳳園賞蝶

另一方面，香港的生態旅遊發展面臨挑戰和約束。其中一些包括：
1. 保護野生動物資源和它們棲息地的現存法例的漏洞；
2. 來自香港和南中國的污染；
3. 政府部門對環境保護和生態旅遊發展的衝突；
4. 缺少受到良好訓練的生態導遊；
5. 不同相關利益者的衝突；以及
6. 基礎設施和城市化的迅速發展。

生態旅遊在中國不是一個新的概念。由於有特別的風景和豐富的物種，中國在近幾十年來大力發展生態旅遊為主要的收入。生態旅遊意味著以自然／文化為主的旅遊活動。儘管很難估計生態旅遊在中國的重要性，不過生態旅遊在內陸地區較沿海地區多，原因是內陸有更多自然和文化為主的旅遊 (Weaver, 2001)。

在1994年3月，中國批准和公佈中國議程21：廿一世紀的人口，環境和發展的白皮書，以促進「綠色旅遊」和利用生態旅保存生物多樣性。中國接受了生態旅遊能夠保護生態系的完整性，以及在財政上提供機會保存自然資源和當地社群。

中國國家旅遊局指定1999年為「中國生態旅遊年」。中國的內陸的地區有豐富的生態旅遊資源。目前成立了超過200個自然保護區，覆蓋七千二百萬公頃，佔中國的領土的7.2%。有全國有14個海洋自然保護區，710個森林公園，60個大型植物公園和512個歷史和景色優美的景點，以及在不久前建立了更被保護的地區 (China National Tourism Administration, 2003)。

結論

生態旅遊涉及遊覽相對保存較好的自然地區。這需要保護環境和促進保育，以提供非消費性的體驗，以及教育參加者。生態旅遊無疑有能力對資源造成有利和不利的變化。

簡言之，生態旅遊應該：
- 提供給參加者一些深入的知識；
- 尊重當地文化，傳統和環境；
- 對環境敏感；以及
- 提供當地人直接的財務支持和環境需要。

最後，這裡有一些香港生態旅遊可持續發展的建議：

1 完整的生態旅遊政策
2 發展任命制
3 生態旅遊的市場
4 跟隨環境影響評估
5 與珠江三角州合作
6 平衡發展和保育
7 意識到共有的悲劇 *(Tragedy of the Commons)*
8 越早越好

Questions for Further Thinking 思考問題

1. Suggest the qualifications of an ecotour guide/interpreter. How is s/he different from a traditional tour guide?
建議一個生態導遊/解說者的資格。他/她與傳統的導遊怎樣不同？

2. Name some of the famous ecotour sites and their features in Hong Kong.
試列一些香港著名的生態旅遊景點和它們的特色。

	Name of Site 景點名稱	Features 特色	Have you been there? 你曾否到過?
1)			Y / N
2)			Y / N
3)			Y / N

References 參考資料

Cater, E. and Lowman, G. (1994), *Ecotourism: a Sustainable Option?* John Wiley & Sons, Chichester, England.

Ceballos-Lascurain, H. (1987), "The Future of Ecotourism", *Mexico Journal*, January, 13–14.

The Ecotourism Society (1991), *The Ecotourism Society Newsletter*, No. 1, Spring.

Fennell, D. (2002), *Ecotourism Programme Planning*, CABI Publishing, New York, NY, p. 44–47.

Fennell, D. (2003), *Ecotourism: an Introduction (2nd edition)*, Routledge, New York, NY, p 35–38, 61–67, 98.

Hong Kong Facts Sheet (2003), *Hong Kong: The Facts: Tourism*. Retrieved January 15, 2004, from website: http://www.info.gov.hk/hkfacts/tourism.pdf

Holden, A. (2000), *Environment and Tourism*, Routledge, London, p.19–22, 195.

The International Ecotourism Society (TIES) (2000), *TIES World Wide Statistical Fact Sheet 2000*. Retrieved January 15, 2005 from TIES website: http://www.ecotourism.org/research/stats/files/stats.txt

Page, S. and Dowling, R. (2002), *Ecotourism*, Pearson Education, Edinburgh Gate, Harlow, p.24–25, 57, 281.

Wearing, S. and Neil, J. (1999), *Ecotourism Impacts, Potentials and Possibilities*, Elsevier Science, Burlington, MA, p.8–9, 24–28

Weaver, D. (2001), *The Encyclopedia of Ecotourism*, CABI Publishing, New York, NY, p.73–82, 123–129.

China National Tourism Administration (2003), *Fact and Figure*. Retrieved January 15, 2003, from CNTA website: http://www.cnta.com/lyen/2fact/index.htm

Advanced Topics
進階題目

Persistent Toxic Substances
持續性有毒物質

Genetically Modified Food
基因改造食物

Solid Waste Management
固體廢物處理

Sewage Treatment
污水處理

E-waste
電子廢物

Environmental Impact Assessment
環境影響評估

Environmental Management System
環境管理系統

Chapter 9. Persistent Toxic Substances

Pollution has been defined as "the adverse effect on the natural environment, including human, animal or plant life, of a harmful substance that does not occur naturally or the concentration to harmful levels of a naturally occurring substance" (Koren and Bisesi, 1996). Usually it is caused by the introduction into a system of materials in excess of normal amounts or concentrations. The level of pollutants exceeds the tolerance limits of the system, and thus leads to physical, chemical and biological changes in the ecosystem.

A group of chemicals known as **persistent toxic substances** is now receiving worldwide attention. These chemicals are toxic, persistent in the environment, bioaccumulate in fatty tissues, biomagnify through the food web and have a tendency for long-range transport (Walker and Livingstone, 1992). They can produce long-term contamination of our environment and they can cause acute and chronic toxicity to humans due to their endocrine disrupting capabilities.

I. What are Persistent Toxic Substances (PTS)?

PTS are natural/anthropogenic organic chemical compounds with the following features:

- TOXIC (acute and chronic)
- PERSISTENT (resist photolytic, chemical & biological degradation)
- BIOACCUMULATIVE (lipid-seeking)
- SEMI-VOLATILE (transports through various media)

The presence of PTS has a significant effect on our quality of life. Due to their unique characteristics, in particular, their ability for long-range transport, they pose a threat to the environment and to human health throughout the world.

A sub-group of PTS is called the persistent organic pollutants (POPs). The **United Nations Environment Programme** (UNEP) has identified twelve POPs to be the initial chemicals targeted for global elimination or restriction of production and use. Under the Stockholm Convention, an international legally binding treaty formally adopted on 22 May 2001, global awareness on the adverse effects of POPs has been raised. The twelve POPs, also known as the **"Dirty Dozen"** (Dudley, 1987), are comprised of pesticides, industrial compounds and unintentional by-products.

The "Dirty Dozen" include:

<table>
<tr><th>Pesticides</th><th>Industrial chemicals</th><th>Unintended by-products</th></tr>
<tr><td>Aldrin, Chlordane
Dieldrin, DDT *
Endrin, Heptachlor
Mirex, Toxaphene</td><td>PCBs</td><td rowspan="2">Dioxins
Furans</td></tr>
<tr><td colspan="2">Hexachlorobenzene</td></tr>
</table>

* Most of the 12 POPs are subject to an immediate ban, however, DDT has been granted a health-related exemption because it is needed in many countries to control malarial mosquitoes. These countries are permitted to use DDT until a suitable alternative is found.

II. Sources and Effects of PTS

The source of PTS into the environment could be from its production, its use, or import. It is indeed important to determine the location of the production facilities and the quantity of chemicals being produced, as well as where and how they are used or stored. Likewise, the source for unintentional by-products is due to the process responsible for its unintentional production and emission. Therefore, it is important to identify how they are produced and how much are produced and emitted.

Emission of PTS →
© Frank Lee

Due to their affinity to lipids, PTS are easily absorbed by the fatty tissue of people and animals. They are bioaccumulated and biomagnified through the food chain. PTS pose adverse health effects, such as reproductive disorders, developmental deformities and cancer in both humans and wildlife. As a consequence to their significant threat to both our environment and to our health, PTS have become a major concern at the local, national and global level (Kamrin, 1997).

PTS affect the environment and human health in several ways. According to UNEP (2002), since many PTS have a high affinity for soil and are hence retained in this environmental medium, the PTS may be taken up by crops and by grazing animals and finally reach the human food chain. The contaminants may also be transported during run-off from land and watercourses. In areas that are characterized by heavy rainfall, soil erosion can be severe and eroded soils may therefore cause significant pollution of waterways. Weak and ineffective control of emissions by industries and lack of stringent control by governments have also contributed to the global pollution of the environment by PTS.

On health, ecology and environment the adverse effects are:

- cancer
- reproductive impairment
- developmental & immune system disorders
- endocrine disruption
- threat of lowered reproductive success
- loss of biological diversity

© Zoe Lee

POPs are chemicals that remain intact in the environment for long periods, become widely distributed geographically, accumulate in the fatty tissue of living organisms and are toxic to humans and wildlife. POPs circulate globally and can cause damage wherever they travel. Three more significant POPs, namely **DDT**, **PCB** and **Dioxin** are explained in the next section.

III. Three Significant POPs – DDT, PCB and Dioxin

1) DDT

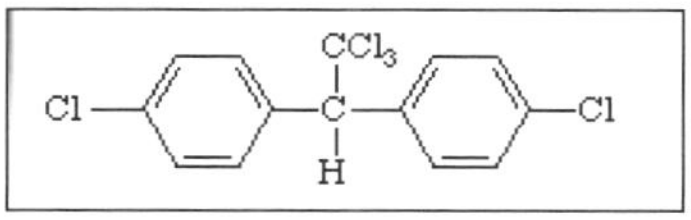

(**d**ichloro-**d**iphenyl-**t**richloroethane)

It was first synthesized in 1874, but its insecticidal properties were discovered only in 1939. It is relatively neither soluble nor readily decomposed by reducer organisms in natural ecosystems, so they are very persistent in the environment. Chlorinated hydrocarbons, however, do readily dissolve in fatty animal tissue, which is one of the factors important in their role as insecticides, since they must be absorbed through the fatty chitin exoskeleton of insects to be effective. They then commonly disperse through an ecosystem as they pass through successive stages of a food chain. DDT may remain unchanged or may be modified to substances of similar chemical structure and activity Derivatives of DDT include DDD and DDE. In 1971, the Environmental Protection Agency banned the use of DDT in the USA.

Effects on humans:

- affects brain development of fetuses and children exposed to DDT contaminated human milk
- suppresses immune system
- causes liver cancer and breast cancer

The term "Pesticide" not only refers to insecticides but also applies to herbicides, fungicides, rodenticides and various other substances used to control pests. In general any substance or mixture of substances intended for preventing destroying, repelling or mitigating any pests is considered to be a pesticide. Pesticides can be products in different forms and for different uses. For instance, they include agricultural pesticides, public health pesticides, wood insect spray, mosquito coil, cockroach and rat baits etc.

Important Facts for Pesticides (Dudley, 1987):

Q: **Are pesticides commonly used?**
A: In an urban area like Hong Kong, the high population density provides an ideal condition for rapid expansion of pests. To control these pests, pesticides are widely used in homes, offices, parks, factories, schools, hospitals, restaurants and public transport.

Q: **Are pesticides hazardous?**
A: While pesticides are very effective for controlling pests, their potential risks should not be overlooked. By their very nature, pesticides may cause serious harm to humans, animals and the environment if they are used improperly.

Q: **How should we use pesticides?**
A: Pesticides should only be used when they are absolute necessary. Whether you apply pesticides yourself or you hire a pest control operator, avoid exposure to the pesticides.

2) PCB

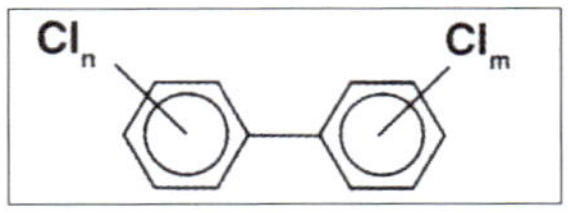

Polychlorinated biphenyl (PCB) sources are thought to be classified into two groups divided by the formation process:

i) one group are PCBs contained within PCB products manufactured in past industrial activities, which are intentional products; ii) the other group are PCBs formed as a byproduct in the process of combustion, such as municipal solid waste incineration (Francis, 1994).

80% of PCBs is released to the atmosphere during the burning of paper, plastic, or paint, and 20% was released to fresh and coastal water by leaks, disposal of industrial wastes, leaching, and atmospheric fallout. This made PCBs the first industrial (non-pesticidal) products to be identified as ubiquitous environmental pollutants.

The effects on humans include probable carcinogen, skin irritations such as acne and rashes, irritation to nose and lungs, as well as possible liver and kidney damage.

3) Dioxin

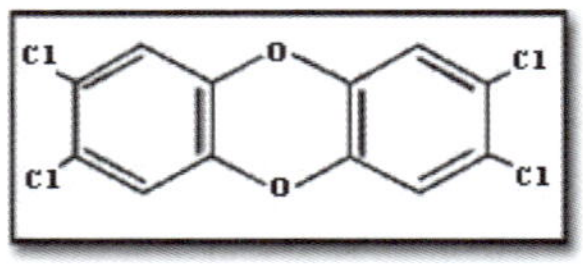

Also known as polychlorodibenzo-dioxin (PCDD), dioxin is colourless solid with no known odour. It is similar with another by-product, Furan (PCDF). The most toxic dioxin: 2,3,7,8-tetrachlorodibenzo-*p*-dioxin (TCDD).

Dioxins are by-products of industrial processes but can also result from natural processes, such as volcanic eruption and forest fires. The incineration of waste containing certain types of plastics, the manufacture and use of certain herbicides, chlorine bleaching of pulp and paper are the major sources of releases of dioxins. Dioxins are also found in vehicle exhaust and cigarette smoke. Yet dioxins can never be totally eliminated. Dioxins are found throughout the world in practically all media including air, soil, water, sediment, and food, especially dairy products, meat, fish and shellfish. The highest levels of these compounds are found in some soils, sediments and animals.

When released into the air, dioxins can be transported over long distances in the world. Unlikely, when released into water, dioxins tend to accumulate in the sediments where they can be further transported or ingested by fish and other aquatic organisms. Since dioxins are broken down in the environment very slowly, they can be deposited on plants and taken up by animals and aquatic organisms. Dioxins tend to bioaccumulate in the food chain and tend to accumulate in fat when ingested by animals. In addition, they can be transferred in human milk (Kamrin, 1997).

Effects on humans:

- chloracne: severe skin disease (most noted health effect)
- carcinogenic
- altered liver function and lipid metabolism
- endocrine and nervous system abnormalities
- reproductive/developmental effects
- weakened immune responses
- interferes with thyroid metabolism of 1 week old babies

IV. Case Studies in Hong Kong and China

(The data of this section mainly comes from UNEP, 2002)

Facts on Usage and Ban of Pesticides:

PTS	Hong Kong	China
Aldrin	Banned from all purposes since 1988	Not used
Chlordane	Banned from all purposes since 1991	Banned from agriculture (exemption for controlling termites)
DDT	Banned from all purposes since 1988	Banned from agriculture since 1983 (exemption for controlling mosquito which transmitted malaria)
Dieldrin	Banned from all purposes since 1988	Not used
Endrin	Not registered	Not used
Heptachlor	Not registered	Not registered
HCB	Not registered	No record
Mirex	Banned from all purposes since 1997	No record (exemption for controlling termites)
Toxaphene	Banned from all purposes since 1984	Banned from all purposes since 1987
HCH	Banned from all purposes since 1991	Banned from all purposes since 1983
PCP	Banned from all purposes since 1994	No record

1) Hong Kong

Elevated levels of PTS including PAHs, PCBs and certain organochlorine pesticides are a long-term significance to the health of HKSAR water's ecosystem. In specific, it is apparent that PAHs and pesticides may pose a risk not only to the marine ecosystem, but also to primary and secondary consumers of marine organisms. Although there are now considerable data on the levels of PTS in the marine sediments and, to a lesser extent, other environmental compartments, such as biota, there is still a general paucity of information of the sources of PTS and their precise effects on biological systems in HKSAR's marine waters. They are the potential ecological effects on the population that are more difficult to measure or predict.

Work has recently been initiated on the environmental management of chemicals in the Hong Kong SAR. The on-going study of toxic substances pollution in Hong Kong, commissioned by the Environmental Protection Department (EPD), aims to identify the trade, usage, production and disposal of priority toxic substances in HKSAR and assess the potential impact that any such substances may pose to local aquatic life and human health.

Major laws and regulations concerning PTS in Hong Kong include:

- Waste Disposal Regulation under the Waste Disposal Ordinance (1992)
- Air Pollution Control Ordinance (1983)
- Water Pollution Control Ordinance (1980)
- The Pesticides Ordinance (1977)
- Dangerous Goods Ordinance (1956)

DDT: DDT was banned from use on 31 December 1987, and currently can be traded only under permit. From 1979 to 1982, more than 5,000 kg of DDT was imported into Hong Kong annually.

PCBs: EPD has commissioned environmental consultants to monitor PTS in the environment. Scientists from local universities have conducted studies related to PTS contamination in different ecological compartments. EPD has been monitoring PCBs and PAHs in marine sediment since 1987 and in 2000, a total of 60 stations (45 in open waters and 15 in typhoon shelters) were routinely monitored. Routine monitoring of total PCBs in ambient air at two urban locations has also been conducted since mid-1997.

Dioxins: Soil on the property of a previously privately-owned shipyard had recently been found to be contaminated with PCDD/PCDF among other contaminants. The estimated volume of the soil is 30,000 m^3. The source of the PCDD/PCDF may be from open burning on the site. Two incinerators may also contribute to sources of PCDD/PCDF.

2) China

Since 1994, the environmental administration of chemicals has begun in China, especially with the development of environmental management controls on the import and export of toxic chemicals. This is a direct result of the management of chemicals being brought into routine government procedures. A number of scientific research institutes and institutions of higher learning have carried out scientific research and investigation on PTS. The Chinese government is now carrying out an inventory of POP pesticides and the PDF-B of national implementation plan (NIP) required by the Stockholm Convention.

Major laws and regulations concerning PTS in China include:

- Code of Occupational Disease Prevention of PRC (2002)
- Regulations on Hazardous Chemicals Safety (2002)
- Regulations on the Management of Pesticides (1997)
- Regulations for Environmental Management on the Import and Export of Toxic Chemicals (1994)

DDT: China has been a significant producer and user of DDT since the 1950s until its ban on production and agricultural use were enforced in 1983. During its thirty years of production, 0.4 million tonnes of DDT were produced (nearly 10,000 tonnes of DDT produced annually). China has requested exemptions from the Stockholm Convention for DDT for the production and use of DDT for vector control.

The pesticides ultimately enter the Pearl River through evaporation (including wind-driven transport of suspended particulates) and surface runoff. The Pearl River carries a considerable load of up to 863 tonnes per year of chlorinated pesticides. There is evidence from the distribution profiles of DDT and its degradation products that current input of fresh DDT as an impurity of other pesticides may still continue in the Pearl River Delta.

PCBs: PCB levels were measured in gas and particulate phase of air sampled from three cities (Guangzhou, Shenzhen and Zhaoqing) in the Pearl River Delta. The study reported PCB concentrations of 0.297 to 0.537 ng/m^3 in gas phase, and 0.048-0.157 ng/m^3 in particulate phase.

Dioxins: A large amount of the pesticide sodium pentachlorophenol (Na-PCP) salt have been sprayed over vast areas in central China to control schistosomiasis, a parasitic disease of epidemic proportions. About 5,000 tonnes of Na-PCP are produced in China annually. PCDD/PCDF are found in impurities in significant contribution to human exposures to PCDD/PCDF in China.

V. International Concern

(The data of this section mainly comes from UNEP, 2002)

Stockholm Convention

Conducted on 22 May 2001, the **Stockholm Convention** was considered the first worldwide collaboration in tackling the problem of persistent toxic substances. Representatives from over 120 countries attended the Convention and signed the international legally binding treaty. It aims to protect human health and the environment from persistent organic pollutants by:

- implementing international action on 12 POPs through coordination and cooperation between national, regional and international levels in countries;
- setting control measures - production, use, import, export, disposal; and
- setting criteria and procedure for adding further POPS.

Regionally Based Assessment of Persistent Toxic Substances Project

Complementary to the Stockholm Convention, the Chemicals Division of United Nations Environment Programme (UNEP) implemented the Regionally Based Assessment of PTS Project. While UNEP plays a pivotal role in developing consensus globally on control of PTS, about 160 countries worldwide are slotted into 12 regions. China (including Hong Kong) is placed in **Region VII** (Central and North East Asia), along with countries such as Japan, Republic of Korea, and Russia.

The Project aims to identify sources of PTS, assess the impact of PTS on human health and the environment, assess transboundary pathways of PTS, assess root causes of problems and capacity to manage regionally, and identify regional and global environmental issues.

After two years of effort, the Project resulted in many outputs, including the establishment of a network of PTS experts, setting up a management and co-ordination structure, production of twelve comprehensive Regional Reports as well as a Global Report published in December 2002. All helped to raising public awareness on PTS management.

Conclusion

Persistent toxic substances are chemical substances that persist in the environment, bioaccumulate through the food web, and pose a risk of causing adverse effects to human health and the environment. Noticing the long-range transport of these substances to regions where they have never been used or produced and the consequent threats they pose to the environment of the whole globe, the international community has now called for urgent global actions to reduce and eliminate the releases of these chemicals.

Among the twelve major PTS known as the "Dirty Dozen", Dioxins, PCBs, and DDT are chemicals of concern in terms of environmental levels and ecotoxicological effects. China and Hong Kong are giving more and more effort on monitoring the levels of PTS, which have posed potential threat to the regions. Both governments have set up responsible authorities and tight measures on regulating the production of the toxic chemicals. However, it is observed that there is certainly more room for improvements in this aspect.

Much attention has been paid to the effects of PTS within the region. The Stockholm Convention is a global treaty to protect human health and the environment from persistent organic pollutants. In implementing the Convention, Governments take measures to eliminate or reduce the release of POPs into the environment.

第九章　持續性有毒物質

污染被定義為「一種不自然的有害物質或一種自然物質達到有害水平，對自然環境，包括人類，動物或植物生命造成不良影響」(Koren and Bisesi, 1996)。污染成因通常是超過正常濃度的物質進入一個系統。污染物質的水平超過系統的限制，因此導致生態系在物質，生物與化學上的變化。

全球正關注一組化學物，名為**持續性有毒物質**。這些化學物對環境是有毒和持久的，並累積於生物脂肪組織，透過食物網而將毒素濃度增加，以及有廣大的傳送範圍(Walker and Livingstone, 1992)。它們對環境造成長期污染，並且由於有擾亂內分泌的能力，可引致人類急性和慢性的毒性。

I. 什麼是持續性有毒物質（PTS）？

PTS是自然/人造的有機化合物，並有以下特徵：

- 有毒（急性和慢性）
- 持久性（抵抗光分解及化學和生物降解）
- 生物累積（尋找脂肪）
- 半揮發性（經各種媒體運送）

PTS 對我們的生活質素有重大的影響。由於他們的獨特的特徵，尤其它們具長遠運輸能力，對全球環境和人類健康構成威脅。

PTS 的一個分類稱為持久性有機污染物（POPs）。聯合國環境計劃 (UNEP) 識別了十二種POPs為消除或限制全球生產和使用。在斯德哥爾摩會議，於2001年5月22日，正式採納國際合法條約，提出全球關注POPs的不良影響。那十二種POPs被稱為「**Dirty Dozen**」(Dudley, 1987)，由殺蟲劑，工業的化合物和副產品組成。

「Dirty Dozen」包括：

<table>
<tr><th>殺蟲劑</th><th>工業化學物品</th><th>非規劃生產的副產品</th></tr>
<tr><td>阿特靈（Aldrin）、
氯丹（Chlordane）、
滴滴涕（DDT）*、
地特靈（Dieldrin）、
安特靈（Endrin）、
飛佈達（Heptachlor）、
米瑞克司（Mirex）、
毒殺芬（Toxaphene）</td><td>多氯聯苯（PCBs）</td><td rowspan="2">二惡英（Dioxins）、
夫喃（Furans）</td></tr>
<tr><td colspan="2">六氯苯（Hexachlorobenzene）</td></tr>
</table>

* 大部分的 POPs 都即時禁止使用。不過 DDT 被授予與健康相關的例外，因為很多國家需要 DDT 控制瘧疾蚊子。這些國家被允許使用 DDT 直到有合適的替代品。

II. PTS的來源和影響

PTS來自生產，使用或輸入。對於生產設施和化學物的產量，以及它們在哪裡和如何使用或儲存，都是十分重要。同樣地，由於非規劃生產的副產品的來源是無意生產和排放，因此識別它們如何生產和生產數量亦是重要的。

由於對脂肪較吸引，PTS輕易地被人類和動物的脂肪組織吸收。它們透過食物鏈被生物累積和生物放大。PTS對健康造成不良影響，如繁殖紊亂，在人類和野生動物發展畸形和癌症。由於對我們的健康和環境構成嚴重的威脅，PTS受到本地，全國和全球的關注 (Kamrin, 1997)。

PTS對環境和人類健康的影響有幾種途徑。根據UNEP (2002) 說法，因為很多PTS對土壤有較高的吸引，因此停留在土壤之中，PTS被農作物和放牧動物攝取，並透過食物鏈而被人類吸收。污染物也可以經陸地和水道運送。在降雨量大的地區，土壤侵蝕是嚴重的，受侵蝕的土壤可以引致水道嚴重的污染。工業控制排放的效率低，以及缺乏政府嚴格的控制造成PTS全球環境的污染。

在健康，生態學和環境上，不良的影響是:

- 癌症
- 生殖損傷
- 發展及免疫系統紊亂
- 內分泌中斷
- 減低成功生殖能力
- 生物多樣性的損失

POPs 長時間保留在環境，廣泛傳送，積聚生物的脂肪組織，以及對人類和野生動物是有毒的。POPs 在全世界循環，並引致損害。在下一節，將會解釋另外三種重要的 POPs，即是 DDT，PCB 和 二惡英。

III. 三種重要的持久性有機污染物 – DDT, PCB 及二惡英

1) DDT

(**d**ichloro-**d**iphenyl-**t**richloroethane)

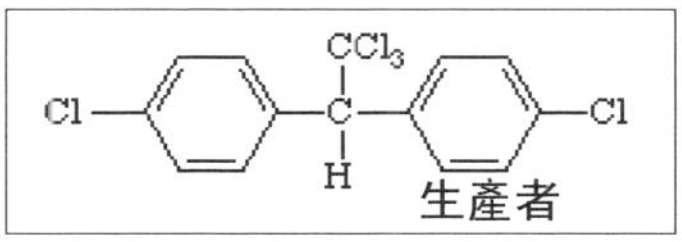

DDT 於 1874 年首次被製造，不過到 1939 年才發現它的殺蟲功效。它不可溶解也不可以在自然生態系中被生物分解，所以它們在環境是非常持久。不過，氯化碳氫化合物溶解於脂肪組織，這是它們作為殺蟲劑的其中一個重要因素，因為它們被昆蟲的脂肪性外骨骼吸收才有效。透過食物鏈，它們在生態系擴散。DDT 可以保持不變或改變成類似的化學結構和活動物質。DDT 的衍生物包括滴滴滴 (DDD) 和滴滴依 (DDE)。在 1971 年，美國環保局禁止使用 DDT 。

對人類的影響:

- 透過受DDT污染的人奶，影響胎兒和小孩的大腦發育
- 抑制免疫系統
- 肝癌和乳癌的成因

「殺蟲劑」一詞不僅指殺蟲劑而且適用於除草劑，殺真菌劑，滅鼠劑和各種其他控制害蟲的物質。一般而言，任何物質或混合物用於阻止，破壞，驅除或減輕任何有害物均被認為是殺蟲劑。殺蟲劑有不同形式的產品和使用。例如，農業殺蟲劑，公共衛生殺蟲劑，林木昆蟲噴霧，蚊香，蟑螂和老鼠誘餌等等。

殺蟲劑的重要事實 (Dudley, 1987)：

Q：殺蟲劑普遍使用嗎？
A：像香港那樣的城市，高度人口密集，提供一種理想環境給害蟲迅速擴張。為了控制這些害蟲，殺蟲劑在家裡，辦公室，公園，工廠，學校，醫院，餐館和公車都廣泛使用。

Q：殺蟲劑有害嗎？
A：當殺蟲劑為了有效控制害蟲的時候，它們的潛在風險不應該被忽視。如果它們不恰當使用時，殺蟲劑對人類，動物和環境造成嚴重的損害。

Q：我們應該如何使用殺蟲劑？
A：絕對必要時才使用殺蟲劑。不論你自己使用殺或你雇用滅蟲公司，避免接觸殺蟲劑。

2) PCB (多氯聯苯)

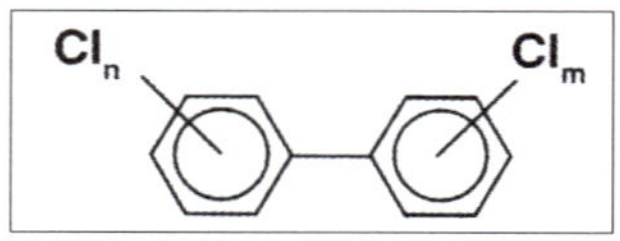

PCB的來源可分為兩類:

i) 一類是在過去工業活動製造的 PCB 產品，屬於規劃性產品；ii）另一類是在燃燒的過程中產生PCB，屬於副產品，如焚化都市固體廢物 (Francis, 1994)。

80% 的PCB在燃燒紙張，塑料或油漆時排放到大氣層，20% 經由滲漏，傾倒工業廢物，過濾而排放到淡水及海岸。這樣令 PCB 由工業（非殺蟲劑）產品變成普遍存在的環境污染物。

對人類的影響包括致癌，刺激皮膚例如粉刺和皮疹，刺激鼻和肺，以及可能令肝臟和腎受損。

3) 二惡英 (Dioxin)

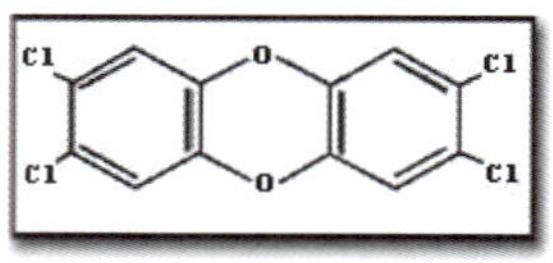

也稱為多氯聯笨二惡英 (PCDD)，二惡英是具無色無味的固體。類似另一個副產品，多氯聯笨夫喃 (PCDF)。最毒的二惡英是TCDD。

二惡英是工業過程的副產品，不過自然過程也會產生，如火山爆發和森林火災。焚化含有塑膠的廢物，製造和使用某類的除草劑，用氯氣漂白紙漿和紙張都是釋放二惡英的主要來源。汽車廢氣和香煙也發現二惡英。不過，二惡英不能夠完全被消除。二惡英在全球媒介均被發現，包括土壤，水，沉澱物和食物，尤其是乳製品，肉，魚和貝類。在一些土壤，沉澱物和動物發現最很多這種化合物。

當釋放到空氣的時候，二惡英能夠擴散世界。不過，當排於到水裏，二惡英會積聚於沉澱物而擴散或被魚類和其他水生生物所攝食。因為二惡英在環境分解的速度非常緩慢，它們能夠積存於植物，及被動物和水生生物所吸收。二惡英在食物鏈而累積，積聚於動物脂肪。此外，它們也可以透過人奶傳送 (Kamrin, 1997)。

對人類的影響：

- 嚴重的皮膚病（最受關注的健康影響）
- 致癌
- 改變肝臟功能和脂肪的新陳代謝
- 內分泌和神經系統異常
- 生殖/發展的影響
- 削弱免疫反應
- 干擾一星期嬰孩的甲狀腺新陳代謝

IV. 香港和中國的案例

(此部分資料主要來自UNEP, 2002年)

殺蟲劑的使用和禁用情況:

PTS	香港	中國
阿特靈 (Aldrin)	自1988年全面禁止	不被使用
氯丹 (Chlordane)	自1991年全面禁止	農業禁用(用作控制白蟻除外)
滴滴涕 (DDT)	自1988年全面禁止	自1983年農業禁用(用作控制虐蚊除外)
地特靈 (Dieldrin)	自1988年全面禁止	不被使用
安特靈 (Endrin)	沒有註冊	不被使用
飛佈達 (Heptachlor)	沒有註冊	沒有註冊
六氯苯 (HCB)	沒有註冊	沒有記錄
米瑞克司 (Mirex)	自1997年全面禁止	沒有記錄(用作控制白蟻除外)
毒殺芬 (Toxaphene)	自1984年全面禁止	自1987年全面禁止
HCH	自1991年全面禁止	自1983年全面禁止
PCP	自1994年全面禁止	沒有記錄

1) 香港

PTS的濃度增加，包括 PAHs，PCBs 和某類有機氯化殺蟲劑對香港特別行政區水生態系的健康有長期的影響。尤其是， PAHs 和殺蟲劑不只對海洋生態，而且對主要和次要消費者的海洋生物也有風險。

儘管有很多關於海洋沉澱物或特定環境的 PTS 資料，不過有關整體 PTS 對香港水域生物的影響的資訊卻很少。它們對人口的潛在生態影響是很難測量或預測。

最近，香港特別行政區進行有關環境化學物的管理工作。在香港，由環境保護署 (EPD) 委托的有毒物質污染研究持續進行，目的是識別有毒物質的貿易，使用，生產和丟棄，以及評估這些物質對於本地水生物的生活和人類健康的潛在影響。

在香港，有關PTS的主要法律和規則包括：

- 在廢物處理條例下的廢物處理規則（1992 年）
- 空氣污染控制條例（1983 年）
- 水污染控制條例（1980 年）
- 殺蟲劑條例（1977 年）
- 危險的貨物條例（1956 年）

DDT：在1987年12月31日起禁用DDT，有許可證才可以進行貿易。從1979年至1982年，每年超過5,000公斤的DDT輸入香港。

PCB：EPD有受委託環境顧問公司監測PTS。來自本地大學的科學家在不同生態系區域進行有關PTS污染的研究。自從1987年，EPD一直監測在海洋沉澱物的PCBs和PAHs，在2000年，共50個監測站（45個位於公開水域和15個位於避風塘）定期進行監測。自從1997年中期，在兩個城市地點進行空氣方面的總PCBs的日常監測。

二惡英：在私有的造船廠的土壤裏，發現受PCDD/PCDF污染。受污染的土壤估計有30,000 m3。PCDD/PCDF來自露天焚燒，以及兩座焚化爐。

2) 中國

自從1994年，在中國開始了化學物的環境管理，尤其發展有關有毒化學物的輸入和出口的環境管理。化學物管理被列入例行的政府程序。很多科學研究組織和高等學府進行有關PTS的科學調查和研究。中國政府如今正執行斯德哥爾摩會議所要求有關POP和PDF-B的全國執行計劃（NIP）。

在中國，有關PTS的主要法律和規則包括：

- 中華人民共和國的職業病預防的法規（2002 年）
- 危險化學物品安全的規則（2002 年）
- 殺蟲劑管理的規則（1997 年）
- 有毒化學物的輸入和出口的環境管理規則（1994 年）

DDT：自五十年代，中國是DDT主要的生產者和使用者，直到1983年，才禁用於生產和農業上。在這三十年期間，生產四十萬公噸的DDT（每年生產將近10,000公噸的DDT）。中國為了控制傳染病，向斯德哥爾摩公約申請豁免生產和使用DDT。最終，殺蟲劑透過蒸發流入珠江（包括風力散播的懸浮粒子）和表土流失。每年珠江流入863公噸的氯化殺蟲劑。有證據顯示DDT和殺蟲劑的降解物繼續流入珠江三角州。

PCB：在珠江三角州的三個城市（廣州，深圳和肇慶)，以氣體和空氣的微粒量度PCB的水平。研究報告指出PCB氣體的濃度由0.2907至0.537 ng/m3，以及微粒的濃度由0.048至0.157 ng/m3。

二惡英：在中國中部地區，噴射大量鈉（Na-PCP）鹽殺蟲劑，以控制血吸蟲病，一種寄生蟲的疾病。中國每年生產大約5,000公噸的Na-PCP。PCDD/PCDF雜質對於中國人民接觸PCDD/PCDF有重大的影響。

V. 國際關注

(此部分資料主要來自UNEP, 2002年)

斯德哥爾摩公約

在2001年5月22日召開的斯德哥爾摩約，是第一次全球合作處理持久性有毒物的問題。來自超過120 個國家的代表參加會議和簽署國際公約。目的是保護人類健康和環境，避免持久性有機污染物質：

- 透過在全國，區域和國際之間的協調和合作，對十二種持久性有機污染物採取國際行動；
- 設定控制措施 — 生產，使用，輸入，出口，棄置；以及
- 為了更多的持久性有機污染物設定標準和程序。

持續性有毒物質區域性評估計劃

為了配合斯德哥爾摩公約，聯合國環境規劃署 (UNEP) 的化學物部門實施持久性有毒物質區域性評估計劃。當 UNEP 在控制 PTS 扮演關鍵角色的時候，大約 160 個國家被劃分為12 個區域。中國（包括香港）在區域 VII 中被放置（中亞和東北亞），與日本，韓國和俄羅斯國家一起。

計劃的目的：

- 識別PTS的來源；
- 評估PTS對人類健康和環境的影響；
- 評估PTS的跨境路線；
- 評估PTS的根本問題和區域性的處理能力；以及
- 識別區域和全球首要的PTS環境問題。

經過兩年的努力，計劃有很多成果，包括成立PTS專家網路，建立管理和協調架構，出版十二份完整區域性報告以及在 2002年12月出版的一份全球報告，以提高公眾對PTS管理的意識。

結論

持久性有毒物質是長存於環境的化學物質，透過食物網而累積，以及對人類健康和環境造成不良影響。由於長途的傳送，經過沒有被使用或生產的區域，因此對全球環境構成威脅，國際社會現要求緊急的全球行動，以減少和消除這些化學物的散佈。

在十二種主要PTS被稱為「Dirty Dozen」當中，二噁英、PCB 和 DDT 對環境和生態毒理學影響最受關注。中國和香港投入更多努力監測PTS的水平，因為它對地區構成威脅。政府成立專責機構和措施，以調控有毒化學物的生產。不過，現時在這範疇有更多改善的空間。

在區域以內，PTS的影響更受關注。斯德哥爾摩公約是一個全球公約，保護人類健康和環境，避免持久性有機污染物質。在實施條約方面，政府設法消除或減少持久性有機污染物排放到環境。

Questions for Further Thinking 思考問題

1. One existing difficulty for regulating the impact of PTS in Hong Kong and China is the low public awareness/participation. Do you agree with this comment?
   在香港和中國要調控PTS的影響的其中一個困難是公眾的意識／參與率低。你同意這個評論嗎？

2. How can we as individuals contribute to reducing the impact of PTS and sustaining environmental health in our daily lives?
   作為個人怎樣去減少PTS的影響，以及維持生活環境的健康?

3. Suggest some ways our government to enhance the coordination with the authorities in the Pearl River Delta in order to control the emission of PTS in the region.
   試建議政府一些方法加強珠江三角洲之間的協調，以控制PTS的排放。

References 參考資料

Dudley, N. (1987), *This Poisoned Earth: the Truth about Pesticides*, London: Judy Piatkus Ltd., p.1–56.

Francis, B.M. (1994), *Toxic Substances in the Environment*, New York: John Wiley & Sons, Inc., p.93–119.

Huang, P.M. and Iskandar, I.K. (2000), *Soils and Groundwater Pollution and Remediation: Asia, Africa, and Oceania*, Boca Raton, Florida: Lewis Publishers.

Kamrin M.A. (1997), *Pesticide Profiles: Toxicity, Environmental Impact, and Fate*, Boca Raton, Florida: CRC Press, p.623–629.

Koren, H. and Bisesi, M (1996), *Handbook of Environmental Health and Safety: Principles and Practices* (vol. II), Boca Raton: Lewis Publishers, p395–463.

Laegreid, M., Bockman, O.C. and Kaarstad, O. (1999), *Agriculture, Fertilizers and the Environment*, New York: Cabi.

Ragsdale, N. (1999), *Pesticides- Managing Risks and Optimizing Benefits*, Washington, DC: American Chemical Society: Oxford University Press.

United Nations Environment Programme (UNEP) Chemicals (2002), *Regionally Based Assessment of Persistent Toxic Substances — Central and North East Asia Regional Report*, Switzerland: United Nations, p.1–27.

Walker, C.H. and Livingstone, D.R. (1992), *Persistent Pollutants in Marine Ecosystems*, Oxford: Pergamon Press, p.3–34.

Wiebers, U. (1993), *Integrated Pest Management and Pesticide Regulation in Developing Asia*, Washington, DC: World Bank.

Wong, M.H., Cao, Z.H. and Luo Y.M. (2003), *Chemosphere: Soil and Environmental Quality in Modern China*, Oxford: Pergamon.

Chapter 10. Genetically Modified Food

Genetically modified (GM) food is a subject of much controversy. On the one hand, opponents wishing to prevent its use cite reasons ranging from the spiritual possible cause of genetic disease to environmental disaster. On the other hand, its proponents see it as a cure for world hunger, a preventer of disease and crop blight, and even improvement to the environment. In actual fact, the truth of exactly what GM food will bring in the future is yet unknown.

Corn field © Frank Lee

However, to make an educated judgement on how to use and develop this food, we must first understand what GM food is, what its existing and potential benefits and disadvantages are, and, once deciding what risks actually exist, how risk can be reduced to a minimum. This chapter aims to make an introductory discussion of these topics and hopefully sort out some of the myths from the reality.

I. What is GM food?

Human beings have been genetically manipulating crops for centuries. For example, farmers have been cross-breeding various crops, trying to breed a type of corn that is more resistant to a cold climate, a type of rice that is heavy with grain but can still support the weight of the head, or even a type of watermelon without seeds. These are all a subtle form of genetic manipulation, an attempt to enhance a certain characteristic of a plant to achieve a desired result.

The term genetically modified (GM) food commonly refers to plants that have been 'genetically' engineered using techniques specific to molecular biology. They are also sometimes called GMOs (genetically modified organisms) or transgenic foods (Whitman, 2000).

Genetic Engineering

Genetic engineering, in terms of producing GM foods, refers to techniques which modify the genetic characteristics of an edible plant in order to enhance a certain trait or characteristic. Genes that correspond to certain characteristics are isolated, and are then spliced into another plant (Nottingham, 1998). Exactly what can be achieved by genetic engineering techniques is only limited by the different genes that can be isolated from and inserted into different plants. The potential simply in transferring genes from one plant to another is enormous, but the picture becomes more complex as it has now been found possible to transfer genes from non-plant organisms into plants themselves. Recent research has even been working on ways to transplant genes from coldwater fish into potatoes and other crops to help make them frost-resistant.

The development of GM cropping

The first genetically modified food to be approved for consumption was the *Flavr Savr Tomato* which was produced in 1994. This tomato was modified to make it last longer on the shelves by splicing the tomato with genes from an arctic trout. Since then the production of GM crops has become more widespread, sustaining a growth rate of over 10 percent per year. By 2003 reports estimated that over 7 million farmers in 18 different countries planted as much as 67.7 million hectares of genetically manipulated crops. However, 99 percent of these crops were planted in just six countries, the United States, Argentina, Canada, Brazil, China, and South Africa, with 52% grown in the USA and gradually decreasing to 4% in China, and 1% in South Africa (ISAAA, 2003).

But what types of GM crops are grown? We have already mentioned tomatoes, but these are no longer a major crop. According to statistics, the four major crops nowadays are:

Four Main GM Crop Planting Statistics 2003

Crop	Million Hectares
Soybeans	41.4
Maize	15.5
Cotton	7.2
Canola	3.6

Data source: ISAAA 2003 (International Service for the Acquisition of Agri-biotech Applications)

Other GM crops such as chicory, squash, potatoes, and papaya are also produced but on a much smaller scale.

II. Process of Genetic Modification

To get a better understanding of what is involved in making GM food, what potential it has, and what the possible ramifications of the process are, one must know a little of what genes are and how genetic modification is carried out.

DNA, Genes, and Engineering GM Food

DNA

The chemical DNA was first discovered in 1896, but its real significance for inheritance was not discovered until 1953. In that year, two British scientists, James Watson and Francis Crick made what is sometimes called the scientific discovery of the century. They declared that DNA is a 'double helix polymer'; this means that two strands of DNA twist around each other and link together by phosphate bonds from the base of one molecule to the complementary base of another (Watson, 1968).

Genes

Put simply, a gene is a section of DNA containing a sequence of base pairs that determines part of the hereditary characteristics of a biological organism (Saenger, 1984; Nottingham, 1998).

Engineering GM Food

The actual process of genetic modification involves taking a specific gene from one organism and splicing it into the DNA of another organism. The process is quite complex and involved but, for our purposes, we will simply consider that modifying the DNA of an organism with a new gene consists of 4 processes (Saenger, 1984; Nottingham, 1998):

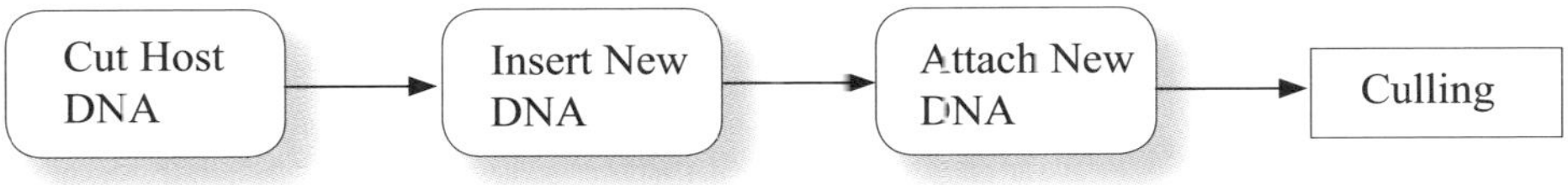

> The DNA is cut at specific points using something called restriction enzymes, which leaves a gap for the new DNA. The new DNA is then inserted into the cells. This process is quite difficult and commonly uses a retrovirus that can carry the DNA through a cell wall and into the cell's nucleus. Once the new DNA is inserted, an enzyme called Ligase is used to attach the sections of DNA together and the cell starts carrying out the instructions of the new DNA. However, the process of attaching DNA does not always work properly, DNA ends up in the wrong place or not attached at all. A special drug is therefore introduced that kills all the unwanted cells, leaving only the cells in which the DNA attached correctly.

Clearly the process of genetic modification allows for very specific changes to be made to the structure of a biological organism or, in the case of GM food, plant. At the beginning of this section we touched upon the possibility of both enormous benefits and enormous disadvantages to our future lives brought about by making such specific changes to plant DNA. In the following sections we will look in much greater detail at those benefits and possible disadvantages, and at how the risks involved, if any, may be reduced.

III. Benefits of Genetic Modification

We previously mentioned that genetic modification has been around for centuries, but what are the advantages of genetic engineering over the traditional techniques of cross-breeding and what can genetic engineering achieve that traditional cross-breeding cannot? Furthermore, what practical achievements has genetic engineering made so far and what may it be able to achieve in the future?

Traditional Breeding

Traditional plant breeding takes plants from within a species and cross-breeds them in an attempt to combine two or more of their desirable traits. In fact, traditional breeding is theoretically capable of achieving many of the things that GM is capable of, but the speed at which it is achieved and the range of possibilities are much greater for genetic engineering. The following table sheds some light on this (FEHD, 2005).

Traditional Breeding	Genetic Engineering
Cross-breeds thousands of genes at a time in a somewhat random process.	Isolates and splices a single gene with a specified characteristic.
Achieves results slowly, may take many generations of cross breeding and selection of plants with particular characteristics.	The desired characteristic is achieved in a single generation.
Usually limited to within a certain species.	Genes are not species sensitive and can be transferred across the species barrier.

Existing Benefits

Economic

As the growing of GM food is a relatively new type of commercial venture, few long-term studies of the economic benefits have been performed. However, short-term studies have shown economic advantages of GM farming resulting in improved yield (during times of high pest and weed pressures), and reduced herbicide and pest control costs.

Agricultural

Apart from the economic benefits GM have, the crops themselves have specific advantages. Some of these advantages may not have immediately apparent benefits, but in time of high biological and environmental pressures these advantages become more evident. Current GM crops have been bred with tolerance of cold, herbicides, and pests.

Health

Nutritional and Medical Benefits

Genetic modification of crops may, in future, be able to make them more nutritionally valuable and may have medical benefits. However, no crops, that have been modified in this way, have yet been licensed for production. We will discuss the potential benefits of these later.

However, there may be existing medical advantages to the GM crops currently used. When plants compete with weeds they often produce some of the most powerful toxins know, which may damage human health when eaten. The ability to control weeds when using GM herbicide tolerant crops reduces the level of these toxins (Nottingham, 1988). Another possible advantage comes from pest tolerant crops. When pests attack plants they damage the plant's defensive layer, which in turn allows fungus infection which results in the production of chemicals called mycotoxins. Mycotoxins are known causes of cancer, so pest tolerant crops may bring about a reduction of cancer, especially liver cancer.

Environmental

Exactly what the benefits GM crops have on the environment requires long-term studies, but some current environmental benefits are relatively obvious, even at present. Reduced use of both pesticides and herbicides cannot but have advantages to the environment and will inevitably result in reduced groundwater contamination and reduced impact on non-target species such as butterflies. Increased crop productivity will give rise to a reduction of the land area needed for cropping, which can be reforested, resulting in an improved environment. Reduced levels of tilling can also help stabilize soils which reduce erosion and water loss. Other GM plants, not grown as crops, also have benefits to the environment. Certain types of trees have been genetically engineered to clean heavy metal pollution out of contaminated soil.

Potential Benefits

In the previous section we discussed the existing benefits of GM foods. However, there is a great deal of active research that may bring a much wider range of benefits, especially in the areas of agriculture and health.

Economic

The development of future economic benefits for GM crops depends on the development of the public's attitude toward them. At the moment GM crops are stringently controlled and monitored. If the level of control drops in future and people accept the use of GM crops as everyday food, the economic advantages of growing these crops will increase, especially as new breeds of greater yield are produced.

Agricultural

Apart from the current benefits of GM crops, further developments in those areas are likely to bring other significant benefits in the future. But the benefits of GM food are not just restricted to these areas. Research work is being made in other areas, in an attempt to improve the agricultural properties of various plants. The majority of research is concentrated in the following areas:

<u>Drought/Salinity Tolerance</u>

Although not a huge priority in research by economically driven genetic engineering companies, drought and salinity resistance could bring great benefits to farming in certain poorer areas of the world. Areas such as Africa and parts of Asia, susceptible to drought and high salt levels in soil, may be the biggest benefactors, especially in areas where increased utilization of irrigation is difficult (Zhang and Blumwald, 2001).

<u>Disease Resistance</u>

Various viruses, fungi and bacteria affect plants. Scientists are working on producing plants resistant to diseases that commonly destroy crops. This is usually done by adding a gene from the virus itself into the plant so it produces its own antiviral/antibiotic chemicals (Moellenbeck *et al.*, 2001).

B.t. genes can let corn crop persist to insect larvae © Frank Lee

Health

GM food may have future benefits for human health by improving the nutrition of food and by introducing medicines into food itself. Certain nutritionally altered food has already been developed such as "Golden rice", genetically altered rice developed with added vitamin A to help fight the worldwide deficiency of this vitamin. However, as with most GM developments in these areas, it is yet to receive regulatory approval in most countries. Of course, there are various other benefits in the areas of nutrition and medicine; some of these are introduced below:

Nutrition

Around 850 million people around the world are malnourished. Much of the cause of malnourishment is dependence on a single staple food, such as rice, potatoes and manioc. However, these stable foods rarely contain all the nutrients essential for a healthy diet. GM foods offer a possible solution to these problems by altering the nutritional content of various foods to add new nutrients, such as the vitamin A described above (**Swiss Federal Institute of Technology Institute for Plant Sciences**). By adding a variety of different nutrients staple foods can be made much healthier and may be of huge benefit to poorer countries in the future.

Medicine

Manufacture, transport, and storage of pharmaceuticals and medicines often cause problems, especially in poorer areas of the world. Scientists are attempting to develop GM food with edible vaccines, thus making production, transportation and administering of these vaccines much easier (Daniell *et al*., 2001).

IV Disadvantages of Genetic Modification

In many countries throughout the world there have been huge outcries against using genetically modified foods. Many people fear the dangers of this new, and in some ways disturbing, technology. Some say that scientists' reasoning that one gene is coded for one trait is flawed and may have possible consequences, because it is unknown what may actually result from gene splicing. It was commonly thought that one gene was coded for a single protein, but recently it has been discovered that a single gene may produce as many as 38,000 proteins. Clearly unless the action of a single gene is fully understood before splicing, random and possibly harmful results may occur. In this section we will look at the disadvantages of genetic modification, as we did with the benefits, by looking at the current disadvantages and the potential disadvantages in the future.

Existing Disadvantages

Economic

Opponents of GM food stress that the economic benefits of GM crops are not so marked as many of the biotech companies try to make out. In poorer countries, having to buy new seeds each year, as needed for most GM crops, is not a viable option for most farmers. They also point out that although savings are made in some areas, the costs needed for controlling monitoring and testing GM foods counteract their other economic advantages.

Agricultural

Although many potential problems with GM foods have been identified, to date no study has been able to identify any significant agricultural disadvantages to growing GM food.

© Frank Lee

Health

Few health problems with GM foods have been identified to date. However, one or two cases have indicated the potential for problems occurring in the future. For example, a genetically bred type of soybean had a gene from a brazil nut inserted (Nordlee et al., 1996). Insertion of this gene resulted in the new hybrid containing an allergen, but this allergen was identified during testing and the plant was not used.

Environmental

As with other areas few current significant environmental effects of growing genetically modified crops have been identified. Certain issues were recognized for pest resistant crops when it was found that the anti-pest toxin in the crop also killed butterflies and insects in the vicinity.

Potential Disadvantages and Dangers

Numerous potential disadvantages and dangers for GM foods have been identified. Many of these result from the indeterminate nature of natural processes whereby genes transfer between species. A discussion of the actual degree of danger these represent is beyond the scope of this introductory material, so we will limit the following discussion to identifying what these risks actually are.

Economic

It is difficult to predict what future economic disadvantages will result from GM foods. However, if any of the disadvantages mentioned below actually occur, the potential for economic loss is great. There is also a risk of economic disadvantages for developing countries depending on the policies taken by bioengineering companies when selling GM products and on policies and restrictions for their exportation of GM food.

Agricultural and Environmental

Numerous possible scenarios of disadvantage and danger have been identified in relation to agriculture itself and the environment. Most of the disadvantages relating to these two areas are mutually connected; hence the two have been grouped together.

<u>Herbicide Tolerance</u>

A number of potential risks from plants that have been modified for herbicide tolerance have been identified, including:

- Weed problems. The crop may actually become a weed itself by spreading to other fields or by appearing in the following crop. It is also possible that the GM herbicide tolerance trait will transfer to weeds close in species, thereby creating 'super weeds' that are themselves tolerant to herbicides
- The broad spectrum herbicides used with GM herbicide resistant crops do effectively clear weeds, but clearing this deprives many small animals of a valuable food source.
- Although the overall use of herbicides will reduce, the amount of the specific 'tolerant' herbicides will increase dramatically leading to possible unforeseen consequences or the possible evolution of resistant weed strains.
- Nearby crops that are organic may become pollinated by the GM crop resulting in contamination of organic crop.

Pest Tolerance

As for herbicide tolerance a number of potential disadvantages have been identified for plants modified for pest tolerance, including:

- Pest resistance: Insects are known for adapting to various pesticides, although the process of adapting to pest tolerant plants is likely to be slower, many scientists believe that insects will become resistant to these plants eventually causing crop failure.
- As with herbicide resistance, it is feared that genes may transfer to other species, creating pest tolerant 'super weeds'
- The toxin killing pests may also kill harmless non-target insects (Losey *et al*., 1999).

Disease Resistance

Plants bred for disease resistance will also have the potential cross pollination and 'super weed' disadvantages described above. But, a few other areas of potential problem have also been identified, including:

- Bacteria and fungi may develop resistance and new disease resistant GM crops must be developed.
- Viral DNA in the plant may possibly combine with an infecting virus and produce a new virus. The new virus may be even more virulent than the original virus which the GM plant was bred against

Health

Possibly the biggest bone of contention for opponents of GM food is the potential risk to public health. However there is little evidence to support either its safety or danger. Opponents say that the lack of scientific research into the potential risks to health is due to the protectionist behaviour of large biotech companies who do not wish to find evidence of possible health hazards in the new technology they have spent huge amounts of money developing, while supporters of GM food say there is little need for extensive long-term testing when GM food has already been proven to be one of the safest developments ever introduced as so far not a single death can be attributed to it. Concerns about public health are usually levelled at possible allergic reactions, creation of new antibiotic resistant viruses and bacteria, antibiotic resistance, and cite laboratory tests already made on GM food.

Allergic Reactions

The possibility of allergic reactions may be divided into two categories; the introduction of known allergens and the unforeseen creation of new allergens (Nordlee et al., 1996).

During the splicing of genes from one species to another, allergens from one gene may get accidentally transferred into the other. An example of this occurred when an allergen from Brazil nuts was discovered in a GM soybean. This meant that people allergic to Brazil nuts may have the same reaction to the soybean (Nordlee et al., 1996). This particular allergen was found before the soybean was marketed and so was scrapped. However, the possibility exists of allergens being transferred and not discovered.

<u>Creation of new antibiotic resistant viruses and bacteria</u>
There is a possibility that genes may transfer between species, especially seeing as genes do not always break up in the alimentary tract and intestinal bacteria have been known to assimilate genes and GM plasmids. With this occurring, it is possible that these genes may create new bacteria viruses that present drugs are unable to counteract. If, for example, an antibiotic resistant gene was transferred to a bacterium such as streptococcus, an antibiotic resistant bug would be formed.

<u>Laboratory Tests</u>
Although hard to substantiate, claims have been made that laboratory tests used to determine the safety of certain approved GM foods were inaccurate, poorly designed and that important procedures were not carried out, such as histology of the rats' digestive tracts.

V. Reducing Risks from Genetic Modification

Genetically modified food has become one of the most controlled and monitored substances ever. This is for two main reasons: governments do not wish to take any risks with the health of their citizens, the country's economy, and the environment; and biotech companies, apart from also being concerned with the same risks mentioned above, do not wish to be open to huge lawsuits in the future.

It goes without saying that most governments are taking all the measures they deem necessary to both protect the interests and safety of their citizens while ensuring the ability to take advantage of this new technology and also protect the interests of biotech corporations.

Currently, the main method of reducing agricultural risk is to segregate GM crops from other nearby similar crops in order to prevent cross contamination. However, crop segregation is not always strictly controlled and has been proven to be ineffective. Recent guidelines have been made that suggest further procedures to prevent risk. These suggest plants should be engineered that are genetically incompatible with local species to prevent gene contamination, the minimum amount of genes necessary to produce the desired trait should be used, the use of antibiotic resistant genes should be reduced, and that genes should only be added when absolutely necessary. However, such guidelines are only recommendations, and are not enforced by law.

To reduce risks, various countries have taken up policies by restricting the use of GM foods, and labeling foods that contain them. These policies vary from country to country and depend on various factors such as the country's economy, the political climate, and the amount of social opposition voiced to GM foods.

Policies on restriction and testing

Various stances have been taken throughout the world on the restriction of GM foods from having no policy, through to restricting its use, and in some places completely banning it. Testing for GM food content also varies from country to country where some countries perform no testing, some have voluntary testing, and some have mandatory testing (Whitman, 2000).The majority of countries currently have no policy on restriction, including Hong Kong, India, and many third world countries. Some of these, like Hong Kong, have no policy because GM crops are not grown locally. And third world countries are unlikely to make any restrictions partly through political inability to instate procedures, and partly because GM foods may offer a way to solve endemic food shortages. Areas like the European Community restrict the amount of GM food 'contamination' in other unmodified foods to 1%, while in certain areas of Brazil GM food is completely banned.

Labeling

Labeling is probably one of the most contentious issues related to GM foods, both politically, socially and economically. The standpoint taken on labeling has particular significance to international trade and on consumer usage of GM products. However, labeling is also a way of giving consumers an 'informed choice' of whether to use GM products or not. GM corporations believe that labeling should be voluntary and that the market should regulate itself, whereas consumer organizations believe that labeling should be mandatory as consumers have a right to know what they are eating (Wong, 2003).

In January 2000, an international trade agreement on the labelling of GM foods was made. Over 130 countries signed the agreement, including the United States, who is the world's largest producer of GM foods. In this agreement, each country agreed that exporters must label all GM foods and that importing countries may judge for themselves the potential risks, and be free to use or reject GM foods as they choose (Wong, 2003).

At present, labeling is carried out in two ways: the selective approach and the pan-labeling approach. The selective approach simply requires that foods not substantially equivalent to the same traditional type of food require labeling. This approach has been adopted mainly in those countries that are large producers of GM food such as the United States and Canada (Wong, 2003).

Pan-labeling is a requirement for when a foodstuff contains over a certain amount of GM foods, or has a significantly different characteristic such as an allergen, which must be labeled. Some countries, like China, have proposed mandatory labeling of all food containing any GM foodstuffs; however this is not practicable as extremely low levels are undetectable. Other areas like the European Community require products to be labeled if they have over 1% GM foods, while countries like Japan and Korea adopt a limited labeling approach (Wong, 2003).

Conclusion

Since their first appearance over ten years ago, GM foods and their consumption, have been a topic of furious debate. Obviously, GM foods have the potential to bring great benefits, such as improving nutrition, but equally obviously they have the potential for causing a variety of agricultural, environmental, and health problems. However, neither the benefits nor disadvantages foreseen have yet to be fully displayed.

With any new technology, there is a period of balancing and regulation when first introduced. Yet, for GM foods this process shows little or no signs of universal agreement. Clearly the use of GM food and its consequences deserve as much research as the development of new GM foods, yet economics make this unlikely. Of course, GM foods and the their potential benefits cannot be dismissed out of hand for fear of possible repercussions, so the sensible conclusion is that GM food research should continue, while taking a cautious approach with more unified regulation of this provocative new technology.

Questions for Further Thinking

1. Have you found some GM food in your daily life? Please give an example.

2. The GM industry has wasted no time to solve starvation problems. It will be necessary to feed the growing world population. Are there other sustainable solutions to help farmers and people from developing countries to fight poverty and hunger?

第十章　基因改造食物

基因改造食物是一個具爭議性的主題。一方面，反對者的理由是基因改造食物引致遺傳病和環境災難。另一方面，建議者認為基因改造食物可以解決饑餓，預防疾病和農作物枯萎，甚至改善環境。事實是，基因改造食物不知會帶來什麼後果。不過，為了對使用和發展這種食物作出判斷，我們首先必須理解基因改造食物是甚麼，現有和潛在的好處和不利的條件是甚麼，以及存在的風險，如何把風險減至最少。這一章會討論這些主題。

1. 基因改造食物是甚麼？

人們運用遺傳方法於農作物已有數百年歷史。例如，農夫一直雜交繁殖各種農作物，嘗試繁殖一種抗寒的粟米，一種具有大顆的穀粒但不會下垂的稻米，或沒有種子的西瓜。這全部需要精巧的遺傳處理，加強植物某個特點，以達到要求的結果。

基因改造 (GM) 食物一詞普遍指利用分子生物學技術改造植物的基因。它們有時也稱為 GMOs（基因改造生物）或跨基因食物 (Whitman, 2000)。

基因工程

基因工程，生產基因改造食物，是指一種修改可食用植物的基因特點，以增加某個特性或特點的技術。相關的基因被抽離，然後連接到另一種植物。基因工程技術達到的僅僅是抽離不同的基因，並插入到不同的植物。基因由一種植物轉移到另外一種植物的潛力是巨大的，不過當基因從非植物轉移到植物時，就變得複雜了。最近的研究甚至把生存在冷水的魚類基因轉移到馬鈴薯和其他農作物，以抵抗嚴寒。

基因改造食物的發展

1994年生產的 Flavr Savr 番茄是第一種在市面出售的基因改造食物。透過從一條北極的鰈魚基因連接到這種番茄，以加長存放的時間。自此，基因改造食物生產變得更廣泛的，每年有超過10% 的增長率。2003 年報告預計在 18 個國家，超過七百萬個農夫，栽種多達六千七百七十萬公頃的基因改造農作物。不過，其中 99% 的農作物僅僅六個國家種植，美國，阿根廷，加拿大，巴西，中國和南非，美國佔 63%，而中國和南非則分別減少至4% 和 1%（ISAAA, 2003）。

但是種植甚麼類型的基因改造農作物？我們已經提及了番茄，但是這些不再是一種主要的農作物。根據統計數字，四種主要的農作物如今是：

四種主要農作物的統計2003

農作物	百萬公頃
大豆	41.4
粟米	15.5
棉花	7.2
油菜	3.6

Data source: ISAAA (International Service for the Acquisition of Agri-biotech Applications)

其他基因改造農作物，例如菊苣，南瓜，馬鈴薯和木瓜也被生產，但是規模較小。

II. 基因改造過程

為了理解基因改造食物，它的潛力和過程，就必須知道基因是什麼和怎樣改造基因。

DNA，基因，和基因改造食物

DNA

化學的DNA於1896年首先被發現，但是它對遺傳的重要性要直到1953年才被發現。那年，兩名英國科學家，James Watson和Francis Crick創造了被稱為世紀科學的發現-就是宣佈DNA的結構是“雙螺旋聚合體”；這意味著兩條DNA鏈纏繞一起，並由磷酸鍵連接。

基因

簡單地說，一個基因是DNA的一部分，包含鹼基對的序列，以決定生物的遺傳特徵。

設計基因改造食物

基因改造的實際過程涉及從一個生物取出基因，剪接到另一個生物的DNA。過程是相當複雜，不過我們會只考慮四個基因改造步驟：

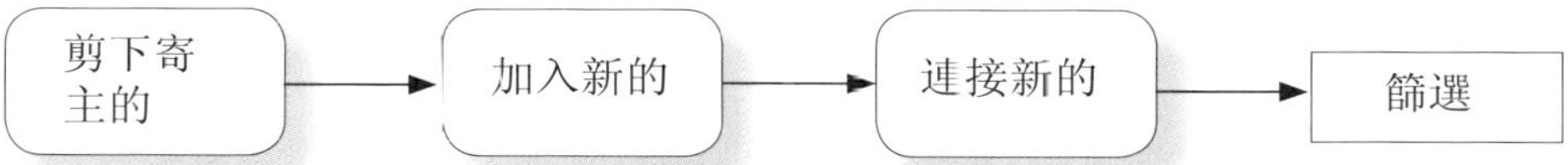

使用限制性內切酶切開DNA某一點，留下一個缺口給新的DNA插入。然後把新的DNA插入細胞。這個過程是相當困難的，通常會使用逆轉錄病毒攜帶DNA，穿過細胞壁到細胞核。一旦插入新的DNA，連接酶會將DNA 連接起來，細胞開始執行新的DNA指令。不過，連接DNA的過程不一定順利，DNA連接錯誤的地方或不完全連接起來。因此利用一種特別的藥物殺死所有不想要的細胞，只留下正確連接的DNA細胞。

很明顯，基因改造過程考慮一個生物或受基因改造的植物結構的特定改變。這個部分一開始，我們簡單提及透過對植物的DNA進行特定的改變，帶來的好處和不利條件。隨後，我們會深入討論這些好處和不利條件，及如何減少風險。

III. 基因改造的好處

我們曾提及基因改造已有數百年歷史，但是基因工程比傳統雜交繁殖有什麼優勝之處，以及基因工程可以做到傳統雜交繁殖不能做到的地方呢？此外，基因工程目前的成就和前景如何？

傳統繁殖

傳統的植物繁殖是同種植物互相雜交，結合它們兩個或以上的預期特性。事實上，傳統繁殖理論上有能力達到基因改造能做到的事情，不過基因工程達到的速度較高和可能性更大。下表可以幫助理解（FEHD, 2005）。

傳統繁殖	基因改造
數以萬計的基因同時混合	可選擇、分辨並轉移個別基因
需要較長時間的觀察和自然進化，預期的特性才會彰顯	預期的特性可在一代之內彰顯
通常只限於同一生物品種	可把特定的基因引進另一生物品種之中

現存的得益

經濟

基因改造食物是新類型的商業，沒有經濟利益的長期研究。可是，短期研究顯示基因改造食物改善產量（在高蟲害和雜草壓力期間），減少除草劑和控制害蟲的成本。

農業

除了帶來經濟利益外，農作物也有特定的好處。這些好處並不是即時顯現，不過當在高生物和環境的壓力下，這些好處就變得明顯。目前基因改造食物可以抵受寒冷、除草劑和蟲害。

健康

營養和醫藥的得益

在未來，農作物的基因改造可以有更好的營養價值和醫療好處。不過，暫時沒有以這種改良方式的農作物得到批准生產。我們往後會討論這些潛在的好處。

不過，對於目前正在使用基因改造食物的醫療好處。當植物與雜草競爭時，會釋放毒素，危害人類的健康。利用基因改造食物忍耐除草劑的能力，以控制雜草，減少毒素。當害蟲攻擊植物時，它們會損壞植物的防衛層，引致真菌感染，產生真菌毒。真菌毒素可導致癌症，因此有抗害蟲能力的農作物可減少患上癌症，尤其是肝癌。

環境

基因改造食物對環境的好處需要作長期的研究，不過目前有一些對環境的好處是相當明顯的。減少使用殺蟲劑和除草劑可以減少地下水污染和對非目標物種，例如蝴蝶的影響。提高農作物的產量可減少種植的土地，讓森林再生，以改善環境。減少種植可幫助穩定土壤，減少侵蝕和水土流失。

其他基因改造植物，不被培植為農作物，也對環境帶來好處。利用基因改造某品種植物，清除受污染土壤的重金屬。

潛在的得益

前一節，我們討論基因改造食物的現存好處。不過，有大量研究顯示基因改造食物可以帶來更廣泛的好處，尤其在農業和健康方面。

經濟

基因改造食物的未來經濟得益收取決於公眾態度的發展。現時，基因改造食物受到嚴格控制和監視。如果將來可減少控制，及大眾接受基因改造食物如日常的食物，種植這類農作物的經濟得益會增加，新的品種可以大量生產。

農業

除了目前基因改造的農作物的得益外，還有為未來帶來更重要的得益。不過，基因改造食物的好處不僅局限於這些領域。其他領域的研究工作正在進行，嘗試改善各種植物的農業特性。大部的研究集中於：

耐旱／耐鹽

儘管沒有基因工程公司在經濟上推動大量的研究，不過抗旱和抗鹽的研究為貧窮地區的農業帶來莫大的益處。例如非洲和亞洲多處，受到乾旱和土壤高鹽水平的影響，而且灌溉困難的地區，是最大的得益者。

抗病

各種的病毒，真菌和細菌影響著植物。科學家正在生產抗病的農作物。通常做法是透過從病毒取出基因，加到植物，以產生抗病毒/抗菌的化學物。

健康

基因改造食物透過改良食物的營養及加入藥物，為人類的健康帶來好處。發展營養改造的食物，如「Golden Rice」，用基因方法將維生素A加入稻米，以抵抗缺乏該種維他命。不過，就像大部分基因改造食物的發展一樣，尚未在大部分國家獲得批准。然，在營養學和醫學的領域裡有其他的好處；其中一些如下：

營養

全球有大約八億五千萬人營養不良。原因是依賴單一的主要糧食，如稻米，馬鈴薯和木薯。不過，這些食物很少包含所有健康飲食的營養。基因改造食物透過改變營養，加入新的營養如維生素A，以解決問題。透過加入不同的營養，加強健康，為貧窮國家的未來帶來很多的好處。

醫藥

醫藥的製造，運輸和儲存製經常引發問題，尤其在窮困的地區。科學家在嘗試發展可食用疫苗的基因改造食物，因此更容易生產，運輸和管理這些疫苗。

IV. 基因改造的不利條件

全球很多國家強烈抗議使用基因改造食物。很多人害怕這種新興，具干擾性技術所帶來的危險。有部分人認為科學家的一個基因代表一個特性的推論是有缺陷，因為基因剪接後的後果是未知的。普遍認為一個基因代表一個蛋白質，不過最近發現一個基因可以生產多達38,000種蛋白質。除非一個基因的行動在連接前被完全理解，否則可能出現有害的結果。在這個部分，我們會討論基因改造的目前和潛在的不利條件。

現存不利條件

經濟

基因改造食物的反對者強調基因改造食物帶來的經濟好處並不如生物科技公司所提出的那樣明顯。在窮困國家，對大部分農民來說，每年購買新的基因改造農作物的種子不是一個可行的選擇。他們也指出儘管在某些方面得到節省，不過監察和測試基因改造食物抵消了經濟益處。

農業
儘管識別了很多基因改造食物的潛在問題，不過到目前為止卻沒有研究基因改造食物對農業的重大的不利條件。

健康
至今，沒有很多基因改造食物的健康問題。不過，一兩個案例表明了未來出現的潛在問題。例如，基因改造的大豆被插入一個巴西果仁的基因。插入這個基因引致產生致敏原，但是這致敏原在測試期間被識別，並沒有使用該植物。

環境
如其他領域一樣，種植基因改造的農作物沒有幾個重大的環境影響。無疑，農作物裏的抗蟲毒素也會殺死蝴蝶和昆蟲。

潛在的不利條件和危險
很多關於基因改造食物的潛在不利條件和危險均被識別。很多源於物種之間基因轉移的自然過程。討論這些危險的確實程度已超出這套教材的範圍，因此以下討論只限於基因改造的風險。

經濟
基因改造食物的未來經濟不利條件是很難預測的。不過，如果發生以下提及的任何不利條件，經濟損失是很重大的。在發展中國家的經濟不利條件取決於生物工程公司的政策，售賣基因改造產品和出口的限制。

農業和環境
很多與農業本身和環境的不利條件和危險均被識別。大部分關於農業和環境的不利條件是相互關連。

耐除草劑
很多被改造成耐除草劑的植物有潛在的風險，包括：
- 雜草問題。當農作物傳播到其他地方時，它本身也成為雜草。基因改造的耐除草劑特性會轉移到其他品種，因此成為“超級雜草”。
- 基因改造的耐除草劑的範圍很大，有效清除雜草，不過會破壞很多小的動物和珍貴的食物資源。
- 儘管除草劑的總體使用量會減少，不過特定的除草劑卻急劇增加，引致無法預料的結果或抗雜草品種的進化。
- 基因改造的農作物可能傳粉到附近的農作物，造成有機農作物的污染。

耐蟲害
正如耐除草劑，基因改造植物的潛在不利條件為：
- 抵抗害蟲：儘管適應耐蟲害的過程可能較緩慢，不過昆蟲可以適應各種的殺蟲劑，很多科學家相信昆蟲會抵抗這類植物。
- 正如抗除草劑一樣，基因可以轉移到其他物種，造成超級耐蟲害的品種。
- 毒素可殺死害蟲，也可以殺死非目標和無害的昆蟲。

<u>抗病</u>
抗病植物也會有互傳花粉和“超級雜草”的潛在不利條件。不過還有其他的問題：

- 細菌和真菌可作出抵抗，那麼要發展抵抗新疾病的基因改造農作物。
- 植物的病毒DNA與傳染病毒結合，產生一個新的病毒。這個新的病毒或許較原先的病毒更為致命。

健康
對於基因改造食物的反對者，最大的爭論是對於公共衛生的潛在風險。不過，沒有證據顯示基因改造食物是安全或危險。反對者說那缺少對健康的潛在風險的科學研究，是因為生物科技的公司投下大量資金發展基因改造食物，基於保護的心態，不希望找到新科技影響健康的證據，而基因改造食物的支持者認為不需要作長期的測試，因為到目前為止基因改造食物被證實為其中一個最安全的發展，並沒有引致死亡的個案。公共健康的關注通常是過敏反應，產生耐抗生素的病毒和細菌，以及在基因改造食物上的測試。

<u>過敏反應</u>
過敏症的反應可分為兩類：可知的致敏原和無法預料的新致敏原。

當在基因從一個物種連接到另外一個物種時，致敏原從一個基因偶然傳到另一個基因。例子是基因改造大豆發現了巴西果仁的致敏原。這意味對巴西果仁致敏的人也同樣對大豆過敏。這個致敏原在銷售前被發展，因此掉棄這批大豆。不過，過敏原轉移的可能性依然存在而且不被發現。

<u>創造耐抗生素的病毒和細菌</u>
當基因在物種之間轉移時，基因可能在食道不被分解，腸細菌將會同化基因和基因改造食物質體。因此，可能產生藥物不能夠對抗的新細菌病毒。例如，如果一個耐抗生素基因被轉移到鏈球菌時，就會形成一隻耐抗生素的細菌。

實驗室測試
儘管很難證明，不過有部分聲稱實驗室測試基因改造食物的安全性是不準確的。不良的設計及沒有進行重要的程序，如老鼠的消化道的組織學。

V. 減低基因改造的風險

基因改造食物受到控制和監視。有兩個主要的原因：政府不希望以公眾的健康，國家的經濟和環境承擔任何風險；此外，生物科技的公司不希望面對大量的訴訟。

大部分政府所推行的措施是同時保護公眾的利益和安全，以及確保利用這種新技術保護生物科技公司的利益。

目前，減少農業風險的主要的方法是基因改造農作物與其他類似的農作物隔離，以防止交叉感染。不過，農作物隔離不一定嚴格控制和證明是沒有效的。最近的指引建議更多的程序以防止風險。建議植物被設計為在基因方面與本地品種不相容，以阻止基因污染，使用最小基因數目生產預期特性，減少使耐抗生素基因，而在必要時才加入基因。不過這些指引僅僅是建議，沒有法律效力。

為了減少風險，各國透過政策限制基因改造食物，及標籤化。這些政策各國不同，取決於該國的經濟，政治氣候和反對基因改造食物的程度。

限制和測試的政策

不同國家對限制基因改造食物有不同立場，從沒有政策到限制使用，甚至一些地方徹底禁止。不同國家測試基因改造食物的內容也不同，一些國家不實施測試，一些進行自願性測試和一些進行強制性的測試。

大部分的國家目前沒有限制政策，包括香港，印度和很多第三世界國家。其中一些，如香港，沒有政策，是因為沒有種植基因改造的農作物。第三世界國家不願作出任何限制，部分原因是政治上沒有能力任命程序，部分是因為基因改造食物可以提供一種解決地方的糧食短缺。像歐洲共同體那樣的地區限制基因改造食物為1%，以防“污染”其他不被改造食物，在巴西甚至徹底禁止基因改造食物。

標籤化

標籤化大概是與基因改造食物有關的其中一個在政治，經濟和社會上最有爭議的議題。標籤化對基因改造食物的國際貿易和消費者都有特別的意義。不過，標籤化也是給消費者的一種“非正式的選擇”以表示使用基因改造產品與否。基因改造食物公司相信標籤化應該是自願性及市場自我調節，而消費者組織相信標籤化應該是強制性，消費者有權利知道他們吃的是什麼。

在2000年1月中，為基因改造食物標籤化達成國際貿易協議。超過130個國家簽署協議，包括美國，世界最大生產基因改造食物的國家。在這個協議中，每個國家同意出口商必須在所有基因改造食物貼上標籤，而輸入國可衡量對自己潛在的風險，自主地使用或拒絕基因改造食物（Wong, 2003）。

目前，標籤化有兩種：選擇性的方法和全面性的方法。

選擇性的方法只是要求那食物不同於傳統食物就需要標籤。基因改造食物的大生產國主要採用這種方法，例如美國和加拿大。

全面性的的方法是當食品含一定程度的基因改造，或有明顯不同的特徵，例如致敏原，就需要標籤。一些國家，像中國，擬議的將所有基因改造食物標籤化；不過這是不可行的，因為低程度的基因改造是測不出來的。像歐洲共同體要求產品有超過1%的基因改造，就需要貼上標籤，而日本和韓國則採用限制標籤方法。

結論

自從十幾年前第一次出現，基因改造食物和消費是激烈爭論的一個主題。顯而易見，基因改造食物帶來潛在的好處，如改善營養，同樣它們亦引致各種農業，環境和健康問題。不過，好處和不利條件都尚未完全顯示。

任何新技術，當第一提出的時候，需要時間去平衡和制定法規。不過，對基因改造食物沒有得到普遍的協議。很明顯，基因改造食物和使用結果需要大量研究，不過經濟使這變得不大可能。當然，不能因為害怕基因改造食物的結果，以不理會它的潛在好處，所以基因改造食物的研究應該繼續進行，而這種新的技術需要採用謹慎的方式和統一的法規。

思考問題

1. 在日常生活，你能夠找到一些基因改造食物嗎？試舉例.

2. 基因改造工業尋求解決飢餓的方法，以應付增長的人口。有沒有其他可持續的方案去幫助發展中國家的農民和人民解決貧窮和飢餓?

References 參考資料

Daniell H, Streatfield SJ and Wycoff K (2001). Medical molecular farming: production of antibodies, biopharmaceuticals and edible vaccines in plants. *Trends in Plant Science*, Vol 6, No 5, pp 219–226.

FEHD (2005). *New technology – genetic modification (GM) foods* (Chinese / English) Website: http://www.fehd.gov.hk/safefood/gmf/index1.html

Gene Translation: RNA -> Protein (2005). Website: users.rcn.com/jkimball.ma.ultranet/biologyPages/T/Translation.html

ISAAA (2003). *Global status of commercial transgenic crops 2003.* Website: http://www.isaaa.org/Resources/Publications/briefs/isaaa_briefs.htm

Losey, Rayor, and Carter (1999). Transgenic Pollen Harms Monarch Larvae. *Nature*, vol. 399, no. 6733, p. 214.

Moellenbeck DJ, ML Peters ML, Bing JW, Rouse JR, Higgins LS, Sims L, Nevshemal T, Marshall L, Ellis RT, Bystrak PG, Lang BA, Stewart JL, Kouba K, Sondag V, Gustafson V, Nour K, Xu D, Swenson J, Zhang J, Czapla T, Schwab G, Jayne S, Stockhoff BA, Narva K, Schnepf HE, Stelman SJ, Poutre C, Koziel M and Duck N (2001). Insecticidal proteins from *Bacillus thuringiensis* protect corn from corn rootworms. *Nature Biotechnology* 19, 668–672.

Nordlee, J.A. S.L. Taylor, J.A. Townsend, L.A. Thomas and R.K. Bush. (1996). Identification of a Brazil-nut allergen in transgenic soybeans. *New England Journal of Medicine* 334, 688–692.

Nottingham, S. (1998). *Eat your genes: How genetically modified food is entering our diet.* Zed Books Ltd., London.

Swiss Federal Institute of Technology Institute for Plant Sciences. Genetic engineering towards carotene biosynthesis in endosperm.

Whitman. D. B. (2000). *Genetically modified foods: harmful or helpful?* Website: http://www.csa.com/hottopics/gmfood/overview.php

Watson, J.D. (1968). *The double helix*. New York: Atheneum

Wong, D. (2003). *Genetically modified food labelling*, Research and Library Services Division, Legislative Council, Hong Kong. 35pp.
Website: http://www.legco.gov.hk/yr02-03/english/sec/library/0203rp05e.pdf

Zhang H and Blumwald E (2001). Transgenic salt tolerant tomato plants accumulate. salt in foliage but not in fruit. *Nature Biotechnology*, Vol 19; Aug 2001; pp. 765–768.

Chapter 11. Solid Waste Management

Waste is a common problem of most developed societies. It is particularly when people can afford greater convenience and have more purchasing power, that they tend to throw away more garbage. Hong Kong is no exception to this. According to the data from Environmental Protection Department (EPD, 2005a), every year the people of Hong Kong dispose of over six million tonnes of refuse, enough to make a 13-floor building of the area of Victoria Park. In 2003, 17,760 tonnes of solid waste on average a day were landfilled. Domestic waste generation is twice as much as 20 years ago. It is expected that waste arising will be doubled in 20 years.

I. Types of Waste

Operation of landfill © Frank Lee

Hong Kong generates several different types of waste, and each has its own requirements for handling (EPD, 2000). **Municipal solid waste (MSW)** is solid waste from households commercial and industrial sources. MSW is disposed of at landfills. **Construction and demolition (C&D)** waste includes waste arising from activities like construction, renovation, demolition, land excavation and road works. Ideally, the waste is separated and inert material is used as fill in reclamation sites, when available. However, a significant portion of waste still goes to landfills.

Chemical waste comprises substances which pose a possible risk to health and/or the environment. **Special wastes** include clinical waste, animal carcasses, livestock waste, radioactive waste, grease trap waste and sewage sludge. These wastes need to be treated separately. Some livestock waste is treated at the Sha Ling Composting Plant.

In 2003, the distribution of total waste in Hong Kong is:

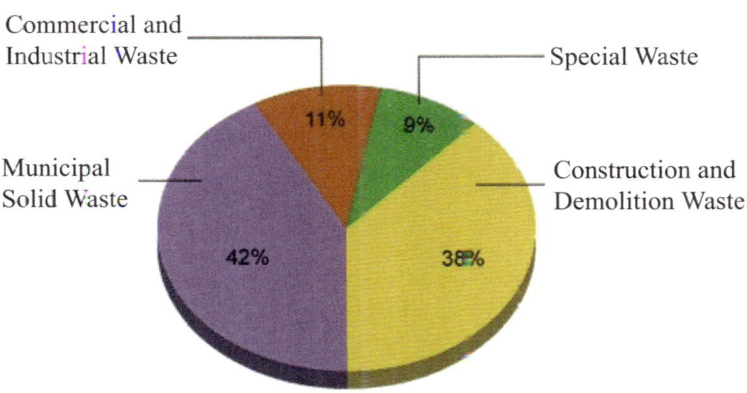

(source: EPD, 2005a)

II. Waste Management and Disposal

The amount of waste increases along with the population. Our government has put effort in research and developed new methods in waste treatment. To collect, transport and dump 1 tonne of municipal solid waste, the Government is currently paying (EPD, 2000):

- HK$850 for treatment cost
- HK$125 for landfill construction and operation

Using solid waste as an example, we are disposing of 9,300 tonnes of refuse per day, representing the treatment cost of HK$7.9 million. In one year it will add up to HK$2.9 billion, which is primarily paid by taxpayers.

The fates of municipal solid waste (modified from EPD, 2000):

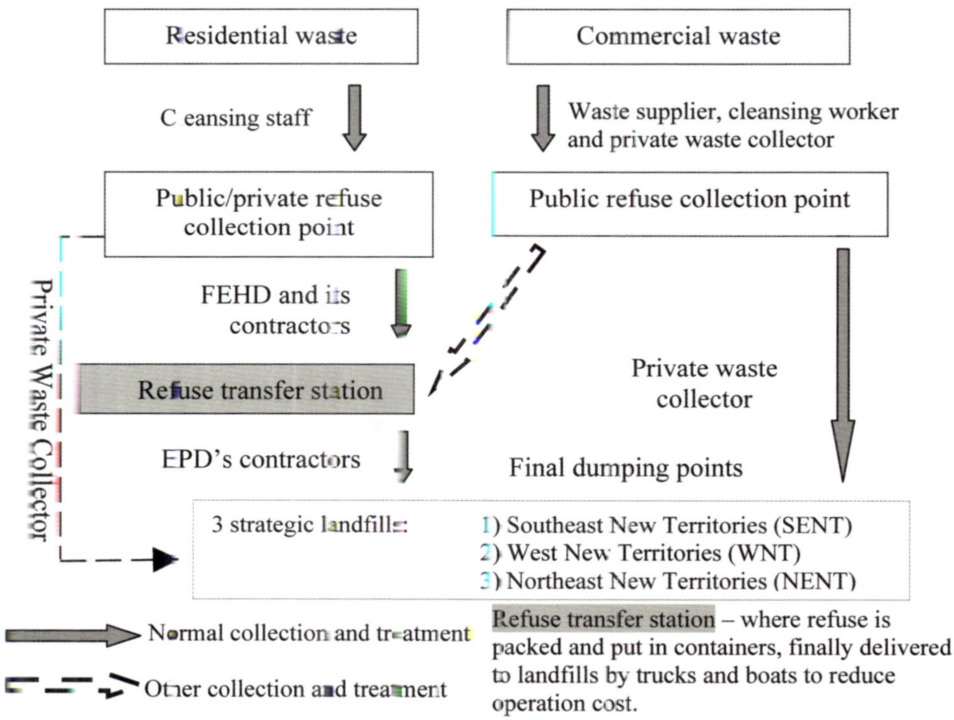

Waste is also managed through legislations. The **Waste Disposal Ordinance** is used to enforce controls on waste disposal, including collection and disposal and the import and export of waste. The **Dumping at Sea Ordinance** is enforced to control disposal of dredged mud and excavated materials at designated marine disposal sites. **Livestock Waste Control Scheme** has been fully implemented in the management and disposal of livestock waste.

Waste Management Facilities in Hong Kong:

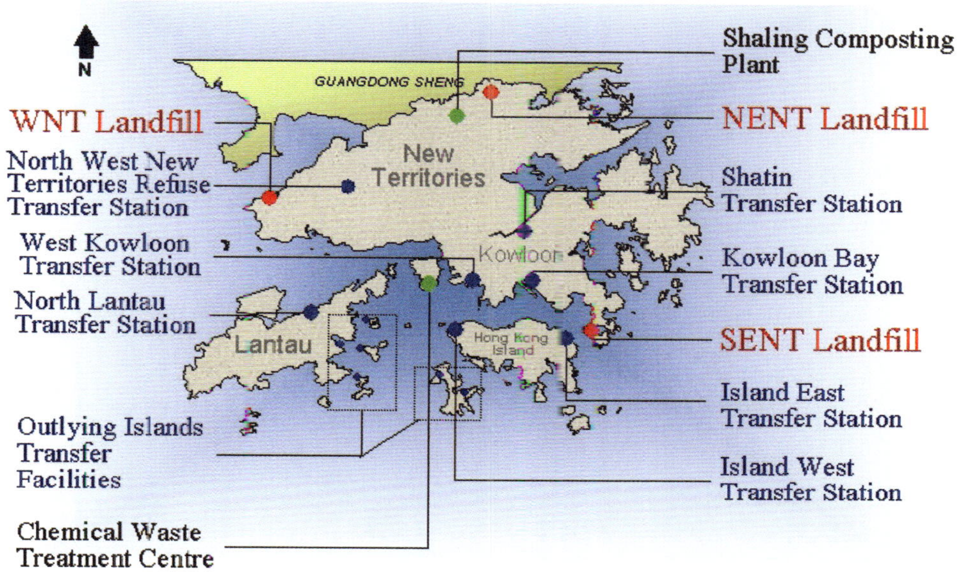

(modified from EPD, 2005b)

Landfilling

Landfilling consists of adding a soil layer on top of solid waste (Ng, 1993). Waste decomposes inside the soil, destroys organic materials, and releasing methane and carbon dioxide as by-products. Dioxin gas from high concentrated methane can be used as power source for generating electricity.

It is estimated that the three strategic landfills will be completely filled up in 10 to 15 years. Not only will developing new landfills be costly, this is not the best solution to waste treatment, considering the city's crowdedness and incredibly small size. The landfills are:

- NENT (in Ta Kwu Ling 打鼓嶺)
- SENT (in Tseung Kwan O 將軍澳)
- WNT (in Nim Wan 稔灣)

SENT landfill in Tseung Kwan O
© Frank Lee

Also, the special waste facilities include **Chemical Waste Treatment Centre** in Tsing Yi and **Shaling Composting Plant** in Sheung Shui.

Ng (1993) illustrated that there are some drawbacks of landfilling to be considered. The most obvious feature from landfill refuse is the bad odor, which pollutes the environment and is harmful to our health. Secondly, landfilling pollutes groundwater. The decomposed methane gas could also penetrate into households, posing a risk of sudden explosion. Another problem is that the refuse-water composite is difficult decompose into water and carbon dioxide. Finally, industrial waste is harmful to human health with the toxic chemicals and metals.

III. Recycling of Waste

Waste recycling has indeed a long history for mankind. In the old days, manures from poultry were recycled as fertilizers for crop production. Today, waste paper, aluminium cans and plastic bottles are the major items for recycling. In particular:

- Waste paper will become recycled paper and cardboard paper.
- Aluminium will be turned into new aluminium can or other materials.
- The recycled products of plastic bottles have a wide range of usage. They will be produced as benches for public parks, clothes, toys and other plastic materials.
- Other products such as compact discs, batteries, printing toners are also favorite recycling items.

Waste reduction and recycling are very important elements of the local waste management framework. They help both to conserve natural resources, and to reduce demand for valuable landfill space.

Through the existing waste recovery system, about 2.4 million tonnes of municipal solid waste were recovered in Hong Kong in 2003. Of that total, 9% was recycled locally and 91% was exported to the Mainland and other countries for recycling, with an export earning of more than HK$2.5 billion for Hong Kong (EPD, 2005a).

Major materials recovered/recycled in 2003:

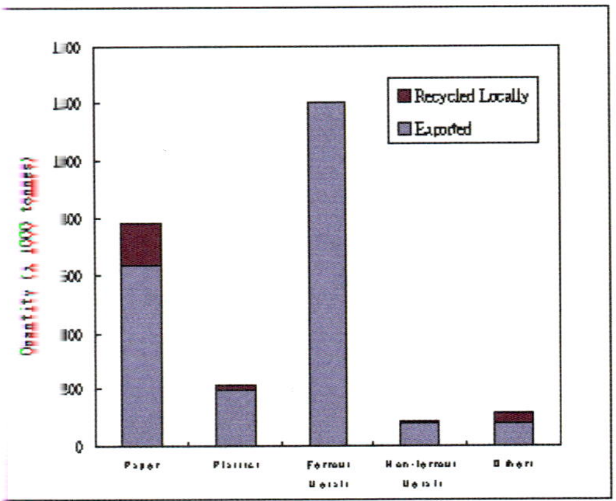

(source: EPD, 2005a)

Estate management of most public and private housing estates has strived to enhance the public's participation rate in waste reduction and recycling. At present, over 60% of the population in Hong Kong has a chance to participate in the waste recovery programmes.

The Recycling Industry

Examples in other parts of the world (extracted from Barron, 1989; Buclet and Godard, 2000; Cheremisinoff, 1989):

Europe:
- "Take-back" system
- European Union landfill directive
- Reduce biodegradable waste

The United States:
- 28% recycling rate in 1999
- Similar concept; more emphasis on market development

Asia:
- Singapore: 40% recycling rate in 2000
- Japan: 25% recycling rate in 1998
- Taipei: "zero waste to landfill" by 2012
- Shanghai: composting

In **Hong Kong**, the recycling industry has been established for more than 20 years. The characteristics of these companies are export-oriented and limited to a few materials. Not only the markets are dominated by small businesses, many of them also have to compete with international competitors from Southeast Asia and overseas countries as well.

So what are the potentials of recycling industry? Ng (1993) and Cheremisinoff (1989) analyzed as follows:

Strengths:

- Entrepreneurial style
- Market-driven approach
- Establish collection and disposal system
- Close to large export market of the Mainland – short delivery period upon order of buyer
- Small cultural difference between China and local exporters

Recycle bins © Frank Lee

Weaknesses

- High labour and land costs
- Low recovery rate
- Insufficient legal support and government assistance

Recycling depot © Frank Lee

Opportunities

- Takes advantage of low labour and land cost of the Mainland in establishing treatment plants in nearby cities
- Concentration of population provides opportunities for mass collection
- Chances of re-introducing processing industry into the economy

Threats

- Export-oriented recycling is subject to up-and-downs of the international market. Hong Kong's recycling programme relies heavily on external factors hardly under our control.
- Recycling programmes in Europe and North America operate under heavy government subsidy. They export high quality materials at competitive prices.

Recycled products © Frank Lee

IV. Waste Reduction and Other Measures

Waste reduction is often mentioned in waste management, but it is not a simple issue either. Nonetheless, avoiding the generation of waste in the first place and minimizing waste are both crucial in any waste reduction strategy.

The Government announced six major initiatives to enhance the prevention, separation and recycling of domestic waste in Hong Kong on 11 September 2001. These include the injection of HK$100 million into the *Environment and Conservation Fund* and making available long-term land to set up a Recovery Park. These new initiatives were put forward following the early completion of a review of the implementation of the 1998 **Waste Reduction Framework Plan (WRFP)**.

Residential waste © Frank Lee

The six major initiatives include (EPD, 1998):

1. Setting up a 20-ha **Recovery Park** at Tuen Mun.
2. Adding of HK$100 million into the *Environment and Conservation Fund.*
3. Enhancing collection of separated waste and adding 8,000 newly designed separation bins.
4. Enhancing public education and community work.
5. Setting examples on waste reduction by Government.
6. Reinforcing producers' responsibility.

However, the review of WRFP expected that Hong Kong's overall waste recycling rate would increase from 32% in 1998 to 40% in 2009 (EPD, 1998). Obviously, much more work has to be done, especially on the prevention and recovery of domestic waste to better manage the increasingly serious waste problem in Hong Kong.

Since 1 April 2001, the Environmental Protection Department has been running a **plastic bottle collection trial scheme**. As plastic bottles are light and bulky, their transportation is uneconomic and many estates have faced difficulties in finding a large and secure recycling outlet for plastic bottles. This trial scheme has been initiated in order to evaluate the cost-effectiveness of a centralized collection system. Two recycling contractors are responsible for collecting plastic bottles twice every month at about 250 housing estates throughout Hong Kong, Kowloon, and the New Territories. All the plastic bottles collected are sent for recycling. Since the commencement of the scheme, many housing estates, especially those on Hong Kong Island, have shown significant improvement in their plastic bottle recovery rates.

27 million plastic bags are disposed of in Hong Kong every day. © Frank Lee

Other latest campaigns include:

- ***"No Plastic Bag, Please"* Campaign**
- **Waste Bulk Reduction Programme** - consider the treatment methods of incineration and composting, in order to reduce the ultimate volume of waste.
- New waste charging scheme for construction waste disposal at landfills – aims to provide an incentive for waste producers to reduce waste and carry out sorting to facilitate reuse/recycling of waste, in order to slow down the depletion of limited landfill capacity.

The diagram on the right is called the "**Waste Management Principle**". It aims to illustrate that avoiding consumerism gives the greatest desirability, or benefits towards waste management, followed by minimizing the amount of waste, and then recycling and treatment (Cheremisinoff, 1989; EPD, 1998). Finally, disposal of waste is the least desirable because it has the greatest impact on our health as well as the natural environment.

Waste Management Principle

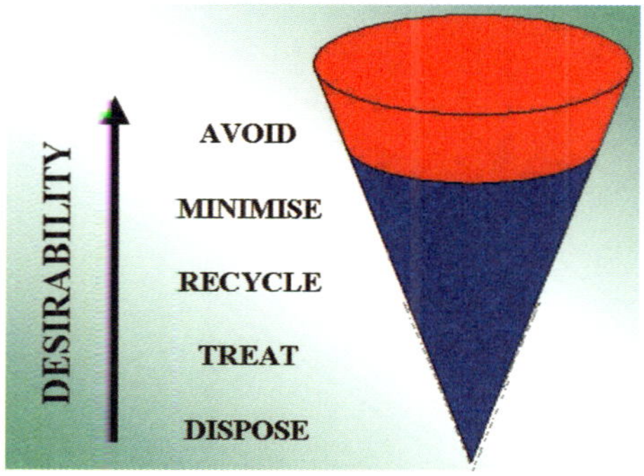

(modified from EPD, 1998)

Visits for Further Study

The management of the Southeast New Territories (SENT) landfill at Tseung Kwan O welcomes students and the public to conduct group visits at the site. For details, please refer to the introduction and guidelines described in Chapter 4.4 (Applying Resources and Waste Management) of the **Field Trip Guide**.

Questions for Further Thinking

1. What would be the best practical environmental options (technologies, management systems, etc.) to tackle the waste problem?

2. Will we eventually solve Hong Kong's waste problem by more recycling?

3. What should the Government do to increase reduction of waste? Any specific measures?

Let's keep our environment clean! © Frank Lee

第十一章　固體廢物管理

廢物是多數已開發的社會的一個共同問題。尤其當人們為了方便及購買力增加時，他們就丟掉更多垃圾。香港亦都如此。根據環境保護署的資料 (EPD, 2005a)，每年香港人棄置六百萬噸廢物，足夠做十三層維多利亞公園面積大小的大廈。於2003年，平均每日有17,760噸固體廢物棄置堆填區。家居廢物是二十年前的兩倍。預計在二十年內廢物數量倍增。

I. 廢物的種類

香港每天產生多種廢物，這些廢物必須以不同方式處置 (EPD, 2000)。**都市固體廢料**是來自家庭，商業和工業的固體廢物，最終被棄置於堆填區。**建築廢物**包括來自建築、裝修、拆卸、土地挖掘及修路等工程所產生的惰性廢物。這類廢物最好是分類處置，惰性廢物可用作填海物料，但通常大部份拆建廢物都會不當地被運往堆填區。

化學廢物指對人類健康及/或環境構成潛在危險的物質。特殊廢物包括**醫療廢物**、**動物屍體**、**禽畜廢物**、**輻射性廢物**、**隔油池廢物**及濾水廠/污水廠污泥。這類廢物需要獨立處理。其中禽畜廢物會運往沙嶺堆肥廠。

2003年香港總廢物量的分佈：

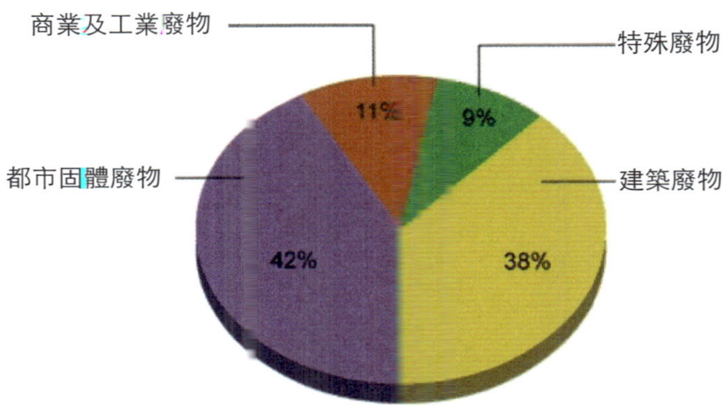

(來源：EPD, 2005a)

II 廢物管理和處置

廢物數量隨著人口增加。政府正研究和開發新方法去處理廢物。要收集，運輸和傾銷1噸的都市固體廢物，政府要支付 (EPD, 2000):

- HK$850 作為處置費用
- HK$125 作為興建和運作堆填區

以固體廢物為例，我們每日處理9,300噸廢物，處置費用是HK$790萬。一年就是HK$29億，主要由納稅人支付。

固體廢物的處理 (EPD, 2000):

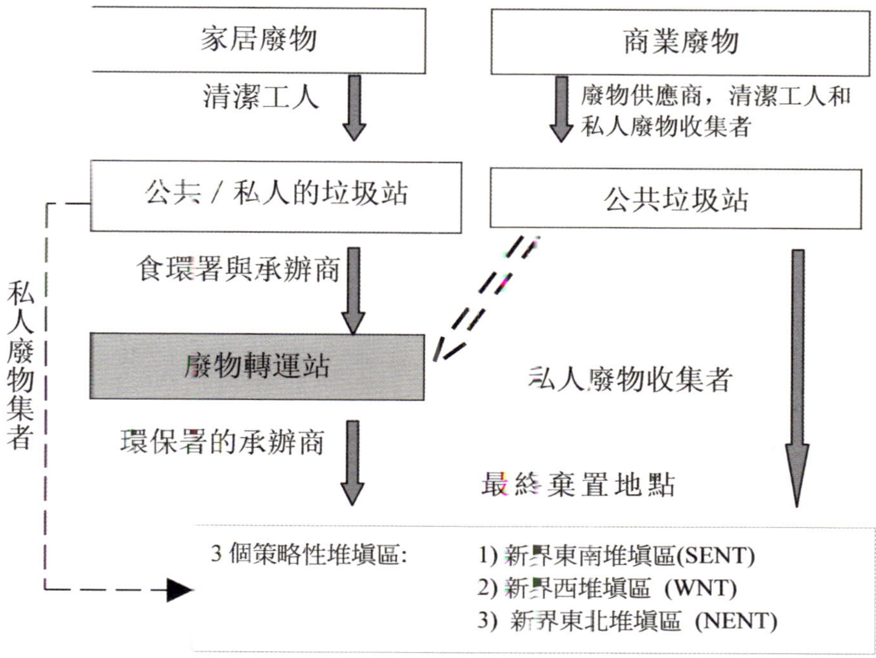

正常收集及處置程序

其他收集及處置程序

廢物轉運站—所收集的廢物會被壓縮，然後裝入特製的貨櫃內，再運往策略性堆填區處置，以降低運費。

政府主要透過立法管制廢物。**廢物處置條例**是管制廢物處置的法例，監管範圍包括廢物的收集、處置及**廢物進出口**。**海上傾倒物料條例**則是管制在指定海泥傾卸區棄置疏浚海泥及挖掘物料的法例。我們已全面實施**禽畜廢物管制計劃**，以管理及處置禽畜廢物。

堆填

堆填是固體廢物上增加一層土壤 (Ng, 1993)。廢物在土壤裡面分解，破壞有機物質，釋放甲烷和二氧化碳，作為副產物。來自高濃度甲烷的二噁英氣體可用作發電。

三戰略垃圾堆填區在10 到15 年內將被填滿。考慮到城市面積小及擠迫，興建新的堆填區不單昂貴，而且不是一個最佳解決廢物處理的方法。現有的堆填區分別位於打鼓嶺、將軍澳及稔灣。

再者，特殊廢物設施包括青衣的化學廢物處理中心及上水的沙嶺堆肥廠。

Ng (1993) 認為堆填有一些缺點。最明顯是堆填廢物有惡臭，污染環境和對健康有害。第二，堆填污染地下水。堆填沼氣滲入民居，造成突然爆炸的風險。另一個問題是，滲漏污水是很難分解成水和二氧化碳。最後，工業廢料含有毒的化學品和金屬，對人類健康是有害的。

III. 廢物回收

回收廢物已有悠久的歷史。以前，禽畜糞便作為種植的肥料。今天，廢紙、鋁罐和膠樽是主要的回收項目。尤其是：

- 廢紙成為再造紙和紙板
- 鋁成為新鋁罐或其它材料
- 回收後的膠樽有很多用途。它們再造成公園的長凳、衣物、玩具和其它塑料材料
- 其它產品，如：光碟，電池，列印調色劑亦是受歡迎的回收物品

減少廢物和回收對於廢物管理是很重要的。他們既保存自然資源，亦減少對堆填區的需求。

透過現有的回收廢物系統，2003年回收了大約240萬噸都市固體廢物。其中有9%於本地回收，另外91%運往大陸和其它國家作回收，為香港取得25億元的收益 (EPD, 2005)。

公共屋邨和私人屋苑的管理加強公眾的參與減少和回收廢物。目前，超過60% 的香港人可以參與廢物回收計劃。

回收工業在世界各地的例子 (Barron, 1989; Euclet and Godard, 2000; Cheremisinoff, 1989)：

歐洲：
- 歐洲共同體的堆填指令
- 減少生物可降解的廢物

美國：
- 於1999年，有28% 回收的率
- 相似的概念；注重對市場的發展

亞洲：
- 新加坡：於2000年，有40% 回收的率
- 日本：於1998年，有25% 回收的率
- 臺北：在2012年，達到「零廢物到堆填區」
- 上海：堆肥

在香港，回收業已有超過20年的歷史。這些公司的特點是只出口某幾種物料。回收市場裏，大部分都是小型公司。這些公司還要與東南亞和海外國家競爭。

然而，回收業有什麼潛質呢? Ng (1993) 和 Cheremisinoff (1989) 有以下分析：

優勢:	不足
• 企業模式	• 勞動力高和土地費用昂貴
• 市場推動的方式	• 低退稅率
• 建立收集和處置系統	• 政府協助不足
• 接近中國的出口市場- 縮短運載時間	
• 中國和本地出口商之間的文化差異	

機遇	威脅
• 利用低工資和土地費用在附近的城市建立處理廠 • 人口集中可收集更多回收物料 • 把加工業重新加入經濟	• 以出口為主的回收會受到國際市場的影響 香港的回收計劃依重外在因素，很難受到控制 • 在歐洲和北美洲的回收計劃得到政府很多資助。他們以較有競爭力的價格出口高質量的物料。

IV. 減少廢物和其它措施

減少廢物涉及廢物管理，不過這不是一個簡單的問題。誠然，減少廢物的最佳辦法是避免製造廢物和減少廢物。

在2001年9月11日，政府宣佈六項建議，以減少，分類和循環使用家居廢物。這包括一億元的"環境和保護基金"和利用長期土地作回收園。這些新動議在1998減少廢物綱要計劃 (WRFP) 實施回顧後提出。

六項建議包括 (EPD, 1998)：

1. 在屯門設立 20 公頃的回收園。
2. 增加一億元作「環境和保育基金」。
3. 加強廢物分類和增加 8,000 個新設計的回收箱。
4. 加強公眾教育和社區工作。
5. 由政府作減少廢物的榜樣。
6. 增加生產者的責任。

不過，WRFP 的回顧期待香港的整體回收廢物率會由1998年的32%增加年至2009年的40% (EPD, 1998) 。明顯地，要投入更多的努力，尤其是減少和回收家居廢物，以處理越來越嚴峻的廢物問題。

從2001年4月1日，環境保護署推行膠樽回收試驗計劃。因為膠樽輕身和體積大，運輸不便宜，而且很多屋苑發現膠樽回收很困難。這個試驗計劃是為了評估一個具成本效益的中央收集系統。兩間回收商，每月兩次，在香港，九龍及新界，約250個屋苑收集膠樽，以作回收。計劃開始後，很多屋苑，特別是香港島，膠樽的回收率大大提升。

近期的活動包括：

- **「唔駛膠袋，唔該」活動**
- **大型減少廢物計劃** — 考慮焚化和堆肥的處置方案，以減少廢物量。
- 向棄置堆填區的建築廢物徵收費用，令廢物製造者減少產生廢物，促進再用和回收，以減低堆填區填滿的速度。

「廢物管理原則」說明減少消費和製造廢物，及促進回收和處置是最好的廢物管理方法 (Cheremisinoff, 1989; EPD, 1998)。最終，可以減少棄置廢物，因為這樣對我們的健康和自然環境有很大的影響。

考察

位於將軍澳的新界東南堆填區歡迎學生和公眾人仕參觀。詳請可參閱野外研習指引章節4.4 (應用資源與廢物管理)。

思考問題

1. 處理廢物問題的最好方法是什麼(如：科技，管理系統等)?
2. 透過多些回收，我們最終可以解決香港的廢物問題嗎?
3. 政府可以做什麼去減少廢物? 試舉具體的措施?

References 參考資料

Barron, J. (1989). *An Introduction to Wastes Management*, Institution of Water and Environmental Management, London, p.46.

Buclet, N. and Godard, O. (2000), *Municipal Waste Management in Europe: A Comparative Study in Building Regimes*, Kluwer Academic Publishers, Boston, p.207.

Cheremisinoff, P. (1992), *Waste Reduction for Pollution Prevention*, Butterworth-Heinemann, Boston, p.9, 164, 191–193, 199.

EPD (1998), *Waste Reduction Framework Plan 1998-2007*, Printing Department, Hong Kong, p.4, 10.

EPD (2000), *Monitoring of Solid Waste in Hong Kong*, Facilities Planning Group, EPD, HK Government, p.5.

EPD (2005a), *EPD – Waste Recycling Statistics* and *Waste Treatment & Disposal Statistics*. Retrieved January 20, 2005, from Environmental Protection Department website: http://www.epd.gov.hk/epd/english/environmentinhk/waste/data/stat_recycle.html

EPD (2005b), *EPD – Waste*. Retrieved January 20, 2005, from Environmental Protection Department website: http://www.epd.gov.hk/epd/english/environmentinhk/waste/waste_maincontent.html

Ng, G. (1993), *Decision Makers' Guide to Solid Waste Management in Hong Kong: A Sustainable Approach*, Conservancy Association, Hong Kong, p.30-34, 48, 58, 77–80.

Chapter 12. Sewage Treatment

Sewage refers to the "used water supply" of a community, or a combination of the waterborne wastes from homes, businesses, institutions and industries. It may contain groundwater, surface water, and storm water.

The composition of sewage is a complex mixture of natural inorganic and organic material mixed with man-made substances. Coming from sewer or run-off, sewage is 99.9% water and 0.1% material (in which 70% organic like protein and 30% inorganic like metals) to be removed (Ramalho, 1983). Municipal sewage could be domestic or industrial, and both the strength and composition change all the time.

Sewage © Frank Lee

I. Sewage and Health

Drinking water © Frank Lee

Untreated sewage contains a dangerous load of infectious bacteria, viruses, parasites and toxic chemicals. When it ends up in our recreational and drinking water, it takes a severe toll on human health and the environment. Every year, millions of diseases are caused by swimming in water polluted by sewage outflows and by drinking water as well.

In addition, increases in population density and per capita water consumption have caused sewage systems to become overloaded. Most of the people in developing countries live in conditions of poor sanitation. Thus, the environmental impact on water bodies involving poor hygiene could be quite tremendous.

Environmental Effects

Death of fish - result of untreated sewage to the sea © Dennis Kong

Fresh raw sewage carries the types and large numbers of microorganisms found in our intestinal tract. Microorganisms like pathogens that can be transmitted from humans through sewage to water or food and back to humans again. Examples include bacteria, protozoa and a wide range of resistant viruses.

BOD testing © Frank Lee

Haller (1995) and Ramalho (1983) pointed out that biological oxygen demand (BOD) is a measurement of the oxygen requirement of the bacteria and other organisms as they decompose the organic material contained in sewage. Effluent with high biological demand depletes oxygen in the receiving body of water and is harmful to aquatic life. As sewage decomposes, it releases many volatile chemicals such as methane, hydrogen sulphide and carbon dioxide.

II. Sewage Treatment

Sewage treatment is a process in which the solids in sewage are partially removed and partially changed from complex organic solids to minerals or relatively stable organic solids by decomposition brought about by microbial action. The major objective is to convert the solid portion to a manageable sludge (2% dry solids), leaving a small portion in the final effluent (0.003% dry solids).

Processes involved in sewage treatment plants for treating wastewater:

- **Pretreatment**: screening, grit separation, flotation, equalization
- **Primary sedimentation**
- **Secondary treatment:** fixed film systems, suspended growth systems
- **Tertiary treatment**: effluent polishing, nitrogen and phosphorus removal
- **Disinfection**: chlorination, using ultraviolet, ozone
- **Sludge treatment**
- **Discharge effluent:** wastewater reuse

Overview of Sewage Treatment

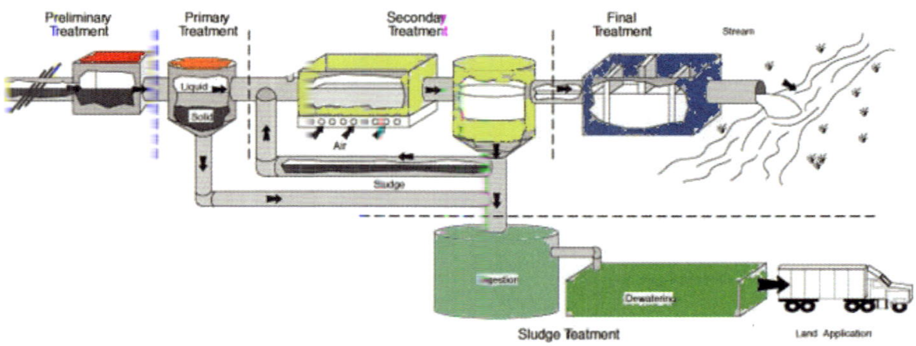

(modified from Biotopics, 2005)

Preliminary Treatment
Not only it damages or clogs pumps in the later treatment processes, it also screens out of debris with bar screens and a settling of grit through the grit-setting tank (Arundel, 1995; Haller, 1995).

Primary Treatment

It refers to the physical removal of particulate organic material. About 30–50% of total organic material settles at the bottom. Fatty or oily materials float on the top where it is skimmed from the surface. Both particulate organic materials and fatty materials become raw sludge (Arundel, 1995; Haller, 1995).

Settling Tanks of Primary Treatment ➔

© Frank Lee

Secondary Treatment

It is the biological removal of colloidal and dissolved organic materials (Grady, 1999; Ramalho, 1983). It makes use of organisms (like natural decomposers and detritus feeders). The colloidal and dissolved organic materials breakdown to carbon dioxide and water via cell respiration. Oxygen is also added to facilitate and enhance organism respiration and growth.

🡐 Aeration Tanks of Secondary Treatment

Tertiary (Anaerobic) Treatment of Solids

Grady (1999) and Malina (1992) mentioned that this treatment is solely for high-strength wastewater treatment such as industrial, agricultural wastewaters, turning organic materials into biomass and methane. Anaerobic digesters are used to treat solids produced in an aerobic treatment process.

Other Treatments include (Ramalho, 1983; Haller, 1995):

Sedimentation:
- Primary clarifier: to remove solid before biological treating it, less expensive than aeration treatment
- Secondary clarifier: to remove biomass formed during biological treatment and other solids
- In physical-chemical treatment: follow coagulation and flocculation units

Sludge concentration and dewatering – reduces the volume of sludge for disposal in a landfill, incinerator, on land.

Sludge digestion occurs in aerobic or anaerobic sludge digesters.

Sludge thickening:
- A settling process where the sludge concentrates in a basin by the action of gravity
- Supernatant returned to the primary clarifier to pass through the treatment process

Stabilization ponds – a series of ponds that settle and biologically treat wastewater, in which aquaculture occurs in the final ponds.

III. Sewage Sludge

As mudlike product of sewage treatment, sludge is the soft mud or dirty matter which settles at the bottom of sewage. The compound has these properties (Arundel, 1995):
- high moisture content;
- dewatering is needed for disposal; and
- chemical properties: high levels of nitrogen (N), phosphorus (P), potassium (K), sulfur (S) and heavy metals such as manganese, calcium, zinc, copper, etc.

According to Vesilind (1986), **municipal sewage sludge** contains over 200 different substances, like organic chemicals, pathogenic organisms and toxic metals. The five major disposal and reuse of municipal sludge are land application, distribution and sale of sludge products, landfill, incineration and ocean disposal.

Recycling/utilization of Sewage Sludge

The main objective of sludge recycling is to minimize potentially detrimental environmental effects whilst maintaining agriculture as a recognized outlet of value for the recycling of sewage sludge. It is essential to minimize potential environmental problems like (Arundel, 1995):

- potentially toxic elements (PTEs);
- infection from pathogenic organisms that present in sludge; and
- other adverse effects on human health arising from the contamination of sludge

Agricultural Application of Sewage Sludge

Benefits (Vesilind, 1986):

- It provides a cost-effective method of sludge disposal
- It becomes the principal way of deriving a beneficial use for crop production

Crop fields in Yuen Long © Frank Lee

Stabilized sludge as fertilizer © Frank Lee

Problems:

- Heavy metals
- Pathogens
- Metal contamination of soil and uptake by plants
- Not suitable for grasses used for grazing, and the crops for human consumption

The regulatory system for agricultural use includes defining soil concentration limits and proper sludge treatment and management practices. It may have significant economic implications if regulations are too precautionary.

IV. Wastewater Management in Hong Kong

Hong Kong sets standards for wastewater discharges into drainage and sewerage systems, inland waters and coastal waters (EPD, 1995). Some of the prohibited substances in discharges into foul sewers, inland and coastal waters include: polychlorinated biphenyls, polycyclic aromatic hydrocarbons, pesticides, radioactive substances, chlorinated hydrocarbons, toxic solvents, petroleum oil and calcium carbide. The effects of some of these toxic substances are also discussed in **Chapter 9** (Persistent Toxic Substances).

In the government, the **Planning, Environment and Lands Bureau** is responsible for wastewater policy formulation, while the **Drainage Services Department** (DSD) executes the treatment of sewage. *Water Pollution Control Ordinance (WPCO)* is passed in 1980 and applied by EPD in 1986. This Ordinance and its subsidiary legislation of WPCO are general laws to regulate the water quality and wastewater discharge standards in Hong Kong. WPCO declares ten Water Control Zones to cover the entire Hong Kong area.

A sewerage network of over 1,320 km in total length and around 200 plants ranging from preliminary treatment (screening) plants to secondary treatment (biological) plants, treating sewage from residential, commercial and industrial premises in the territory prior to disposal to the sea for dilution and dispersion through submarine outfalls (DSD, 2003).

© Frank Lee

Sewerage Master Plans (SMPs)

In 1989, a sewage disposal strategy was adopted by the Government (DSD, 1995). SMPs provide a blueprint of the sewage infrastructure required to collect sewage on a catchment-by-catchment basis and direct it to treatment facilities. The entire territory has been divided into 16 areas, and SMPs have been produced for all of them. The recommendations of these SMPs are implemented progressively to cater for the present and future development demands of Hong Kong.

Construction work under the Chai Wan and Shau Kei Wan sewerage master plan was completed, in 1997, followed by the Tsuen Wan, Kwai Chung and Tsing Yi sewerage master plan in 1998, the Hong Kong Island South sewerage master plan in 2000 and the East Kowloon sewerage master plan in 2001. The Government aims to implement the remaining plans by around 2009.

Under the Strategic Sewage Disposal Strategy (SSDS), sewage is collected from around the harbour, transported by deep tunnels to a sewage treatment plant on Stonecutters Island for Chemically-enhanced Primary Treatment (CEPT) and discharged through an outfall to the South China Sea.

Sewage Charges in Hong Kong

All water users who discharge their sewage to public sewers are required to pay a basic sewage charge (DSD, 2003). In addition, a number of trades and industries (such as the catering industry) whose effluent strength exceeds that of domestic sewage need to pay a trade effluent surcharge to reflect the additional cost for treating their stronger effluent. These charges aim at partially recovering the operation and maintenance costs of sewage collection, treatment, and disposal facilities.

In addition to the effluent surcharge, the level of treatment, the impacts of effluent disposal, and the engineering problems during the construction of tunnel networks, all are subject to debate by the public.

第十二章　污水處理

污水是指社區曾使用的水源，或由住宅、商業、及工業所產生帶水的廢物，其中包括地下水、表面水及雨水。

污水由複雜地混合的天然有機物和無機物，及人造物質混和組成。來自污水渠或雨水渠污水含99.9%的水，其餘需要除去的物質佔0.1% (其中百分之七十為有機物，如蛋白質，另外百分之三十為無機物，如金屬) (Ramalho, 1983)。都市污水由住宅或工業產生，無論其濃度和成分都會不時改變

I. 污水與健康

未經處理的污水所含的傳播疾病的細菌、病毒、寄生蟲及有毒化學物量達危險水平。若這些污水用來作康樂活動及飲用用途，將嚴重危害人類的健康及環境。每年因在受污水污染的水域游水及飲用食水所引致的疾病數以百萬計。

隨着人口密度及人均耗水量日增長，污水處理系統已不勝負荷。許多發展中國家的人民都是生活在惡劣的衛生環境中，因環境問題導致水體不潔的影響可以很大。

對環境的影響

未經處理的污水含有大量可在人類腸胃內找到的微生物，其中一些如病原體可由人類經污水傳送至食水或食物，再傳送回給人類，其中包細菌、單細胞生物及各類抵抗性的病毒。

Haller (1995) and Ramalho (1983) 指出生化需氧量測試 (BOD) 可用作量度細菌和其他生物分解污水中的有機物質時所需的氧氣。若排放污水的生物需氧量高，便會消耗被排放水體的氧氣，並危害水中生物。在污水分解的過程中，會釋放許多揮發性化學物如甲烷、硫化氫及二氧化碳。

II. 污水處理

污水處理會先將污水中部分固體分隔開來，並以微生物的作用將一些有機複合物分解為礦物或相對較穩定的有機物，這是為了將污水中的固體轉變為可處理的污泥 (百分之二的乾固體)，以將排放污水中的成分減至最低 (0.003% 乾的固體)。

污水處理廠處理廢水所牽涉的步驟：

- 初級處理：分隔、沙礫分隔、浮選、均衡混合
- 一級處理：沉澱
- 二級處理：固定隔層系統、防止生長系統
- 三級處理：淨化排放流質、除去氮氣和磷酸
- 消毒：加氯消毒、利用紫外線、臭氧
- 污泥的處理
- 排放污水：廢水循環再用

基本處理

初級處理不但破壞或堵塞其後處理程序的泵水管，這程序亦會利用隔篩將殘渣分隔開來，並在沉砂池將沙礫沉澱 (Arundel, 1995; Haller, 1995)。

一級處理

這是一個利用物理工序來移除微粒般的有機物，約有30-50%的有機物會沉澱於底部。浮於表面油脂會被撇除。微粒般的有機物和油脂物質便轉化為污泥 (Arundel, 1995; Haller, 1995)。

二級處理

利用微生物去淨化膠質和溶解性的有機物 (Grady, 1999; Ramalho, 1983)，利用生物的細胞呼吸作用(如天然分解者和進食碎屑的生物)將膠狀和溶解性的有機物會分解為二氧化碳和水，這過程需要提供氣氧給生物作呼吸和生長。

其他處理包括 (Ramalho, 1983; Haller, 1995)

沉澱:
- 一級淨化：在生物處理前先將固體移除，這方法比有氧處理便宜
- 二級淨化：移除生物處理所產生的生物量及其他固體
- 物理及化學處理：用於凝結沉選及膠凝後

污泥濃縮和脫水：將污泥體積減少以便棄置，如運往堆填、焚化等
污泥消化：以有氧或厭氧消化污泥。
污泥調厚：
- 這是一個沉澱過程，以重力將池中的污泥濃縮化；
- 上清液再經一級淨化處理

穩定池 — 污水經過一連串的沉澱及生物處理池，最終產生水產養殖、溶液培養、水耕種植法

III. 污泥

污水經過處理後產生如泥漿般的污泥，這些軟泥或骯髒物沉澱於污水下層，其特生如下 (Arundel, 1995):
- 含大量水份；
- 需經脫水處理才可棄置；及
- 化學成分含大量氮(N)、磷 (P)、鉀(K)、硫 (S) 和鎂、鈣、鋅、銅等重金屬

Vesilind (1986) 指出都市污泥含有超過200種不同物質，包括有機化學物、致病生物及有毒金屬。棄置和循環再用都市污泥的主要方法共有五種，土地運用，售賣和分發污泥產品、堆填、焚化和海上傾倒。

循環再用/利用污泥

循環再用污泥的主要目的是要減低污泥對環境潛在的不良影響，此外，亦要維持污泥可循環應用於農業上的價值。整個過程必須減低可能引致的環境問題，如 (Arundel, 1995):

- 潛在的有毒物質 (PTEs)
- 被污泥中的致病生物所污染；及
- 其他由污泥污染對人類健康構成的不良影響

污泥的農業應用

好處 (Vesilind, 1986):	問題：
• 這個棄置污泥的方法具有經濟效益。 • 這已成為有利農作物生產的主要方法	• 重金屬 • 病原體 • 泥土受金屬污染並由植物吸收 • 不適合種植用作放牧的草及讓人消費的農作物。

農業用途的管制措施包括規定泥土濃度限制、合適的污水處理和管理措施。若條例太過嚴緊，可能會造成很大的經濟損失。

IV. 香港的污水管理

香港制定了排入去水渠及污水渠系統、內陸及海岸水域的污水標準 (DSD, 1995)。部分禁止排入污水渠、內陸及海岸水域的物質包括多氯聯苯、聚芳烴、殺蟲劑、放射性物質、氯化烴、有毒的溶劑、石油及碳化鈣。這些有毒物質的影響已於第九章(持續性有毒物質)討論了。

在政府架構中**環境運輸及工務局**擔當策劃污水處理的職責，而渠務署則負責執行污水的處理。水污染管制條例 (*WPCO*) 於1980年通過及於1986年由環境保護署執行。該條例及其附屬條例是本港管制水質及污水排放的準則。水污染管制條例將全港水域劃為十個水質管制區。

現時污水收集系統包括總長度超過1,320公里的污水系統網絡及約 200 座提供基本(隔篩) 至二級 (生物) 污水處理的污水處理廠，負責處理來自住宅、商業和工業區的污水，經深海排放管排放到大海稀釋和擴散 (DSD, 2003)。

污水收集整體計劃 (SMPs)

香港政府於一九八九年採取一套污水整體排放策略 (DSD, 1995)，「污水收集整體計劃」制訂逐一擴展及改善收集及處理污水所需的設施藍圖，全港被分為十六區，訂下污水收集整體計劃，其建議逐步實施以應付目前和日後的需要。

在1997年，柴灣和筲箕灣污水收集整體計劃的建造工作經已完成，接著是1998年荃灣、葵涌及青衣污水收集整體計劃，2000年港島南污水收集整體計劃，以及2001年東九龍污水收集整體計劃。政府打算在2009年實施其餘計劃。

在策略性排污計劃 (SSDS) 下，從海港收集污水，經過深層隧道運送往昂船州的處理廠作化學輔助一級污水處理 (CEFT)，然後排放到南中國海。

香港的污水處理服務費

所有將污水排放到公眾污水渠的使用者都需要支付基本的污水處理服務費 (DSD, 2003)。此外，徵收工商業（如飲食業）污水附加費則旨在收回處理污染程度超出住宅廢水的工商業污水的額外費用。這些費用覆蓋部分污水收集，處理和棄置設施的操作和保養成本。

除了排污費，公眾可以參與討論有關污水處理的程度，排污的影響和建造隧道期間的工程問題。

Visits for Further Study 野外研習

Both the Drainage Services Department and the Water Supplies Department of the Government welcome students and the public to conduct group visits at their treatment stations. For details, please refer to the introduction and guidelines described in Chapter 4.4 (Applying Resources and Waste Management) of the **Field Trip Guide**.

渠務署和水務署歡迎學生和公眾團體參觀它們的濾水廠。詳細請參考野外研習指引內第4.4章的介紹（應用資源和廢物管理）。

參觀沙田濾水廠 © Frank Lee

Questions for Further Thinking 思考問題

1. Should Hong Kong follow the practice of other countries and conduct tertiary treatment of sewage in order to further purify it before discharge?
   香港應否效法其他國家實施三級污水處理，以作更徹底的淨化?

2. The Sewerage Master Plans propose to discharge sewage through an outfall to the South China Sea. Will the sewage released eventually affect our marine environment when it is moved back to the waters of Hong Kong by currents and climatic changes?
   污水收集整體計劃旨在將污水排放到南中國海。當水流和氣候轉變時，污水會否回流香港，影響海洋環境?

References 參考資料

Arundel, J. (1995), *Sewage and Industrial Effluent Treatment: A Practical Guide*, Blackwell Science, London, p.2–3, 111–115.

Biotopics (2005), *Sewage*. Retrieved March 20, 2005, from Biotopics website: http://www.biotopics.co.uk/microbes/sewag.html

DSD (1995), *Sewerage Manual*, Government Printer, Hong Kong, p.13, 23.

DSD (2003), 環境報告 *2003*, Drainage Services Department, Hong Kong, p.25, 53.

Grady, C. (1999), *Biological Wastewater Treatment*, Marcel Dekker, New York, p.3, 599.

Haller, E. (1995), *Simplified Wastewater Treatment Plant Operations*, Technomic Pub. Co., Lancaster, Pa., p.11, 21, 29, 119, 139.

Malina, J. (1992), *Design of Anaerobic Processes for the Treatment of Industrial and Municipal Wastes*, Technomic Pub. Co., Lancaster, p.2.

Ramalho, R. (1983), *Introduction to Wastewater Treatment Processes*, Academic Press, New York, p.9, 31, 68, 80, 341, 383.

Vesilind, P. (1986), *Sludge Management & Disposal for the Practicing Engineer*, Lewis Publishers, Chelsea, Michigan, p.4, 63–64.

Chapter 13. E-waste

Chapter 11 has discussed in general the waste problem and management in Hong Kong. Now we look more deeply into a mounting topic – electronic waste (or simply "E-waste).

I. What is E-waste?

Electronic products © Frank Lee

Computers, televisions, VCRs, stereos, copiers, and fax machines are common electronic products. E-waste refers to the used electronic equipment that is thrown away. It contains all types of hazardous waste that are not natural to the environment. In today's society people seldom repair electronic equipment and but simply throw it away. Although many of these products can be reused, refurbished, or recycled, many western countries simply export them to developing countries, especially to the poor communities of Asia. To the people of these receiving countries, this is indeed an export of real harm to them.

Let us take personal computers (PC) as an example. One may be surprised to learn that a typical PC actually consists of at least 20 different metal and plastic materials. Many of these contain toxic chemicals if left untreated. Just imagine the impact generated by the amount of waste if millions of PCs are constantly imported to our land.

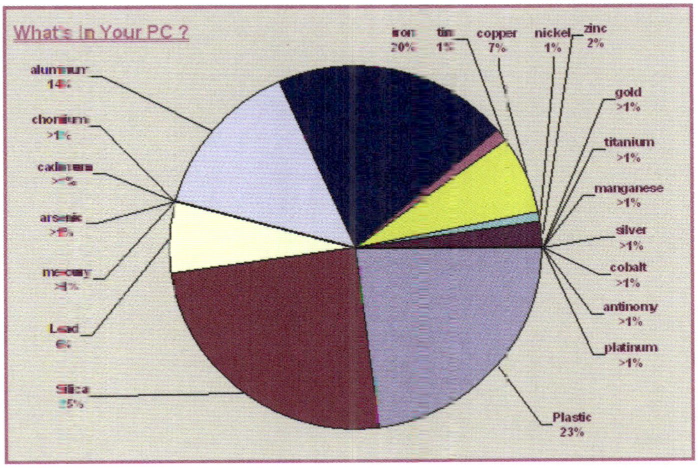

(modified from Eiler, 2003)

In the USA, only 20% of electrical waste is recycled, while the other 80% is exported elsewhere, such as China and India. It is estimated that between 50–80% of the E-waste collected for recycling in the U.S. is not recycled domestically, but is very quickly placed on container ships bound for destinations like China. In 2002, Americans throw out about 10 million old computers (O'Neill, 2000).

E-waste exports to Asia are motivated entirely by global economics. Market forces make toxic waste always run "downhill" on an economic path of least resistance. The toxic effluent of the affluent will flood towards the world's poorest countries where labor is cheap, however occupational and environmental protections are inadequate.

In short, a free trade in hazardous wastes leaves the poorer people of the world with the result of living in poison. Listed below are reasons (Asante-Duah, 1998):

Reasons for transboundary shipment of wastes:

- Overall stringent environmental laws in most industrialized countries.
- Tightening of specific laws, regulations and policies concerning disposal of certain types of wastes.
- High cost of hazardous waste disposal in waste-generating country.
- Diminishing domestic capacity for disposal of certain types of wastes.
- Potential future liability for any damages caused by domestic wastes.
- Market opportunities elsewhere for materials which can be recovered, reclaimed or recycled from wastes otherwise destined for "final" disposal.
- General economic growth which may result in more generation of wastes.
- Economies of scales involving treatment, storage and disposal facilities (TSDF).
- Existence of TSDFs which may serve several countries.
- Existence of an appropriate TSDF in a foreign country which is closer than a similar facility in the home country.
- Lack of environmentally sound TSDFs in some waste-producing countries.

After E-waste is exported, a number of recycling operations for treating different types of waste materials are conducted. Yet, no matter whether the method is open burning, acid baths or toxic dumping, they all cause pollution of the land, air and water, where, in turn, people are exposed to the toxins.

Features of E-waste Recycling Operations (Moore, 2000)**:**

Open Burning	Copper is recovered, but a large amount of black ash residues are generated.
Toner Sweeping	Creates constant clouds of toner that billows around workers and is routinely inhaled; blackening skin and clothing.
CRT Cracking and Dumping	Cathode-ray tubes (CRT) are NOT refurbished into new TV sets or computers. Instead, the CRT is cracked and discarded with the yokes sold to copper recovery operations.
Circuit Board Recycling	The most environmentally destructive; riverbanks are full of charred circuit boards and working areas are severely polluted, emitting harmful heavy metals, dioxins, beryllium and polycyclic aromatic hydrocarbons (PAHs).
Acid Stripping of Chips	Often done by a very primitive process using acid baths (mixture of 25% nitric acid and 75% hydrochloric acid); emits huge clouds of steamy acid gases (toxic chlorine and sulphur dioxide gases) and sludges into river banks where the acidity could go as high as pH level of 0.
Plastic Chipping and Melting	Often conducted by children, the melting is done in rooms with little ventilation and no respiratory protection.

As a result, the following materials are separated for further recycling:

- Material containing copper – from printer and motors, wires and cables, CRT yokes
- Steel – from internal computer frames, power supply housings, printer parts
- Plastic – from housings of computers, printers, faxes, phones, monitors, keyboards
- Aluminum – from printer parts
- Printer toner – from spent toner cartridges
- Circuit boards – from computers, phones, disc drives, printers, monitors, etc.

II. Environmental and Health Impact of E-waste

Not only is E-waste harmful to human health, but it also destroys the environment. A tremendous amount of imported E-waste material and process residues are simply dumped into open fields, along riverbanks, ponds, wetlands and rivers. It pollutes ground water, and when burned it pollutes the air we breathe. Toxic ingredients such as lead, beryllium, mercury, cadmium, and brominated flame retardants pose a threat to both occupational and environmental health. It is affecting many countries in the world. Even young children are being exposed to the harmful materials everyday.

Parts separated from monitors
© Anna Leung

The computers contain materials such as arsenic, mercury, lead, and cadmium. All of these materials pose health risks not only to humans, but also to wildlife. Among various toxic substances, lead and mercury are mostly known to cause brain damage; and are known to cause different kinds of cancer.

Waste of PC circuit boards
© Anna Leung

Lead can be found in computers in the glass panel of monitors (1 to 4 kg per monitor) and as solder in printed circuit boards and other components. Lead causes damage to the central nervous system, the blood system, the kidneys and the reproductive system. It also has serious negative effects on children's brain development.

Mercury can be found in thermostats, sensors, relays and switches (e.g. on printed circuit boards and in measuring equipment). It is also found in great quantities of flat panel displays, and will increase as these become more popular. Mercury can cause brain damage and kidney problems. The developing fetus is highly susceptible through maternal exposure to mercury. Mercury is easily accumulated in living organisms and concentrates through the food chain, particularly fish.

Summary of E-waste Hazards and Their Impacts

Toxics	Sources	Health Impact
Arsenic (As)	Incinerator ash	Lung and tissue damage, liver and kidney injury, potentially fatal
Barium (Ba)	CRT of computers	Brain swelling, muscle weakness, damage to the heart, liver and spleen
Beryllium (Be)	PC (connectors and tiny plugs of mother-boards)	Chronic beryllium disease and skin disease, even lung cancer
Cadmium (Cd)	Chip resistors, infra-red detectors, semiconductor chips, CRT	Kidney damage, cancer, respiratory problems, irritates digestive tract
Chromium (Cr)	Untreated and galvanized steel plates; Incinerator ash	Liver and respiratory problems, causes allergic responses to skin; passes through cell membrane and damage DNA
Copper (Cu)	PC circuit boards, metal waste and scrap, cable scrap	Liver damage, immune system suppression
Dioxins	Burning of PVC, furnace steel dust	Reproductive disorders, immune system suppression; stored in human milk
Lead (Pb)	PC (glass panels and gasket in monitors, circuit boards), incinerator ash, batteries, metal waste and scrap	Neurological disorders, blood, kidney, reproductive and nervous system damage
Mercury (Hg)	Thermostats, sensors, relays, switches, lamps, mobile phones, batteries and flat panel displays; incinerator ash, honeycomb baking coal	Neurological, liver, and kidney damage, coma and death, dangerous during pregnancy; bioaccumulates through food chain in aquatic species
Nickel (Ni)	Electric arc furnace steel dust	Respiratory problems, lung cancer from dust inhalation
PAHs	Air/atmosphere, ash emission	Skin, lung, stomach and liver cancer
Polybrominated Diphenyl Ether (PBDE)	Burning circuit boards and wire insulation	Bioaccumulates in human milk; causes neurological effects on infants
Polychlorinated Biphenyls (PCBs)	Toxic fumes; transformers and capacitators	Reproductive dysfunction, immune system suppression
Zinc (Zn)	PC circuit boards, furnace steel dust	Respiratory problems, lung cancer from dust inhalation

(Asante-Duah, 1998 and Clapp, 2001)

Childhood with E-waste © Janet Chan

Mountains of chipsets in a backyard
© Anna Leung

III. A Case Study in Guiyu, China

Since 1995, Guiyu, a coastal town in Guangdong Province in South China, has transformed from a rural community to a booming E-waste processing center. While rice is still grown in the fields, virtually all of the existing building space has given way to providing many small and specialized E-waste recycling shelters and yards. The entire Guiyu district comprises four small villages along the Lianjiang River: Huamei Village, Longgang Village, Xianpeng Village and Beilin Village. The region is full of workers who spend each day stripping the computers by hand without any special protections. They know little about electronics, not mentioning how to dismantle the computers and where to put the discarded materials.

Location of Guiyu © Frank Lee

Once a beautiful town without E-waste ➔
© Anna Leung

Working at the Expense of Health

E-waste labour © Janet Chan

Most of the labor force working in the recycling industry comes from outlying regions. Farmers from provinces such as Hunan and Anhui move to Guiyu to take the unskilled jobs of dismantling and processing the imported E-waste for an average wage of HK$10 per day. For those collecting metals in rivers, disregarding the harmful toxic chemicals they could earn about HK$400 per month. The most profitable task is the removal of integrated circuits (IC) from PC circuit boards, for which as much as HK$700 per month can be earned. (Eiler 2003).

Earning little but sacrificing health.
© Anna Leung

🡐 Women pick through the wires torn out of computers. The wires are sorted by day and burned by night. Without any respiratory protection, workers inhale acid fumes, chlorine and sulphur dioxide gas all day as they swirl computer chips from circuit boards in acid to collect tiny amounts of gold and other valuable metals.

Ash Town

Lousy workplace © Anna Leung

Most of the activities like the separation of metals from plastic, and the retrieval of copper from burning wires, involve dismantling by hammer, chisel, screwdriver and even bare hands. Workers smash the cathode ray tube from a computer monitor in order to remove the copper laden yoke at the end of the funnel. The glass is laden with lead but the biggest hazard from this is the inhalation of the highly toxic phosphor dust coating inside the tube. Monitor glass is later dumped in irrigation canals and along the river where large amounts of lead are discharged.

After the workers have dismantled the computers, they burn the useless parts. In fact **open burning** of plastics and other materials is common, so to extract the waste from metals. Burning these wires releases dioxins and other toxics gases into the air, from where they then fall back down to earth in the form of ash.

Open burning of plastics and other material is conducted to reduce the waste to metals ➔

© Janet Chan

The whole village is easily covered by this ashen residue, with the children playing in the toxic ash. The ash is made up of residues from melted plastics and heavy metals. Moreover, carcinogenic polycyclic aromatic hydrocarbons (PAH) and dioxins are largely emitted from the burning of wires made from PVC and brominated flame retardants. Every day villagers inhale this dirty air, so their bodies have become weak. Many of them have developed respiratory and skin diseases, too.

Whole village covered with ash
© Anna Leung

Drinking Water Problem

As mentioned previously, toxins are dumped into canals that eventually flow into rivers. These rivers have always been a major source of drinking water for the villagers. Unfortunately, most rivers close to recycling operations have been known to contain **lead** (Pb) levels (1.9 mg/L) that are as much as 190 times higher than levels considered safe according to the World Health Organization (Eiler 2003).

Each dumped computer contains about 2 kg of lead and much of the leaches into the ground and the water supply. Drinking, cooking and washing are usually conducted with local ash-contaminated surface waters. There have been reports of villagers suffering from stomach and breathing problems. Hundreds of people, including pregnant women and small children live among the ash heaps.

Liangjiang is heavily polluted © Janet Chan

Because of the E-waste industry, the groundwater in Guiyu has been completely contaminated. Local residents claim that the water has become foul tasting and the government has warned the public not to drink it.

As a result, a new business has developed with a constant parade of tractors carrying large plastic tanks of fresh drinking water every day from the town of Ninjing, 30 km away. Nowadays, almost 80% of the drinking water in Guiyu has to be imported because of all the contamination in the rivers.

Fresh water transport – a new business in Guiyu © Anna Leung

But why China?

The main reasons for the import of e-waste to China are:

- Some of them are NOT waste, but they are actually reusable goods!
- Some enter the country by smuggling
- Most importantly, the goods have high market value. The benefits of extracting the reusable parts from the goods and refining them into valuable products are good enough to have the e-waste import to be continued.

IV. Basel Convention

Convention © Martin Fung

An international treaty known as the **Basel Convention** was created in 1989 as an effort to counter the unsustainable and unjust effects of free trade in toxic wastes. On 22 March 1989, ***the Basel Convention on the Control of Transboundary Movements of Hazardous Wastes and their Disposal*** was adopted in Basel, Switzerland. The Convention came into force on 5 May 1992 and today has its Secretariat in Geneva, Switzerland.

The Basel Convention calls on all countries to reduce their exports of hazardous wastes to a minimum and deal with their waste problems within the country in order to protect human health and the environment. In 2002, 36 nations including all 15 European Union countries have signed the **Basel Ban Amendment**, which aims to unite most developed countries to prevent the export of waste to developing countries. According to Krueger in 1999, the United States, Afghanistan, and Haiti did not sign this treaty. Bear in mind that the United States is one of the largest producers of E-waste, but Afghanistan and Haiti produce none.

Uncontrolled E-waste disposal
©Anna Leung

Trade Restrictions of the Basel Convention

Transboundary movement of hazardous waste is prohibited in the following cases (Krueger, 1999):

- To states that are neither parties to the Basel Convention nor to an agreement that is less environmentally sound than the Convention (the "limited ban");
- To Antarctica;
- To states who have prohibited such imports or belong to a regional organization whose members have prohibited such imports by national legislation;
- If there is "reason to believe" that environmentally sound management or disposal options are not available in the prospective state of destination.

The reduction of hazardous waste generation, and the movement towards cleaner production, is a key objective of the Basel Convention. In the long run, identifying opportunities to increase the use of low waste technologies and methods of cleaner production will be the basis of the Convention's work, and the role and use of economic instruments for this purpose are also expected to be clarified and pursued.

Conclusion

E-waste is a problem that will be sustained if it is not taken care of. While it is an easy solution for the country exporting the waste, what about the countries that are receiving it? People in those receiving countries have no liability to deal with waste that they did not create. If our appetite for better computers is going to continue to grow, then we need to think of something to eliminate E-waste. Otherwise, we may find that there is no longer space to put it.

Wires of E-waste © Anna Leung

The answer to this E-waste problem lies not in finding new solutions in the downstream for treating the wastes, but actually in moving upstream to prevent the problem from happening at manufacturing level. It is time to strengthen the call for sustainable production, environmental justice, and corporate and government accountability, in order to achieve this goal.

Recommendations for reducing E-waste production

1) Ban hazardous waste exports	Accept the **Basel Convention** and ban all exports of hazardous waste from OECD to non-OECD countries; developing nations should be given the tools and training to develop preventative waste management strategies.
2) Get the poisons out	**Clean production** – producing less quality of waste and less hazardous waste in the first place; must find a way to eliminate hazardous materials in their products, and educate consumers on the potential threat to public health and the environment.
3) "Don't let new poisons in"	Similar to the concept of the "**Precautionary Principle**", this calls for the prevention of a chemical from existing before it has done significant damage to the public.
4) Make the producer responsible	Encourages producers to prevent pollution and reduce resources and energy in each stage of the production cycle through changes in product design and process technology; make sure they are financially responsible for end-of-life waste management.
5) Require producers to "take it back!"	The ultimate aim is to close the loop of the product life cycle, so that producers assume full responsibility for life cycle costs; the free-of-charge "take-back" from customers should apply to products that are obsolete or broken. Many large manufacturers like Hewlett Packard have computer take-back programs.
6) Design for longevity, ability to upgrade, repair and re-use	Rapid technological advances should provide more flexible software and hardware systems that are capable of being upgraded and compatible through time.
7) Design for recycling	The electronic device should be designed to ensure clear, safe and efficient mechanisms for recovering its raw materials.

(Moore, 2000)

It is crucial that the developed countries adopt the policies above, or any other policy or action that will get rid of E-waste. These policies may be very expensive at first, but in the long run they will pay off and make the world a much cleaner place to live.

Undoubtedly, E-waste has evolved into a complex social and global problem with deep-rooted issues. However, data on the impacts of E-waste is still limited and the long-term effects on environmental health and human health are unknown. It is believed that increased awareness and education are crucial.

第十三章　電子廢物

第11章討論了香港的一般廢物問題和管理。現在我們更深入地探討一個主題 — 電子廢物。

I. 甚麼是電子廢物？

電腦，電視，錄影機，音響，影印機，和傳真機是普及的電子產品。電子廢物是指被扔掉的陳舊電子設備。這些有害的廢物並不天然的。在今天的社會，很少人會維修電子設備，只棄掉它們便算。儘管這些產品可以再利用，整修或循環，不過很多西方國家只是把它們出口到發展中的國家，尤其是亞洲窮國。對接收國家的人民，造成損害。

以個人電腦為例。其實，一部個人電腦其實由至少二十種不同的金屬和塑膠材料組成，確是令人驚訝。這些未經處理的化學物是有毒的。試想像如果不斷輸入成千上萬的個人電腦，造成影響有多少。

在美國，只有20%的電子廢物再循環，而其他80%則出口到其他地區，如中國和印度。據估計，50–80%的回收電子廢物不是在美國再循環，而是非常迅速被放到貨櫃裏，運往像中國那樣的目的地。在2002年，美國人棄置大約一千萬部舊電腦 (O'Neill, 2000)。

全球經濟推動電子廢物出口到亞洲。市場力量使有毒的廢物運往經濟較差的地區。有毒的污水流向世界最窮的國家，原因是工資低，不過當地的職業和環境保護不足。

簡而言之，有毒廢物的自由貿易令窮國人民生活在有毒的環境。原因如下 (Asante-Duah, 1998)：

廢物跨境轉運的理由：

- 大部分工業國嚴格的環境法律。
- 收緊關於某類廢物的法律，規則和政策。
- 在生產廢物的國家，有害廢物棄置的代價高昂。
- 縮減棄置某類廢物於國內。
- 國內廢物造成損害的潛在責任。
- 在其他地區，物質有被改造或再循環的市場機會。
- 經濟增長產生更多廢物。
- 涉及處理，儲存和棄置設施的經濟規模(TSDF)。
- TSDFs 的存在可以服務多個國家。
- 鄰近的國家有合適的 TSDF 。
- 一些廢物生產國缺少符合環保要求的 TSDFs。

在電子廢物出口後，會經過不同方法處理。不過，無論是露天焚燒，酸浸或有毒傾銷，它們造成陸地，空氣和水質污染，而且人們亦會接觸到這些有毒物質。

電子廢物回收的運作 (Moore, 2000):

露天焚燒	銅，但是產生大量黑色灰燼。
打印機墨處理	產生煙霧，工人長期吸入；使皮膚和衣服變黑。
陰極射線管 (CRT)粉碎及傾倒	陰極射線管(CRT) 不到新的電視機或電腦中整修。反之，陰極射線管粉碎後，把焊料賣給銅的回收商。
回收電路板	對環境最具破壞力；河岸充滿被燒焦的電路板和嚴重被污染，產生有害的重金屬，二噁英，鈹和多環芳烴(PAHs)。
晶片酸處理	經常使用簡單的酸浸程序（25% 的硝酸和 75% 鹽酸的混合）；產生大量酸霧氣體（有毒的氯和硫磺二氧化硫氣體），滲入河岸，達到0 pH 值水平。
拆掉和溶解塑膠	經常由小孩負責，而且在通風不足和沒有呼吸保護的房間進行熔化。

結果，以下物質會被分開，以作回收：

- 含銅的物質 — 影印機和馬達，電線和電纜，陰極射線管的焊料
- 鋼 — 電腦內部框架，電源箱，影印機零件
- 塑料 — 電腦，影印機，傳真機，電話，顯示屏，鍵盤的遮蓋
- 鋁 — 影印機零件
- 影印機調色劑 — 墨盒
- 電路板 — 電腦，電話，磁片驅動器，影印機，顯出屏等

II. 電子廢物對環境和健康的影響

電子廢物不只對人類健康有害，亦會破壞環境。大量入口的電子廢物和殘餘物傾倒於河岸，池塘，沼澤地和河流。並且污染地下水，燃燒時會污染空氣。有毒的物質有：鉛，鈹，水銀，鎘，溴系阻燃劑等，對職業和環境的健康構成威脅，而且影響很多國家。甚至小孩子經常接觸這些有害物質。

電腦含有很多物質，如：砷，水銀，鉛和鎘。這些物質影響人類和野生動物的健康。在各種有毒的物質之中，鉛和水銀損害腦部是眾所周知；而且引發不同的癌症。

鉛主要在顯示屏的陰極射線管的玻璃（1–4千克/每個顯示屏），電路板和其他組件中找到。鉛會破壞中樞神經系統，血液系統，腎和生殖系統。也對小孩子的大腦發育有嚴重的影響。

水銀在恆溫器，感測器，繼電器和開關制（例如在測量設備和電路板）找到。而且在大很多的平板顯示，並將作為這些的增加變得更受歡迎的。水銀導致腦部損傷和腎的問題。發展胎兒透過母親而接觸到水銀。水銀輕易透過食物鏈集在有機體積聚，尤其魚類。

電子廢物的危害及影響摘要

毒素	來源	健康影響
砷 (As)	焚化灰燼	破壞肺，肝，腎組織，具致命性
鋇 (Ba)	電腦顯示屏	腦部腫脹，肌肉萎縮，破壞心，肝及脾臟
鈹 (Be)	個人電腦（主板連接部份）	慢性皮膚病，甚至肺癌
鎘 (Cd)	晶片電阻器，紅外線探測器，晶片半導體，顯示屏	破壞腎臟，癌，呼吸問題，刺激消化管道
鉻 (Cr)	未經處理及塗鋅的鋼片;焚化爐的灰燼	肝及呼吸問題，皮膚過敏; 穿過細胞膜，破壞
銅 (Cu)	個人電腦電路板，金屬廢料及碎片，電線碎片	肝破壞，抑制免疫系統
二噁英	燃燒聚氯乙烯，熔鋼的灰塵	生殖紊亂，抑制免疫系統，儲存於人奶
鉛 (Pb)	電腦（顯示屏，電路板），焚化爐灰燼，金屬廢料及碎片	神經紊亂，血液，腎臟，生殖及神經系統破壞
水銀 (Hg)	恆溫器，感應器，繼電器，開關制，電燈，手提電話，電池，平面螢幕顯示，焚化爐灰燼，燒煤	神經，肝臟及腎臟破壞，昏迷，死亡，對孕婦危險，經水生物種的生物鏈而累積
鎳 (Ni)	鋼塵	呼吸問題，肺癌
多環芳烴 (PAH)	空氣，灰燼	皮膚，肺，胃，及肝癌
多溴聯苯 (PBDE)	燃燒電路及電線	於人乳累積; 影響嬰兒神經
聚氯聯二苯 (PCBs)	毒煙; 變壓器，電容器	生殖障礙，抑制免疫系統
鋅(Zn)	個人電腦電路板，熔鋼灰塵	呼吸問題，肺癌

(Asante-Duah, 1998 and Clapp, 2001)

III. 中國貴嶼的案例

自1995年，貴嶼，一個廣東省沿海市鎮，從農村轉變成一個迅速增長的電子廢物處理中心。當稻米還在田野生長，很多地方已被作為回收電子廢物用途。整個貴嶼地區沿著練江包含四個小村莊，每天，工人在沒有任何特別保護下，拆除電腦。他們對電子產品所知甚少，不知如何拆除電腦和把廢物安置。

影響健康的工作

從事回收工作的工人，大部分來自偏遠地區。湖南和安徽的農夫為了每日$10港元的薪酬，拆除和處理入口的電子廢物。如果肯不顧有毒的化學物，收集河流的金屬，他們能夠賺取每月大約 $400港元。最賺錢的是從個人電腦拆除集成電路 (IC)，每月可賺取多達700 港元 (Eiler, 2003)。

女工們從電腦收集殘舊的電線。白天把電線分類，晚間燃燒。工人整天把電腦晶片浸在酸溶液中，收集小量金和其他珍貴金屬。在沒有任何呼吸裝置的保護下，工人吸入氯氣和二氧硫磺氣體。

灰燼市鎮

只利用鐵鎚，鑿子，螺絲刀，甚至徒手，燒塑料和燃燒電線提取銅和其他金屬。工人打碎顯示屏的陰極射線管，以抽取含銅的焊料。玻璃管不只含有鉛，而且更有劇毒的磷光劑灰塵。顯示器的玻璃傾倒在用於灌溉的河流，排放大量的鉛。

工人拆除電腦後，他們燃燒無用的部分。事實上，露天燃燒塑料和其他物質，以抽取金屬，是很普遍的。燃燒這些電線，釋放二噁英和其他有毒的氣體，然後以灰燼的形式掉在地上。

整個村莊輕被這個灰白的殘餘物覆蓋著，小孩玩這些有毒的灰燼一起。灰燼是由熔化的塑料和重金屬組成的。此外，燃燒聚氯乙烯製成的電線和溴系阻燃劑，釋放致癌的多環芳烴 (PAH) 和二噁英。每天村民吸入這些空氣，因此他們的身體變得薄弱，大部分人患有呼吸和皮膚病。

食水問題

曾提及，毒素傾倒於水道，最終流入河流。這些河流是村民飲用食水的來源。不幸地，大部分河流鄰近這些回收地區，鉛的水平 (1.9 mg/L) 比世界衛生組織的安全標準高出一百九十倍 (Exler, 2003)。

每台電腦含有大約2公斤的鉛，大部分滲透到地面和食水。受灰燼污染的表面水被用作飲用，煮食和清洗。有報告指村民有胃病和呼吸問題。數以百計的人，包括懷孕婦和小孩在灰堆中居住。

因為電子廢物的二美，貴嶼的地下水徹底被污染。本地居民聲稱水質有惡劣的味道，政府亦警告公眾不要飲用。

結果，新的商業營運而生，每天拖拉機從30公里外的市鎮運載食水。如今，因為河流污染，貴嶼 80% 的食水要從外地輸入。

為什麼在中國?
把電子廢物運往中國的主要原因是:
- 其中一些不是廢物，它們其實可以再用!
- 其中一些透過走私運入
- 最重要的是，貨物有很高的市場價值。從貨物抽取可再用的部分，提煉為珍貴的產品。

IV. 巴爾賽公約

在1989年，創立巴爾賽公約，反對有關有毒廢物不可持續和不公平的自由貿易。在1989年3月22日，瑞士巴爾賽採納巴爾賽公約－控制有害物質跨境運輸及棄置。公約在1992 年5月5日生效，今天在瑞士的日內瓦設有秘書長職位。

巴塞爾賽公約號召所有國家減少輸出有害廢物，以保護人類健康和環境。在2002年，36個國家，包括15個歐盟國家的簽署巴爾賽禁令修訂條約 (Basel Ban Amendment)，以阻止發展國家把廢物運往發展中國家。根據 Krueger (1999)，美國，阿富汗，和海地沒有簽署這條條約。美國是其中一個電子廢物的最大生產者，阿富汗和海地卻沒有生產。

巴塞爾賽公約的貿易限制

在以下情況禁止電子廢物跨境運輸 (Krueger, 1999)：

- 往非公約國；
- 到南極洲；
- 該國禁止輸入，或棣屬於一個有禁止輸入法規的區域機構的成員；
- 有理由相信該目的地缺乏合適管理或棄置方法。

巴塞爾賽公約的主要目標是減少產生廢物，和更潔淨的生產。長遠而言，使用低浪費的技術和更潔淨的生產的方法是公約的基礎，並且闡明經濟工具的角色和運用。

結論

如果不加理會，電子廢物會是一個持久的問題。對於出口廢物的國家，問題是很容易解決，不過接收廢物的國家又如何？在那些接收廢物國家的人民，沒有責任處理不是他們製造的廢物。如果我們對電腦的需求繼續增長，我們需要考慮一些方法去消除電子廢物。另外，我們要知道沒有更多的空間墊存放這些廢物。

電子廢物問題的答案不在於發現新的解決方法去處理廢物，而是在製造層面去預防問題發生。為了達到這個目標，應該加強可持續生產，環境公義，公司和政府職責。

減少電子發物的建議:

1) 禁止出口有害廢物	接受巴爾賽公約和禁止經濟合作發展組織出口有害廢物到非經濟合作發展組織國家；發展中國家應該獲得技術和訓練，以發展廢物管理。
2) 趕走毒物	潔淨生產 – 首先減少產生有害廢物；必須消除產品中的有害物質，教育消費者關於健康和環境的潛在威脅。
3) 「不讓新毒物入口」	類似「預防的原則」的概念，阻止化合物的出現，避免對公眾造成損害。
4) 製造商應負責任	鼓勵製造商在產品設計和過程技術方面，預防污染和減少資源和能源；確保他們負責廢物管理的費用。
5) 要求製造商回收	最終的目標是完成產品生命週期的循環，以便製造商承擔全部週期費用的責任；免費「回收」顧客的廢棄或破損的產品。像惠普那樣的很多大製造商有電腦回收計劃。
6) 耐用，可作升級，維修及再用的設計	迅速的技術發展提供可作升級，更靈活及耐用的軟體和硬體
7) 回收設計	電子儀器的設計應該是清楚，安全和高效率的。

(Moore, 2000)

最重要的是那些已發展國家採納以上或其他政策，以解決電子廢。這些政策或許非常昂貴，但是長遠而言他們會獲得回報和令世界更潔淨。

無疑地，電子廢物發展為一個複雜的社會和全球問題。不過，關於電子廢物影響的資料有限，而且對環境健康和人類健康的長期的效應仍是未知的。加強瞭解和教育是至關重要的。

Questions for Further Thinking 思考問題

1. Apart from the case in Guiyu, China, do you know any other places that are also suffering from the pollution of imported E-waste? How do they combat the problem?
除了貴嶼的案例，你知道其他地方同樣有電子廢物入口的問題嗎？他們該怎樣應付？

2. As mentioned, E-waste recycling operations include open burning, toner sweeping, CRT cracking and dumping, circuit board recycling, acid stripping of chips, as well as plastic chipping and melting. Which of these do you think is the most toxic?
曾提及回收電子廢物的方法有：露天焚燒，打印機墨處理，陰極射線管粉碎及傾倒，電路板回收，晶片酸處理，及拆卸和溶解塑膠，你認為哪種方法毒害最大?

3. Among the seven recommendations dealing with E-waste listed above, which do you think is the most effective? Please give reasons to explain.
在以上七種處理電子廢物的方法，哪種最有效？試解釋。

References 參考資料

Asante-Duah, D. and Nagy, I. (1998), *International Trade in Hazardous Waste*, E & FN Spon, London, p.71, 122–123.

Clapp, J. (2001), *Toxic Exports: the Transfer of Hazardous Wastes from Rich to Poor Countries*, Cornell University Press, Ithaca, p.26–27.

Eiler, S. (2003), *Foreign E-waste Problems and Policies*. Retrieved February 20, 2005, from Ball State University website: http://www.bsu.edu/web/smeiler/ewaste/ewastehome.htm

Krueger, J. (1999), *International Trade and the Basel Convention*, the Royal Institute of International Affairs, London, p.44.

Moore, E. (2000), *An Introduction to the Management and Regulation of Hazardous Waste*, Battelle Press, Columbus, Ohio, USA.

O'Neill, K. (2000), *Waste Trading among Rich Nations: Building a New Theory of Environmental Regulation*, MIT Press, Cambridge.

Chapter 14. Environmental Impact Assessment

Building the Chep Lap Kok Airport and the Tsing Ma Bridge created tremendous impact on the local environment

© Frank Lee

An **environmental impact** is an on-site or off-site change that the project may cause in the environment. It could be an effect of the change on the well-being of people, flora, fauna and ecosystems, on physical and cultural heritage, or even a structure, site or other thing that is of historical or archaeological significance. John Glasson (1994) defined that the impact is a change to the project that the environment may cause, whether the change or effect occurs within or outside the site of the project.

According to Erickson (1994) and Kreske (1996), an environmental impact can be categorized by:

Type:	biophysical, social, health, economic
Nature:	direct, indirect, cumulative, synergistic
Magnitude:	insignificant (no remedial action required) and significant (remedial action and monitoring required)
Extent:	local effect vs. regional, transboundary, global effect
Timing:	immediate vs. future
Duration:	short term vs. long term
Uncertainty:	likelihood, consequences
Reversibility:	reversible vs. irreversible
Significance:	can be socially, economically or ecologically related; the significance of an impact is not necessarily related to magnitude of the impact

I. What is Environmental Impact Assessment (EIA)?

Based on the definitions mentioned by Colombo (1992), Weston (1997) and Erickson (1994), EIA is a set of structured and systematic activities to assess and control the possible environmental impacts arising from a project plan and strategy. It can be applied at all stages of planning and implementation from conception and design to construction, operation and decommissioning. In fact, it attempts to predict or measure environmental effects of certain human activities and to investigate and propose means of mitigation those effects. This can lead to an outcome that balances the interests of the development operations and the environment.

Generally speaking, EIA is the systematic identification and evaluation of the potential impacts (effects) of proposed projects/plans/programs/legislative actions relative to the physical-chemical, biological, cultural, and socioeconomic components of the total environment (Glasson, 1994).

According to Lawrence (2003) and Wood (1995), the **main objective** of EIA is to ensure that development proposals, activities and programs are environmentally sound and sustainable. Specifically, EIA aims to:

- identify, control, contain, minimize and remove potential negative impacts on the environment;
- modify and improve the design of a proposal;
- ensure that resources are used efficiently;
- disclose to decision makers and the public the significant environmental effects of proposed activities;
- disclose to the public reasons for agency approvals of projects with significant environmental effects; and
- enhance public participation.

Origins and History of EIA

The concept of environmentalism emerged along with the publication of *Silent Spring* by Rachel Carson in 1962. Later on, the first formal system of EIA established in the US following the **National Environmental Policy Act (NEPA)** of 1969 (Barrow, 1997; Colombo, 1992; Gilpin, 1995). NEPA ensured that environmental concerns were considered in the decision-making of Federal Government agencies, which were required to prepare a detailed statement on the environmental impact of "*proposals for legislation and other major Federal actions significantly affecting the quality of the human environment*". The statement, also known as **Environmental Impact Statement (EIS)**, included details on (Bregman, 1992; Ortolano, 1997):

- any adverse environmental effects which cannot be avoided should the proposal be implemented, and
- alternatives to the proposed action.

Since 1969 a host of other countries have adopted EIA legislation. In 1977 the European Commission began drafting a directive on EIA and finally published a proposal in 1980. Adopted in July 1985, this **EIA Directive** stated that the EIA should identify, describe and assess the direct and indirect effects of a project on the following factors (Weston, 1997):

- human beings, fauna and flora;
- soil, water, air, climate and the landscape;
- material assets and cultural heritage;
- the interaction between the above factors.

Since the enactment of NEPA, EIA systems have been established in various forms throughout the world, beginning with more developed countries – e.g. Canada (1973), Australia (1974), West Germany (1975), France (1976) – and later also in the less developed countries (Wood, 1995).

II. How is EIA Conducted?

This section explores the different stages of a typical EIA process. As stated by Morris (2001), the **steps of EIA** in detail:

1. Present environmental elements involved in project or affected by the project processes are investigated and analyzed.
2. Prediction, modeling or projecting of environmental impacts caused by the project is studied and mitigation measures for the impacts caused are considered as well.
3. The considered environmental elements and impacts are documented as an EIA report.
4. The reports are submitted to the decision-makers for evaluation on both the reliability of the report and feasibility of the project.
5. Based on the impact and mitigation measures available, the project may be rejected or admitted.

EIA contents cover description of environment, impact prediction, evaluation and mitigation (Gilpin, 1995). In terms of environmental monitoring and auditing, continuous environmental monitoring and auditing after the permission of the project ensure compliance of the project proceeding and the mitigation measures proposed whereas environmental impacts are carefully monitored.

In general, the overall process of EIA is (Barrow 1997; Wood 1995; Weston 1997)**:**

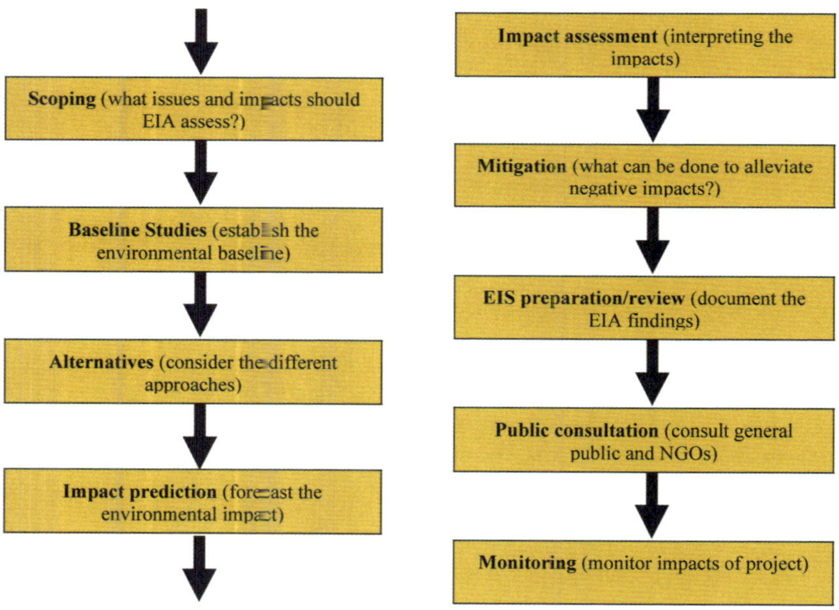

There are three ways to achieve environmentally sound development:

1. Avoidance of adverse environmental impacts;
2. Minimization and control of adverse environmental impacts; and
3. Compensation of adverse environmental impacts.

III. EIA in Hong Kong

Hong Kong now has more than twenty years of EIA experience under the enforcement of Environmental Protection Department (EPD). Effective from 1 April 1998, the **Environmental Impact Assessment Ordinance** has provided for assessing the impact on the environment of many projects and proposals, for protecting the environment and for incidental matters (EPD, 2002). The EIA Ordinance was enacted in 1997 and came into force in 1998. The Ordinance aims to avoid, minimize and control adverse impacts to the environment through the EIA process and environmental permit.

EIA Ordinance in Hong Kong

All specified projects are required to undertake EIAs and properly implement mitigation measures as recommended in EIAs. Both developers and contractors, or whoever carries out the work are responsible for ensuring that mitigation measures are properly followed through. Features include provision of a set of criteria in form of a technical memorandum for undertaking EIA and an explanatory guidebook. These requirements apply to both public and private sector projects.

According to EPD's annual report, "*Environment Hong Kong 2000*", the Ordinance requires designated projects to be proceeded with an environmental permit, which can be acquired once EIA has been carried out and reported to EPD. The EIA report is then reviewed by EPD, and under some circumstances, such as ecological impacts, by Agriculture, Fisheries and Conservation Department (AFCD), to ensure that the project will not cause any major ecological degradation.

EIA is usually carried out by consultant companies by request of the project promoter. The assessment methods may follow the EIA Study Brief and Technical Memorandum. The final report submitted to EPD is reviewed. Once the EIA Report has been approved, an environmental permit is granted to the project promoter or government authority (such as Civil Engineering Department). During the project processes, proper environmental monitoring and auditing are required.

Overview of the EIAO Process:

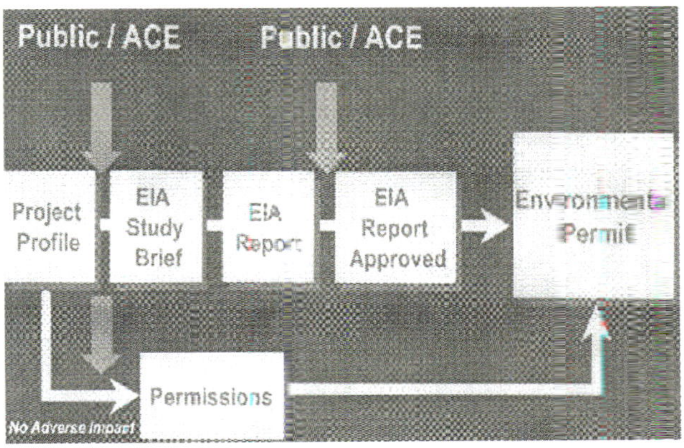

(modified from EPD, 2000)

Parameters evaluated in an EIA:

- air quality
- noise level
- water quality
- wastes
- traffic
- ecological effects
- fisheries
- visual effects
- landscape
- cultural considerations
- socio-economical concerns

Recent EIA Developments

In 1994, a dedicated EIA sub-committee was established under the Advisory Council on the Environment. The sub-committee includes representatives from industry, academics and green organizations.

A further development was the approval of the **Technical Memorandum (TM)** on the EIA Process by the Legislative Council. The TM is a legal instrument, issued under the EIA Ordinance, which seeks to guide the Director of Environmental Protection in decision-making. It provides clear-cut transparent technical guidelines and criteria for EIA and represents best practice in Hong Kong in relation to screening and scoping, conducting assessments, reviewing EIA reports and evaluating impacts. Both the Ordinance and the Technical Memorandum can be obtained from the department's homepage (www.info.gov.hk/epd).

Recently, the first ever transboundary EIA between China and Hong Kong was endorsed by both governments and the Hong Kong Advisory Council on the Environment for the Shenzhen River Regulation Project. This has provided a solid foundation for future cross-border collaboration on EIA.

IV. EIA Applications in Hong Kong

Case 1 – Controversy EIA report "Sheung Shui to Lok Ma Chau Spur Line"

(EIA Project Ref: AEP-067/2000 entitled Sheung Shui to Lok Ma Chau Spur Line by Kowloon-Canton Railway Corporation; application was rejected and permit was not granted on 16/10/2000)

Located in Sheung Shui, New Territories, **Long Valley** is a large piece of flatland known for its richness in bird species (some call it the "Second Mai Po") and the large agricultural wetland irrigated by the local farmers. The area hosts at least 210 migratory bird species, including 29 species of conservation importance or globally endangered (such as Little Egret and Painted Snipe) to stay in the area.

Long Valley © Frank Lee

Valuable resources recorded in Long Valley:

Migratory birds © Frank Lee

Gastropod found in streams © Frank Lee

Dragonfly © Frank Lee

Diagram illustrating the Lok Ma Chau spur line:

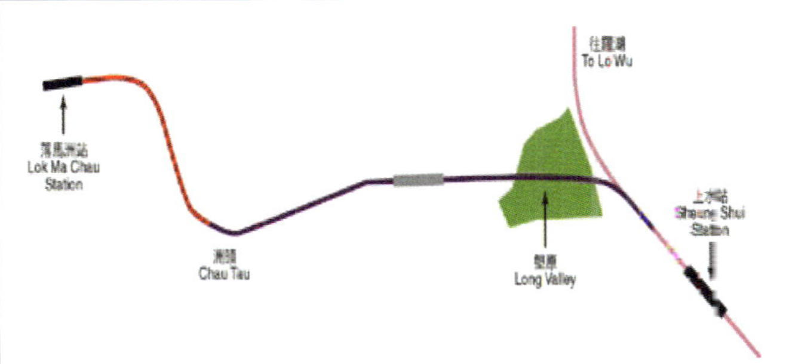

(modified from EPD, 2000)

It was proposed that a railway line would be constructed to pass through this area by Kowloon-Canton Railway Corporation (KCRC). Voices from different parties including local residents, Hong Kong citizens and green groups all raised concerns about the development.

In October 2000, EPD believed the spur line would have adverse environmental impacts on the ecology of Long Valley. The Director of Environmental Protection Department noted "the high ecological value of the areas that would be affected by the Project, and the high diversity of birds in these areas". Finally the government made the decision not to issue a permit for the project. The KCRC appealed to this decision, but the EIA Appeal Board dismissed the appeal.

The most recent news is that KCRC has proposed the construction of an underground railway line to go through the valley instead. Finally, the application was granted a permit by EPD in April 2002. (EIA Project Ref: AEP-129/2002 entitled Sheung Shui to Lok Ma Chau Spur Line by Kowloon-Canton Railway Corporation; permit granted on 6/4/2002)

Case 2 – Impact on Reclamation and Construction in Lantau Island

(EIA Project Ref: AEP-054/2000 entitled Penny's Bay Reclamation by Civil Engineering Department; permit granted on 28/4/2000 and EIA Project Ref: AEP-059/2000 entitled the Theme Park at Penny's Bay, Lantau by Civil Engineering Department; permit granted on 6/5/2000)

The case deals with problems generated from various project developments on Lantau Island including land reclamation and the construction of the Hong Kong Disneyland theme park in Penny's Bay, and other pollution issues in the nearby aquatic environment.

Tung Chung development and North Lantau Expressway
© Frank Lee

Reclamation – construction of the airport at Chep Lap Kok and the North Lantau Expressway have destroyed pristine natural coast and shallow marine habitat. More dolphin habitat will disappear when container terminals are built on reclaimed land at Yam O and Lung Kwu Tan, both areas are frequented by the dolphins.

Boat Traffic – while dolphins are usually adept at sensing and staying out of the way of boats, many show signs of collisions.

Sewage to be treated © Frank Lee

Sewage – some 150,000 m³ of raw sewage (screened, not treated), is dumped into the western harbour area every day. It is expected the volume will reach 700,000 m³ a day by the year 2010. Sewage discharged includes household toxic wastes such as cleaning fluids, bacteria, and viruses that can affect the dolphins directly, and industrial effluent.

Over-fishing – Hong Kong's increasingly sophisticated and virtually unregulated fishing fleet is depleting the dolphins' food supply. Several dead dolphin carcasses have shown signs of having been caught and tangled in fishing nets. In recent years, the government has set up "non-fishing periods" lasting for at least 30 days in the summer to prevent the over-fishing problem.

The consequence is the gradual extinction of Chinese White Dolphins, icon species of Hong Kong. There are now only around 150 dolphins left within the Hong Kong marine boundaries. If these problems are ignored and not dealt with properly, we will definitely lose one of the most beautiful wildlife in Hong Kong.

Dolphin © Frank Lee

Conclusion

Raising public awareness of the environment, the level of public participation in environmental decision-making and the political weight accorded to the environment are essential to increase the influence of EIA on decisions. This is important if the legally binding and well-resourced Hong Kong EIA system is to avoid the fate of other Asian countries, and if the decline in the quality of Hong Kong's environment arising from the cumulative impacts of numerous major projects is to be slowed.

Questions for Further Thinking

1. An EIA application often involves many different related stakeholders. Who should take the role of communicating among these different parties? How can this be efficiently achieved?

2. Apart from the above-mentioned cases of EIA, try to name another current issue with relevance to EIA. Who are the participating stakeholders?

第十四章　環境影響評估

環境影響是指工程對現場或附近環境的轉變。這包括人，植物，動物，生態系，文化遺產，甚至是有歷史性或考古學價值的建築或地區。John Glasson (1994) 將影響定義為工程導致環境的變化，無論變化發生在工程地點以內或以外。

根據Erickson (1994) and Kreske (1996)，環境影響可分為：

種類：	生物物理，社會，健康，經濟
性質：	直接，間接，累積，協同
強度：	不重要 (無須修補行動) 及重要 (須要修補行動)
程度：	當地與地區，跨境，全球影響
時間：	即時與未來
持續性：	短期與長期
不確定性：	可能性，結果
可逆轉性：	可逆轉與不可逆轉
重要性：	與社會，經濟或生態有關；影響的重要性未必與影響的大小相關

I. 什麼是環境影響評估 (EIA)?

根據Colombo (1992)，Weston (1997) 和 Erickson (1994) 的定義，環境影響評估是一套有組織性和系統性的活動以評估和控制因工程帶來的環境影響。它可以應用在工程的概念，設計，建立，運作及拆卸等各階段。事實上，嘗試預測或測量人類活動對環境的影響，及提議緩解的途徑，以平衡發展和環境之間的利益。

一般而言，環境影響評估是系統鑒定和評估有關提議工程／計劃／方案／程序／法律行動對物理化學，生物，文化和社會經濟學的潛在影響 (Glasson, 1994)。

根據 Lawrence (2003) 和 Wood (1995)，環境影響評估的主要的目標是確保發展的提議，活動和程序是符合環保要求和可持續的。環境影響評估目的是：

- 確定，控制，減低和移除潛在的負面影響；
- 修改和改良一個議案的設計；
- 確保資源有效地運用；
- 向決策者和公眾公開擬議活動對環境有關的影響；
- 向公眾解釋批准對環境有影響的工程；以及
- 增加公眾參與。

環境影響評估的起源及歷史

環境主義的概念隨著1962年Rachel Carson的寂靜的春天而出現。後來，第一個正式環境影響評估的系統於1969年在美國的全國環境政策法案(NEPA)後建立(Barrow, 1997; Colombo, 1992; Gilpin, 1995)。NEPA 確保聯邦政府在決策中考慮環境問題，提及任何不能避免，對環境造成不良影響，及替代方案 (Bregman, 1992; Ortolano, 1997)。

自從1969年，很多國家採納了環境影響評估法規。在1977年，歐洲委員會開始起草一個環境影響評估指示，於1980年發表提案。在1985年7月被採納，這個環境影響評估指示聲明環境影響評估應該確定，描述和評估工程帶來的直接和間接的影響，包括以下因素 (Weston, 1997)：

- 人類，動物和植物；
- 土壤，水，空氣，氣候和地貌；
- 文化遺產；
- 上述的因數之間的相互作用。

自從制訂NEPA，環境影響評估系統在世界各地建立各種形式，由已發展國家開始－例如：加拿大（1973 年），澳洲（1974 年），西德（1975 年），法國（1976 年）－以至較少發展的國家 (Wood, 1995)。

II. 如何執行環境影響評估？

這個部分探索一個典型的環境影響評估過程的不同階段。如 Morris (2001) 陳述，環境影響評估的步驟如下：

1. 調查和分析涉及工程或受工程影響的環境元素。
2. 預測工程對環境的影響和考慮緩解措施。
3. 記錄環境因素和影響於環境影響評估報告。
4. 評估提交給決策者，以評估報告的可靠性和工程的可行性。
5. 根據提供的影響和緩解措施，工程或許被拒絕。

環境影響評估包括環境描述，影響預測，評估和緩解 (Gilpin, 1995) 的描寫。在環境監測和審核而言，連續的環境監測和審核可確保工程依據報告的程序和緩解措施。

環境影響評估的程序：

1. **甄選** (工程是否需要環境影響評估?)
2. **範疇** (環境影響評估要評估什麼問題及影響?)
3. **基線研究** (建立環境基線)
4. **替代方案** (考慮其他方案)
5. **影響預測** (預測環境影響)
6. **影響評估** (解釋影響)
7. **緩解措施** (減少負影響)
8. **環境影響報告書** (環境影響評估文件化)
9. **公眾咨詢** (咨詢公眾及非政府組織)
10. **監測** (監測工程影響)

(modified from Barrow 1997; Wood 1995; Weston 1997)

有三種途徑達到符合環保要求的發展：

- 迴避對環境不良的影響；
- 減低和控制不良的環境影響中；以及
- 補償對環境不良的影響。

III. 香港的環境影響評估

在環境保護署 (EPD) 的強制執行下，香港現在有超過二十年的環境影響評估的經驗。環境影響評估條例於1998年4月1日生效，評估工程和議案對環境的影響，以保護環境 (EPD, 2002)。環境影響評估條例於1997年動議，1998年生效。環境影響評估條例的目的在透過環境影響評估程序及環境許可證的機制，以避免、盡量減低或控制指定工程項目對環境所產生的不良影響。

在香港的環境影響評估條例

所有指定的工程必須遵行法定的環境影響評估程序，及恰當地實施環境影響評估建議的緩解措施。建造及營辦，或執行相關工作的人仕確保實施緩解措施。技術備忘錄及指引提供一系列的標準，以執行環境影響評估程序。這些要求均適用於公眾和私營企業工程。

根據環境保護署年報「香港環境2000」，條例要求指定的工程有環境許何證，能夠實行環境影響評且估，並向環境保護署報告。環境保護署審閱環境影響評估報告，在一些情況下，如生態的影響，會由漁農自然護理署(AFCD) 審閱，確保工程不會導致任何重要的生態破壞。

環境影響評估通常由顧問公司負責。評估方法會遵照環境影響評估技術備忘錄。終期報告會交給環境保護署審閱。當環境影響評估報告獲得批准，環境許可證授予工程承辦商或政府當局（例如土木工程署）。在工程進行期間，需要恰當的環境監測和審核。

環境影響評估的因素：

- 空氣質素
- 噪音水平
- 水質
- 廢物
- 交通
- 生態影響
- 漁業
- 視覺影響
- 地貌
- 文化考慮
- 社會經濟

環境影響評估近期發展

1994年，環境咨詢委員會核下，成立環境影響評估小組委員會。小組委員會來自工業，學術和綠色團體的代表。

再進一步發展是在由立法會批准環境影響評估程序上的技術備忘錄 (TM)。技術備忘錄是環境影響評估下的一個法律工具，協助環境保護署長決策。為環境影響評估，提供清晰，透明的技術指引和標準，以作甄選，範疇，進行估價，審閱環境影響評估報告和評估影響。條例和技術備忘錄可從環境保護署網頁 (www.info.gov.hk/epd) 獲得。

最近，就深圳河工程，政府及環境咨詢委員會簽署了第一個中國和香港之間的跨境環境影響評估。為未來環境影響評估的跨境合作，提供穩固的基礎。

IV. 香港環境影響評估應該

案例1—具爭議性的環境影響評估報告「上水至落馬州支線」

(EIA Project Ref: AEP-067/2000九廣鐵路東鐵發展：上水至落馬洲支線；於16/10/2000，拒絕申請及不獲發環境許可證)

位於新界上水，塱原是一大片平地，在鳥類品種豐富（稱為「米埔第二」）和農業沼澤土。該處至少有210種遷移的鳥類品種，包括29種具保育重要性或在全球危害品種（諸如小白鷺）。塱原記錄的珍貴資源包括季侯鳥、溪流的腹足動物及蜻蜓。

九廣鐵路公司 (KCRC) 擬議建造一條鐵路通過該區。不同的團體，包括當地居民，香港市民和綠色團體都關注是項發展。

在2000年10月中，環境保護署相信支線對塱原的生態造成不良的環境影響。環境保護署長認為「工程會影響該區的高生態的價值及豐富的鳥類品種」。最終，政府做決定不發出許可證。九廣鐵路公司上訴，但遭環境影響評估上訴委員會否決。

最新消息是九廣鐵路公司提議建造地下鐵路穿越塱原。終於，環境保護署於2002年4月發出許可證。(AEP-129/2002九廣鐵路東鐵發展：上水至落馬洲支線；於6/4/2002發出許可證)

案例2 – 大嶼山填海及建設的影響

(EIA Project Ref: AEP-054/2000 竹篙灣填海工程 (土木工程署)；於 28/4/2000 發出許可證；及AEP-059/2000位於大嶼山竹篙灣的主題公園 (土木工程署)；於6/5/2000發出許可證)

該案例處理大嶼山不同的工程所產生的問題，包括填海及建造香港迪士尼樂園主題公園，以及附近水域的污染問題。

填海 — 興建赤臘角機場及北大嶼山高速公路摧毀了自然海岸及海洋生境。當陰澳和竹鼓灘建造貨櫃碼頭時，海豚的棲息地將會消失。

航道 — 雖然海豚善於察覺和避開航道，但仍有很多碰撞發生。

污水 — 每日有大約150,000立方米的污水（經過濾，未經處理）排放到海港西面水域。在2010年以前，預料每日會達到700,000立方米。污水包括家居有毒廢物，如清潔劑，細菌，病毒等能夠直接影響海豚，以及工業污水。

過度捕撈 — 香港的捕漁船正減少海豚的食物供應。若干海豚屍體被捕獲及糾纏於漁網中。近年，政府在夏季時實施至少30天的「休漁期」，以阻止過度捕撈的問題。

結果是中華白海豚逐漸絕種。在香港水域，現存大約150隻海豚。如果忽略這些問題和不恰當處理，香港肯定會失去其中一種最美麗的野生動物。

結論

提高大眾的環境意識及鼓勵大眾參環境的決策，以增加環境影響評估的影響力。香港的環境影響評估系統可以避免像其他亞洲國家的遭遇，以及減少源於工程對香港環境的影響。

思考問題

1. 環境影響評估的應用通常涉及很多利害關係者。誰負責與這些不同的關係者溝通?怎樣可以有效地達到目的?

2. 除上述的環境影響評估案例，試舉其他近期的環境影響評估案例。其中的利害關係者是誰?

References 參考資料

Barrow, C. (1997), *Environmental and Social Impact Assessment: an Introduction*, John Wiley, New York, p.105–124, 166–169.

Bregman, J. (1992), *Environmental Impact Statements*, Lewis Publishers, Boca Raton, Florida, p.2–3.

Carroll, B. and Turpin, T. (2002), *Environmental Impact Assessment Handbook: A Practical Guide for Planners, Developers and Communities*, Thomas Telford Publishing, London, p.2–39.

Colombo, A. (1992), *Environmental Impact Assessment*, Kluwer Academic, Dordrecht, Boston, p.1–2.

EPD (2000) *Environment Hong Kong 2000*, EPD, Hong Kong Special Administrative Region, p.85.

EPD (2002), *The Operation of Environmental Impact Assessment Ordinance in Hong Kong: April 1998-December 2001*, Environmental Assessment and Noise Division, EPD.

Erickson, P. (1994), *A Practical Guide to Environmental Impact Assessment*, Academic Press, San Diego, p.3–9

Glasson, J., Therivel, R. and Chadwick, A. (1994), *Introduction to Environmental Impact Assessment: Principles and Procedures, Process*, Practice and Prospects, UCL Press, London, p.2–23.

Gilpin, A. (1995), *Environmental Impact Assessment (EIA): Cutting Edge for the Twenty-first Century*, Cambridge University Press, New York, p.2, 16.

Kreske, D. (1996), *Environmental Impact Statements: A Practical Guide for Agencies, Citizens, and Consultants*, John Wiley & Sons, New York, p.37–40.

Lawrence, D. (2003), *Environmental Impact Assessment: Practical Solutions to Recurrent Problems*, Wiley-Interscience, New Jersey, USA, p.1–22.

Morris, P. and Therivel, R. (2001), *Methods of Environmental Impact Assessment*, 2nd edition, Spon Press, London, p.3–19.

Ortolano, L. (1997), *Environmental Regulation and Impact Assessment*, Wiley, New York, p.326.

Weston J. (1997) *Planning and Environmental Impact Assessment in Practice*, Longman, Harlow, Essex, p.4–9.

Wood, C. (1995), *Environmental Impact Assessment: A Comparative Review*, Longman, Harlow, Essex, p.2, 17, 31–37.

Chapter 15. Environmental Management System

Business activities
© Frank Lee

All business activities have some impact on the natural environment, with the resources used, the processes undertaken and the waste created. For example, using a computer involves consumption of electricity which depletes energy resources. No matter how big or small the impact is, every organization should carry out the business operations in an environmentally responsible manner.

To begin with, we should look at the widely accepted standards known as the International Organization for Standardization (ISO).

I. What is ISO?

The **International Organization for Standardization** is a worldwide federation of national standards bodies from over 100 countries (ISO, 1998; Piper, 2003). Each country can only have one membership. Established in Geneva in 1947, this non-governmental organization aims to promote the development of international standards on manufacturing, trade and communications.

ISO 14000 consists of a series of management standards for helping organizations manage the environmental aspects of their operations (Morris, 2004). The ISO 14000 documents are entitled *Environmental Management Systems-General Guidelines on Principles, Systems and Supporting Techniques*. ISO 14000 Standards is a series of standards, covering (Von Zharen, 1996; Roberts, 1998):

- 14001: Environmental Management Systems
- 14004: EMS general guidelines
- 14010: Guidelines for Environmental Auditing
- 14011: Guidelines for Auditing of an EMS
- 14012: Auditing - Qualification criteria
- Others deal with life cycle assessment, site assessment, labeling and other matters.

Basically, the ISO 14000 standards can be applied to large and small business and industry, service sectors (hospitals, hotels, etc.), government organizations, and all types and sizes of organizations anywhere in the world.

According to Sayre (1996), ISO 14000 states a set of **ten management principles** for organizations considering an environmental management system to:

1. recognize that environmental management is one of the highest priorities;
2. establish and maintain communications with both internal and external interested parties;
3. determine legislative requirements and those environmental aspects associated with activities, products, and services;
4. develop commitment within the organization to environmental protection and clearly assign responsibilities and accountability;
5. promote environmental planning throughout the production cycle;
6. establish a management discipline for achieving targeted performance;
7. provide the right resources and sufficient training to achieve performance targets;
8. evaluate performance against policy, environmental objectives and targets, and make improvements wherever possible;
9. establish a process to review, monitor, and audit the environmental management system to identify opportunities for improvement in performance; and
10. encourage vendors to establish environmental management systems as well.

II. What is Environmental Management System (EMS)?

Environmental management is the process whereby organizations assess, in a methodical way, the impacts of their activities on the natural environment, and take action to minimize these impacts (Whitelaw, 1997). **Environmental Management System** is a management system that allows an organization to control its environmental impacts and reduce such impacts continuously.

EMS is a method of implementing and maintaining environmental management within an organization management by the identification of organizational structures, responsibilities, practices, procedures, processes and resources (Piper, 2003; Roberts, 1998). It advocates sustainable development for each and every nation and sustainable development for each and every person.

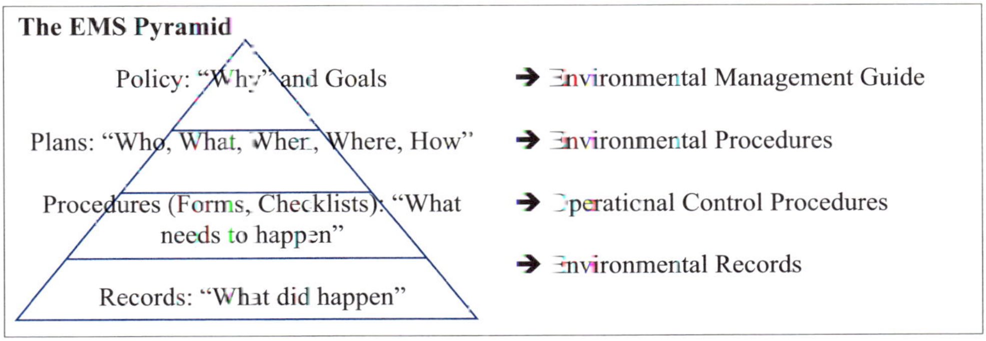

(modified from Lamprecht, 1997; Whitelaw, 1997)

Adopted in 1996, **ISO 14001** is considered the core management document of the entire ISO 14000 series containing the required elements and criteria that must be satisfied by an organization seeking certification (Roberts, 1998). ISO 14001 is the only certification standard that covers processes, not products. The registrar examines EMS for conformity to the ISO 14001 standard while the EMS audit is not a compliance audit. Registration/certification means the organization has a documented EMS that is fully implemented and meets ISO 14001 requirements. Yet, it does not mean that products are more environmentally friendly.

In general, ISO 14001 aims to (ISO 14001, 1996):

- track and satisfy regulatory requirements;
- support effective delegation of responsibilities;
- reduce disruption caused by employee turnover;
- establish a framework to move beyond compliance;
- change a vehicle positively: improved employee morale, enhanced public image;
- identify opportunities for savings; and
- consolidate existing environmental, safety, and health programs into a cohesive system.

Why should companies adopt ISO 14001?

- Improve environmental performance
- Reduce costs
- Attain differentiation advantage
- Competitive necessity
- Public relations
- Regulator relations

ISO 14001/14004 EMS Model

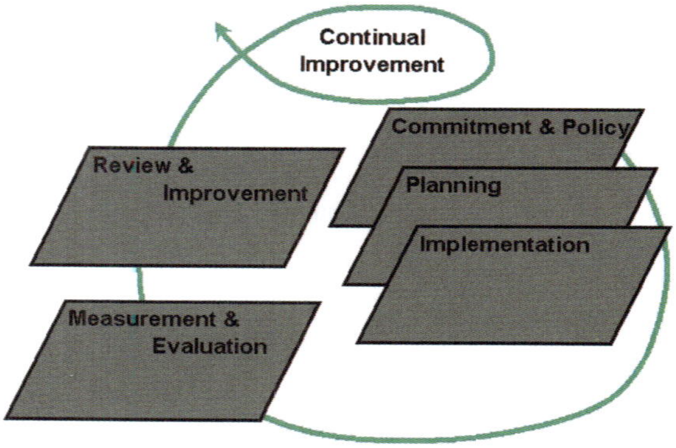

(modified from ISO, 1998)

III. The Five Steps of an ISO 14001 EMS Model

The typical EMS model follows a continual cycle of the five phases, namely (Piper, 2003; Schoffman 2000):

1. Commitments and Environmental Policy

Principle 1: An organization should first focus on what needs to be done-it should ensure commitment to environmental management and define its policy

Top management shall define an organization's environmental policy and ensure that it:

- Is appropriate to the nature, scale, and environmental impacts of its activities, products, and services
- Includes a commitment to continual improvement and prevention of pollution
- Includes a commitment to comply with relevant environmental legislation
- Provides a framework for setting objectives and targets
- Is documented, implemented, maintained, and communicated to all employees
- Is available to the public

2. Planning

Principle 2: An organization should formulate a plan to fulfill its environmental policy

- The organization shall establish and maintain a procedure to
  - identify the environmental aspects of its activities, products, or services
  - identify and have access to legal and other requirements
- It shall establish and maintain documented environmental objectives and targets
- It shall establish and maintain a program for achieving its objectives and targets

3. Implementation

Principle 3: For effective implementation an organization should develop the capabilities and support mechanisms necessary to achieve its policy, objectives, and targets.

- The organization shall define, document, and communicate roles, responsibility, and authority
- It shall identify training needs and ensure appropriate training
- It shall establish procedures for internal and external communication
- It shall document the environmental management system
- It shall establish procedures for emergency response

4. Checking and Corrective Actions

Principle 4: An organization should measure, monitor and evaluate its environmental performance.

- The organization shall establish and maintain procedures to monitor and measure its performance, operational controls, compliance, and conformance with objectives and targets
- It shall establish a procedure for the identification, maintenance and disposal of environmental records
- It shall establish a procedure for auditing its environmental management system

5. Management Review

Principle 5: An organization shall review and continually improve its environmental management system, to insure its suitability, adequacy and effectiveness.

A typical agenda may include questions such as:

- Have the objectives stated in the Environmental Policy been met?
- Does the organization have the continuing capacity to identify the environmental aspects?
- Does the system allow the organization to give a measure of significance to these aspects?
- Have the operational controls that were put in place achieved the desired levels of control?

Fact:

As of March 2000, there were 15,281 ISO 14001 registered worldwide, while 740 ISO 14001 were registered in the United States.

Format of ISO 14001

When writing an ISO 14001 document, the format is usually as follows (Von Zharen, 1996):

0 Introduction
1 Scope
2 References
3 Definitions
4 Environmental management system
- 4.0 General
- 4.1 Environmental Policy
- 4.2 Planning
  - 4.2.1 Environmental aspects
  - 4.2.2 Legal and other requirements
  - 4.2.3 Objectives and targets
  - 4.2.4 Environmental management program
- 4.3 Implementation and operation
  - 4.3.1 Structure and responsibility
  - 4.3.2 Training, awareness and competence
  - 4.3.3 Communication
  - 4.3.4 Environmental management system documentation
  - 4.3.5 Document control
  - 4.3.6 Operational control
- 4.4 Checking and corrective actions
  - 4.4.1 Monitoring and measurement
  - 4.4.2 Nonconformance and corrective and preventive action
  - 4.4.3 Records
  - 4.4.4 Environmental management system audit
- 4.5 Environmental management review

Annexes

A (Informative) Guidance on the use of the specification

B Bibliography

EMS in Hong Kong

In Hong Kong, the ISO 14001 Certified government departments include:

- Architectural Services Department
- Government Laboratory
- Electrical Mechanical Services Department
- Environmental Protection Department
- Drainage Services Department

Certification bodies in Hong Kong include British Standards Institution (BSI), Bureau Veritas (BVQI), Det Norske Veritas (DNV), Hong Kong Quality Assurance Agency (HKQAA), Société Générale de Surveillance (SGS), etc.

Some of the legal requirements in Hong Kong are: 1) Water Pollution Control and Sewage Service Ordinances (Water Pollution Control); 2) Air Pollution Control, Ozone Layer Protection and Road Traffic Ordinances (Air Pollution Control); 3) others include Waste Disposal Ordinance, Noise Control Ordinance, Dangerous Goods Ordinance, Radiation Ordinance, Public Health and Municipal Service Ordinance.

IV. Benefits of EMS

According to Roberts (1998) and Von Zharen (1996), the implementation of EMS indeed results in many benefits. First of all, the **natural resources are more utilised** and utility cost is reduced. The demand for energy consumption can also be decreased. Creditability for loans and insurance is another obvious benefit – lower environmental risks can obtain project financing more easily, at more favourable conditions, and insurance at lower cost.

Compliance to environmental legislation – EMS manages environmental aspects effectively and reduces legal liability to prosecution. EMS can prevent accidents by identifying potential for and response to accidents and emergency situation, by preventing and mitigating the environmental impacts that may be associated with them, and by reducing loss or damage in case of accident.

Increase in public awareness of environmental protection – customers who are environmentally conscious will be more likely to choose and patronise the organisation with sound environmental performance. EMS helps to heighten customer awareness of the organisation's management skills, and serve as a basis for quality service.

Enhancing public image – organisations can distinguish themselves from their competitors, so that their products and services are provided in an environmental friendly manner. Organisation who can exert sound environmental concern can build up a green image and establish a good reputation with the public.

Enhancing company image is indeed one of the potential rewards of certification. In terms of boosting the market share, EMS satisfies both the stockholder's expectation as well as vendor-certification. Financial rewards include improving raw material consumption, reducing waste management costs and down-stream liabilities, and the potential environment-compliance privileges.

Apart from the above benefits, by implementing EMS, organization can also achieve the following (Sayre, 1996):

1. assist in maintaining and improving the quality of the environment;
2. meet customers' environmental expectations;
3. satisfy investor criteria and improve access to capital;
4. provide insurance at a reasonable cost;
5. gain market share;
6. satisfy vendor certification criteria;
7. improve cost control;
8. limit liabilities;
9. provide resource conservation; and
10. provide effective technology development and transfer.

Conclusion

Undoubtedly corporations and organizations are compelled to be more environmentally responsible. The ISO standards are gaining more popularity in the world. Although the requirements of the standards can seem troublesome in the beginning, the implementation is not necessary difficult, once commitment is obtained from top management, along with good planning and an understanding of the concepts.

第十五章　環境管理系統

所有商業活動，包含運用資源和製造廢物，對自然環境有一定的影響。例如：使用電腦消耗電力，減少能量資源。無論影響是大或小，機構應該以環保的方式經營。

首先，我們看看被廣泛接受的國際標準化組織 (ISO) 。

I. ISO是什麼?

國際標準化組織是超過100 個國家標準組織的聯盟 (ISO, 1998; Piper, 2003)。每個國家只有一個會員資格。該組織於1947年在日內瓦成立，這非政府組織目標是發展製造業，貿易及通訊的國際標準。

ISO 14000 包括標準為幫助的組織處理他們的操作的一系列的管理環境方面 (Morris, 2004)。ISO 14000 文件題為*環境管理體系原則、體系和支持技術通用指南*。ISO 14000 標準是一系列的標準，包括 (Von Zharen, 1996; Roberts, 1998)：

- 14001: 環境管理系統(EMS)
- 14004: EMS 通用指南
- 14010: 環境審核指南
- 14011: EMS審核指南
- 14012: 審核 – 資格要求
- 其他是生命週期評估，地點評估，標誌和其他。

基本上，ISO 14000 標準適用於大小企業、工業、服務部門 (醫院、酒店等)、政府機構和全球所有類型和大小的組織。

根據Sayre (1996), ISO 14000 為組織設立一套關於環境管理體系的十個管理原則:

1. 確認環境管理是其中一個優先程序;
2. 建立和維繫組織內部和相關方面的溝通;
3. 確定那些與立法要求和環境方面有聯繫的活動、產品和服務;
4. 對環境保護作出承諾和責任;
5. 在生產週期過程中，促進環境計劃;
6. 建立管理制度，以達到指標;
7. 提供資源和充足的培訓，以達到指標;
8. 評估政策、環境宗旨和目標，以作出改善;
9. 建立評審，監測和審核的環境管理體系程序，以改良表現; 並且
10. 鼓勵賣家建立環境管理體系。

II. 環境管理體系 (EMS) 是什麼?

環境管理是組織透過評估項目對自然環境的影響，並且採取行動把影響減至最低 (Whitelaw, 1997)。**環境管理體系**是一套組織控制和持續減少對環境造成影響的管理系統。

環境管理體系是一套透過組織架構、責任、實踐、程序、過程和資源去實施和維持環境管理的方法 (Piper, 2003; Roberts, 1998)。它主張每個國家及所有人都實行可持續發展。

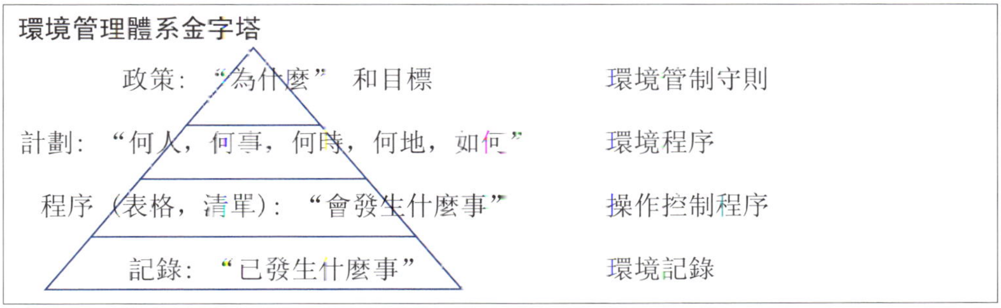

(modified from Lamprecht, 1997; Whitelaw, 1997)

1996年起，ISO 14001 被認為是 ISO 14000 的核心管理文件，而且是組織得到認証的必要條件 (Roberts, 1998)。ISO 14001 是唯一涵蓋生產程序而不是產品本身的認証標準。管理員審查EMS 為整合對ISO 14001標準當EMS 審計不是服從審計。申請／認証是指組織執行EMS的文件，並符合ISO 14001 的要求。不過，它並不代表產品更加環保。

概括而言，ISO 14001目的是 (ISO 14001, 1996)：

- 符合管理的要求；
- 劃分責任；
- 避免因顧員流失而瓦解；
- 根據協意，建立框架；
- 增加員工士氣，加強公眾形象；
- 節省金錢；並且
- 鞏固現存環保，安全和健康計劃為一個體系。

為何公司要採用ISO 14001？

- 改善環境方面表現
- 減少成本
- 取得優勢
- 競爭力
- 公眾形象
- 管理者的聯繫

ISO 14001/14004 EMS 模型

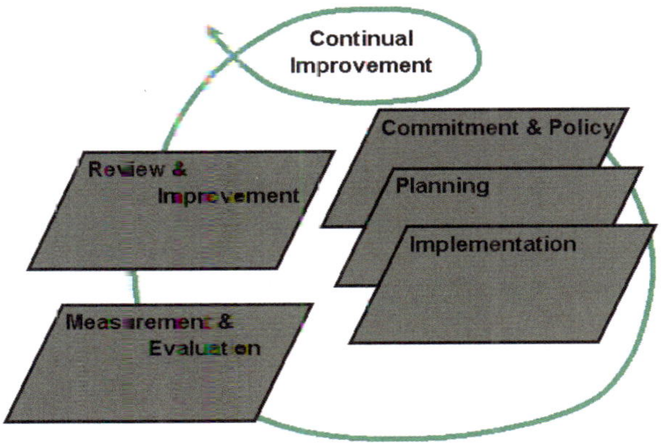

(modified from ISO, 1998)

III. ISO 14001模型的五個步驟

典型環境管理體系從一個循環，其五個階段如下 (Piper, 2003; Schoffman 2000):

1. 承諾和方針
原則 1：一個組織應制定其環境政策並確保其對環境管理體系的承諾
最高管理層應制定組織的環境政策和確保:
- 活動、產品和服務的本質、規模和對環境的影響
- 承諾持續改善和預防污染
- 承諾依從相關的環境法
- 為宗旨和目標設立框架
- 文件化，得到實施和維護，及與所有顧員溝通
- 提供給公眾

2. 策劃
原則 2：一個組織應制定一項實現其環境方針的計劃
- 組織應建立和將維持程序
- 確認活動、產品或服務的環境因素
- 確認法律和其他要求
- 建立和維護環境宗旨和目標的文件
- 建立和維護達到宗旨和目標的計劃

3. 實施
原則 3：為了有效地實施，一個組組應開發其必要能力和支撐機制，以實現其環境方針，目標和指標
- 組織對有關角色、職責和局方作出定義，文件和溝通
- 確認訓練需要和確保有適當訓練
- 建立內部和外部的溝通程序
- 把環境管理系統文件化
- 建立緊急反應的程序

4. 測量和評價
原則 4：一個組織應測量，監測和評價其環境行為
- 組織將建立和維護監測的程序，以測定其表現、運作控制、遵照宗旨和目標的程度
- 建立一個為環境紀錄的標識、維護和處置的程序
- 建立審核環境管理系統的程序

5. 管理評審
原則 5：一個組織以改進環境行為為目標，評審和不斷改進其環境管理體系
一個典型議程應包含以下問題:
- 是否符合環境政策的宗旨?
- 組織有否繼續確認環境因素嗎?
- 體系是否給組織測定這方面的因素?
- 為了達到預期的控制水平，有否作出運作控制?

事實:

直至2000年3月，全球有15,281 個ISO 14001申請，其中740 個ISO 14001來自美國。

ISO 14001 的格式

以下是撰寫ISO 14001 文件格式 (Von Zharen, 1996):

0 引言
1 範圍
2 引用標準
3 定義
4 環境管理體系
  4.0 概述
  4.1 環境方針
  4.2 策劃
    4.2.1 環境因素
    4.2.2 法律與其他要求
    4.2.3 目標和指標
    4.2.4 環境管理方案
  4.3 實施與運作
    4.3.1 組織結構與職責
    4.3.2 培訓，意識與能力
    4.3.3 信息溝通
    4.3.4 環境管理體系文件
    4.3.5 文件控制
    4.3.6 運作控制
  4.4 檢查和糾正措施
    4.4.1 監測和測量
    4.4.2 不一致，糾正與預防措施
    4.4.3 記錄
    4.4.4 環境管理體系審核
  4.5 管理評審
附錄
A (提示的附錄) 規範使用指南
B 參考

香港的環境管理體系

在香港，取得ISO 14001 環境管理體系證書的政府部門有:

- 建築署
- 政府化驗室
- 電子機械供應部門
- 環境保護署
- 渠務署

其他取得證書的機構包括英國標準局 (BSI)，BureauVeritas (BVQI)，Det Norske Veritas (DNV)，香港品質保證局 (HKQAA)，SGS等。

一些香港法律的規定：1) 水污管制條例 (水質污染控制)；2) 空氣污染管制條例、保護臭氧層條例和道路交通條例 (空氣污染控制)；3) 其他包括廢物處置條例、噪音管制條例、危險物品條例、輻射條例、公眾衛生和市政條例。

IV. 環境管理體系的好處

根據Roberts (1998) 和 Von Zharen (1996)，環境管理體系的實施的確有很多好處。首先，有效運用自然資源，節省成本，以及減少能源的消耗量。貸款和保險費亦是其好處 — 環境風險低可以較容易獲得貸款及保險費用較低。

符合環境法規 — 環境管理體系有效地管理環境因素和減少法律起訴。透過意外及緊急事故，預防及緩和與組織有關的環境影響，可減少意外造成的損失或破壞。

加強公眾環保意識 — 關心環保的顧客會選擇較環保的組織。環境管理體系令消費者關注組織的管理技巧，提升服務質素。

提高公眾形象 — 組織較競爭者與別不同，可提供環保產品和服務。組織關注環境問題，加強一個綠色形象和建立良好的聲譽。

提高公司形象是獲得認証的潛在獎勵。環境管理體系滿足了股東期望及賣方的認証要求，而增加市場的佔有率。財政獎勵包括減少原材料消耗量，廢物管理費用，增加成本效益。

除上述好處之外，組織還可以得到 (Sayre, 1996)：

1. 協助維護和改善環境的質素；
2. 滿足顧客對環保的期望；
3. 滿足投資者的標準和改善資金的運用；
4. 提供合理的保險費用；
5. 獲取市場的佔有率；
6. 滿足賣方的認証標準；
7. 改善成本控制；
8. 有限的責任；
9. 提供資源保護；並且
10. 提供有效的技術開發和轉移。

結論

無容置疑，企業和組織不得已對環境負更多的責任。ISO 標準在全球越趨普及。雖然標準的要求起初有些煩瑣，一旦獲得最高管理層的承諾，有好的計劃及理解，實施起來並不是那麼困難的。

Questions for Further Thinking 思考問題

1. What can a company do next, after it has stated all its significant impacts, illustrated improvements and met all its objectives and targets?
   當公司闡明的計劃影響，制訂改善措施及符合標準的目的及指標時，下一步是什麼呢？

2. Does certification indicate that a company's products or services are environmentally superior? 認証是否代表公司的產品或服務是環保的？

References 參考資料

ISO (1998), *Information to Assist Forestry Organizations in the Use of Environmental Management System standards ISO 14001 and ISO 14004*, ISO, Geneva, p.4.

ISO 14001 (1996), *Environmental Management System: Specifications with Guidance for Use*, International Organization for Standards, Geneva.

Lamprecht, J. (1997), *ISO 14000: Issues and Implementation Guidelines for Responsible Environmental Management*, American Management Association, New York, p.43–48, 75.

Morris, A. (2004), ISO 14000 Environmental Management Standards: Engineering and Financial Aspects, Wiley, Hoboken, NJ, p.8.

Piper, L. (2003), Continual Improvement with ISO 14000, IOS Press, Amsterdam, p.21, 44–69.

Roberts, H. (1998), ISO 14001 EMS Implementation Handbook, Butterworth Heinemann, Boston, p.2–15.

Sayre, D (1996), *Inside ISO 14000: the Competitive Advantage of Environmental Management*, St. Lucie Press, Delray Beach, Florida, p.14–25, 155.

Schoffman, A. (2000), ISO 14001: A Practical Approach, Oxford University Press, New York, p.21–55.

Von Zharen, W. (1996), ISO 14000: Understanding the Environmental Standards, Government Institutes, Rockville, Md., p.16–40.

Whitelaw, K. (1997), *ISO 14001 Environmental Systems Handbook*, Butterworth Heinemann, Boston, p.161, preface x.

Abe, T., Levin, S. and Higashi, M. (1997), *Biodiversity: An Ecological Perspective*, Springer, New York, p.59–60.

AFCD, HKSAR (2002), *New Viewpoints – Country Parks in Focus* (2nd ed.), Cosmos Books Ltd, Hong Kong, p.6–7.

AFCD, HKSAR (2005), *Hong Kong Biodiversity Online*. Retrieved February 10, 2005, from Hong Kong Biodiversity Online website: http://www.hkbiodiversity.net/html/en/home.html

Arundel, J. (1995), *Sewage and Industrial Effluent Treatment: A Practical Guide*, Blackwell Science, London, p.2–3, 111–115.

Asante-Duah, D. and Nagy, I. (1998), *International Trade in Hazardous Waste*, E & FN Spon, London, p.71, 122–3.

Barton, J. (1989), *An Introduction to Wastes Management*, Institution of Water and Environmental Management, London, p.46.

Barrow, C. (1997), *Environmental and Social Impact Assessment: an Introduction*, John Wiley, New York, p.105–124, 166–169.

Beeby, A. and Brennan, A. (1997), *First Ecology*, Chapman & Hall, New York, p.23, 26, 86–95.

Bell, P., Greene, T., et al. (1996), *Environmental Psychology* (4th ed.), Harcourt Brace College Publishers, Fort Worth, p.190, 508.

Belshaw, C. (2001), *Environmental Philosophy: Reason, Nature and Human Concern*, McGill-Queen's University Press, Montreal, Canada, p.202–203.

Biotopics (2005), *Sewage*. Retrieved March 20, 2005, from Biotopics website: http://www.biotopics.co.uk/microbes/sewag.html

Bookchin, M. (1980) *Toward an Ecological Society*, Black Rose, Montreal.

Bowen, J. (1994), *Environment Education: Imperatives for the 21st Century*, James Nicholas Publishers, Albert Park, Australia, p.79–103.

Bregman, J. (1992), *Environmental Impact Statements*, Lewis Publishers, Boca Raton, Florida, p.2–3

Buclet, N. and Godard, O. (2000), *Municipal Waste Management in Europe: A Comparative Study in Building Regimes*, Kluwer Academic Publishers, Boston, p.207.

Buell, P. and Girard, J. (1994), *Chemistry: An Environmental Perspective*, Prentice Hall, Englewood Cliffs, N.J.

Bunce, N.J. (1993), *Introduction to Environmental Chemistry*, Wuerz Publishing Ltd., Winnipeg, Canada.

Carroll, B. and Turpin, T. (2002), *Environmental Impact Assessment Handbook: A Practical Guide for Planners, Developers and Communities*, Thomas Telford Publishing, London, p.2–39.

Carson, R. (1994), *Silent Spring*, Houghton Mifflin, Boston, USA (book cover).

Cater, E. and Lowman, G. (1994), *Ecotourism: a Sustainable Option?* John Wiley & Sons, Chichester, England.

Ceballos-Lascurain, H. (1987), "The Future of Ecotourism", *Mexico Journal*, January, 13–14.

Cheremisinoff, P. (1992), *Waste Reduction for Pollution Prevention*, Butterworth-Heinemann, Boston, p.9, 164, 191–193, 199.

China National Tourism Administration (2003), *Fact and Figure*. Retrieved January 15, 2003, from CNTA website: http://www.cnta.com/lyen/2fact/index.htm

Clapp, J. (2001), *Toxic Exports: the Transfer of Hazardous Wastes from Rich to Poor Countries*, Cornell University Press, Ithaca. p.26–27.

Colombo, A. (1992), *Environmental Impact Assessment*, Kluwer Academic, Dordrecht, Boston, p.1–2.

Council for Sustainable Development, HKSAR Government (2004), *Sustainable Development – What is Sustainable Development?* Retrieved January 10, 2005, from website: http://www.susdev.gov.hk/html/en/sd/index.htm

Cox, G. (1997), *Conservation Biology: Concepts and Applications*, Wm. C. Brown, Dubuque, IA, p.2, 6, 33, 44–46, 308, 312.

Dickinson, G. and Murphy K. (1998), *Ecosystems: A Functional Approach*, Routledge, London, p.2, 12, 71.

DSD (1995), *Sewerage Manual*, Government Printer Hong Kong, p.13, 23.

DSD (2003), *環境報告 2003*, Drainage Services Department, Hong Kong, p.25, 53.

Dudley, N. (1987), *This Poisoned Earth: the Truth about Pesticides*, London: Judy Piatkus Ltd., p.1–56.

Economic and Social Commission for Asia and the Pacific (ESCAP) (1998), *Agenda 21 and Challenges for Asia and the Pacific*, New York: United Nations.

Economic and Social Commission for Asia and the Pacific (1999), *Energy, Environment and Sustainable Development III: Energy Infrastructure Policies and Issues*, New York: United Nations.

The Ecotourism Society (1991), *The Ecotourism Society Newsletter*, No. 1, Spring.

Eiler, S. (2003), *Foreign E-waste Problems and Policies*. Retrieved February 20, 2005, from Ball State University website: http://www.bsu.edu/web/smeiler/ewaste/ewastehome.htm

EPD (1998), *Waste Reduction Framework Plan 1998–2007*, Printing Department, Hong Kong, p.4, 10.

EPD (2000a) *Environment Hong Kong 2000*, EPD, Hong Kong Special Administrative Region, p.85.

EPD (2000b), *Monitoring of Solid Waste in Hong Kong*, Facilities Planning Group, EPD, HK Government, p.5.

EPD (2002), *The Operation of Environmental Impact Assessment Ordinance in Hong Kong: April 1998–December 2001*, Environmental Assessment and Noise Division, EPD.

EPD (2005a), *EPD – Waste Recycling Statistics* and *Waste Treatment & Disposal Statistics*. Retrieved January 20, 2005, from Environmental Protection Department website: http://www.epd.gov.hk/epd/english/environmentinhk/waste/data/stat_recycle.html

EPD (2005b), *EPD – Waste*. Retrieved January 20, 2005, from Environmental Protection Department website: http://www.epd.gov.hk/epd/english/environmentinhk/waste/waste_maincontent.html

EPD (2005c), *Environment Hong Kong 2004*. Retrieved February 5, 2005, from Environment Hong Kong website: http://www.epd.gov.hk/epd/misc/ehk04/english/main/contents.html

Erickson, P. (1994), *A Practical Guide to Environmental Impact Assessment*, Academic Press, San Diego, p.3–9.

Exxon Valdez Oil Spill Trustee Council (2005), *Exxon Valdez Oil Spill Trustee Council-Home Page*. Retrieved February 20, 2005, from the website: www.evostc.state.ak.us

FEHD (2005). *New technology – genetic modification (GM) foods* (Chinese / English) Website: http://www.fehd.gov.hk/safefood/gmf/index1.html

Fennell, D. (2002), *Ecotourism Programme Planning*, CABI Publishing, New York, NY, p. 44–47.

Fennell, D. (2003), *Ecotourism: an Introduction (2nd edition)*, Routledge, New York, NY, p.35–8, 61–7, 98.

Footprint (2003), *香港人 — 地球母親的敗家子*, Footprint, vol. 2, p.12–13.

Francis, B.M. (1994), *Toxic Substances in the Environment*, New York: John Wiley & Sons, Inc., p.93–119.

Gene Translation: RNA -> Protein (2005).
Website: users.rcn.com/jkimball.ma.ultranet/biologyPages/T/Translation.html

Glasson, J., Therivel, R. and Chadwick, A. (1994), *Introduction to Environmental Impact Assessment: Principles and Procedures, Process*, Practice and Prospects, UCL Press, London, p.2–23.

Gilpin, A. (1995), *Environmental Impact Assessment (EIA): Cutting Edge for the Twenty-first Century*, Cambridge University Press, New York, p.2, 16.

Grady, C. (1999), *Biological Wastewater Treatment*, Marcel Dekker, New York, p.3, 599.

Haller, E. (1995), *Simplified Wastewater Treatment Plant Operations*, Technomic Pub. Co., Lancaster, Pa., p.11, 21, 29, 119, 139.

Hanks S. (1996), *Ecology and the Biosphere: Principles and Problems*, St. Lucie Press, Delray Beach, p.63–68, 85.

Hardin, G. (1968), "*The Tragedy of the Commons*" Science, 162 December: 1243–1248.

Hong Kong Facts Sheet (2003), *Hong Kong: The Facts: Tourism*. Retrieved January 15, 2004, from website:
http://www.info.gov.hk/hkfacts/tourism.pdf

Holden, A. (2000), *Environment and Tourism*, Routledge, London, p.19–22, 195.

Huang, P.M. and Iskandar, I.K. (2000), *Soils and Groundwater Pollution and Remediation: Asia, Africa, and Oceania*, Boca Raton, Florida: Lewis Publishers.

Information Services Department, HKSAR (2003), *Hong Kong 2003*. Retrieved January 25, 2005, from Hong Kong Yearbook 2003 website:
http://www.info.gov.hk/yearbook/2003/eindex.html

The International Ecotourism Society (TIES) (2000), *TIES World Wide Statistical Fact Sheet 2000*. Retrieved January 15, 2005, from TIES website:
http://www.ecotourism.org/research/stats/files/stats.txt

ISAAA (2003). *Global status of commercial transgenic crops 2003*.
Website: http://www.isaaa.org/Resources/Publications/briefs/isaaa_briefs.htm

ISO (1998), *Information to Assist Forestry Organizations in the Use of Environmental Management System standards ISO 14001 and ISO 14004*, ISO, Geneva, p.4.

ISO 14001 (1996), *Environmental Management System: Specifications with Guidance for Use*, International Organization for Standards, Geneva.

Jackson A. and J. (1996), *Environmental Science: The Natural Environment and Human Impact*, Longman, Harlow, England, p.342–345.

Jarvis, P.J. (2000), *Ecological Principles and Environmental Issues*, Pearson Education Ltd., Harlow, England, p.35–122.

Jeffries, M. (1997), *Biodiversity and Conservation*, Routledge, London, p.59, 139, 150.

Johannesburg Summit (2003), *United Nations: World Summit on Sustainable Development.* Retrieved January 10, 2005, from website: http://www.johannesburgsummit.org/html/basic_info/basicinfo.html

Jones, A. (1997), *Environmental Biology*, Routledge, London, p.3, 58, 163, 169.

Kamrin M.A. (1997), *Pesticide Profiles: Toxicity, Environmental Impact, and Fate*, Boca Raton, Florida: CRC Press, p.623–629.

Kirkby, J., O'Keefe, P. and Timberlake, L. (1995), *The Earthscan Reader in Sustainable Development*, London: Earthscan Publications.

Koren, H. and Bisesi, M (1996), *Handbook of Environmental Health and Safety: Principles and Practices* (vol. II), Boca Raton: Lewis Publishers, p.395–463.

Kreske, D. (1996), *Environmental Impact Statements: A Practical Guide for Agencies, Citizens, and Consultants*, John Wiley & Sons, New York, p.37–40.

Krueger, J. (1999), *International Trade and the Basel Convention*, the Royal Institute of International Affairs, London, p.44.

Lamprecht, J. (1997), *ISO 14000: Issues and Implementation Guidelines for Responsible Environmental Management*, American Management Association, New York, p.43–48, 75.

Lands Department, HKSAR (2003), *Hong Kong and Its Neighboring Region map* (8th ed).

Lawrence, D. (2003), *Environmental Impact Assessment: Practical Solutions to Recurrent Problems*, Wiley-Interscience, New Jersey, USA, p.1–22.

Legreid, M., Bockman, O.C. and Kaarstad, O. (1999), *Agriculture, Fertilizers and the Environment*, New York: Cabi.

Leopold, A. (1949), *A Sand County Almanac*, Oxford University Press, New York.

Lo, F. and Marcotullio. P. (2001), *Globalization and the Sustainability of Cities in the Asia Pacific Region*, New York: United Nations University Press.

Losey, Rayor, and Carter (1999). Transgenic Poller Harms Monarch Larvae. *Nature*, vol. 399, no. 6733, p. 214.

Lovelock, J. (1979), *Gaia: A New Look at Life on Earth*, Oxford University Press, Oxford.

Malina, J. (1992), *Design of Anaerobic Processes for the Treatment of Industrial and Municipal Wastes*, Technomic Pub. Co., Lancaster p.2.

Melchias, G. (2001), *Biodiversity and Conservation* Science Publishers, Inc., Enfield, NH, USA, p.3–5, 84–86.

Moellenbeck DJ, ML Peters ML, Bing JW, Rouse JR Higgins LS, Sims L, Nevshemal T, Marshall L, Ellis RT, Bystrak PG, Lang BA Stewart JL, Kouba K, Sondag V, Gustafson V, Nour K, Xu D, Swenson J, Zhang J, Czapla T Schwab G, Jayne S, Stockhoff BA, Narva K, Schnepf HE, Stelman SJ, Poutre C, Koziel M and Duck N (2001). Insecticidal proteins from *Bacillus thuringiensis* protect corn from corn rootworms. *Nature Biotechnology* 19, 668 – 672.

Moore, E. (2000), *An Introduction to the Management and Regulation of Hazardous Waste*, Battelle Press, Columbus, Ohio, USA.

Morris, A. (2004), ISO 14000 Environmental Management Standards: Engineering and Financial Aspects, Wiley, Hoboken, NJ, p.8.

Morris, P. and Therivel, R. (2001), *Methods of Environmental Impact Assessment*, 2nd edition, Spon Press, London, p.3–19.

Mottershead, T. (2004), Sustainable Development in Hong Kong, Hong Kong: Hong Kong University Press, p.25–35.

Naess, A. (1975), "*The Shallow and the Deep, Long-Range Ecology Movements*" Inquiry 16 (1975) 95–100.

Ng, G. (1993), *Decision Makers' Guide to Solid Waste Management in Hong Kong: A Sustainable Approach*, Conservancy Association, Hong Kong, p.30–34, 48, 58, 77–80.

Nordlee, J.A. S.L. Taylor, J.A. Townsend, L.A. Thomas and R.K. Bush. (1996). Identification of a Brazil-nut allergen in transgenic soybeans. *New England Journal of Medicine* 334, 688–692.

Nottingham, S. (1998). *Eat your genes: How genetically modified food is entering our diet*. Zed Books Ltd., London.

O'Neill, K. (2000), *Waste Trading among Rich Nations: Building a New Theory of Environmental Regulation*, MIT Press, Cambridge.

OECD (2004), *SourceOECD: factbook*. Retrieved February 10, 2005, from OECD Factbook 2005 website:
http://haly.sourceoecd.org/vl=6701579/cl=16/nw=1/rpsv/factbook/11-01-12.htm

Ortolano, L. (1997), *Environmental Regulation and Impact Assessment*, Wiley, New York, p.326.

Page, S. and Dowling, R. (2002), *Ecotourism*, Pearson Education, Edinburgh Gate, Harlow, p.24–25, 57, 281.

Pietra, F. (2002), *Biodiversity and Natural Product Diversity*, Pergamon, Amsterdam, p.3–4.

Piper, L. (2003), Continual Improvement with ISO 14000, IOS Press, Amsterdam, p.21, 44–69.

Planning Department, HKSAR Government (2001), *Study on Sustainable Development in the 21st Century: Final Report*. Retrieved January 10, 2005, from website:
http://www.info.gov.hk/planning/p_study/comp_s/susdev/final_hp/ch1.htm

Pullin, A.S. (2002), *Conservation Biology*, Cambridge University Press, NY, USA, p.15–17, 54–74.

Ragsdale, N. (1999), *Pesticides: Managing Risks and Optimizing Benefits*, Washington, DC: American Chemical Society: Oxford University Press.

Ramalho, R. (1983), *Introduction to Wastewater Treatment Processes*, Academic Press, New York, p.9, 31, 68, 80, 341, 383.

Rees, W. *et al.* (1996), *Our Ecological Footprint — Reducing Human Impact on the Earth*, New Society Publishers, Gabriola Island, BC.

Reiss, M. and Chapman, J. (2000), *Environmental Biology*, Cambridge University Press, Cambridge, p.1–8, 7, 73.

Roberts, H. (1998), ISO 14001 EMS Implementation Handbook, Butterworth Heinemann, Boston, p.2–15.

Sayre, D. (1996), *Inside ISO 14000: the Competitive Advantage of Environmental Management*, St. Lucie Press, Delray Beach, Florida, p.14–25, 155.

Schoffman, A. (2000), ISO 14001: A Practical Approach, Oxford University Press, New York, p.21–55.

Soule, M.E. and Orians, G.H. (2001), *Conservation Biology: Research Priorities for the Next Decade*, Island Press, Washington D.C., USA.

Stokke, O. (1991), *Sustainable Development*, London: Frank Cass & Co. Ltd.

Sustainable Development Unit, HKSAR Government (2002), *Sustainable Development*, Hong Kong: Government Printing.

Swiss Federal Institute of Technology Institute for Plant Sciences. Genetic engineering towards carotene biosynthesis in endosperm.

Tisdell, C. (1999), *Biodiversity, Conservation and Sustainable Development: Principles and Practices with Asian Examples*, Edward Elgar Publishing Ltd, Cheltenham, UK.

Townsend, C., Begon, M. and Harper, J. (2003), *Essentials of Ecology*, Blackwell Pub., Malden, MA, p.8, 55, 96, 317.

United Nations (2005), *Population Growth*. Retrieved February 10, 2005, from Population Reference Bureau website: http://www.prb.org/Content/NavigationMenu/PRB/Educators/Human_Population/Population_Growth/Population_Growth.htm

United Nations Environment Programme (UNEP) Chemicals (2002), *Regionally Based Assessment of Persistent Toxic Substances — Central and North East Asia Regional Report*, Switzerland: United Nations, p.1–27.

United Nations Environmental Programme (2005), *Sustaining life on earth.* Retrieved January 20, 2005, from the Convention on Biological Diversity website: http://www.biodiv.org/doc/publications/guide.asp

Vesilind, P. (1985), *Sludge Management & Disposal for the Practicing Engineer*, Lewis Publishers, Chelsea, Michigan, p.4, 63–64

Von Zharen, W. (1996), ISO 14000: Understanding the Environmental Standards, Government Institutes, Rockville, Md., p.16–40.

Walker, C.H. and Livingstone, D.R. (1992), *Persistent Pollutants in Marine Ecosystems*, Oxford: Pergamon Press, p.3–34.

Watson, J.D. (1968). *The double helix*. New York: Atheneum

Wearing, S. and Neil, J. (1999), *Ecotourism: Impacts, Potentials and Possibilities*, Elsevier Science, Burlington, MA, p.8–9, 24–28.

Weaver, D. (2001), *The Encyclopedia of Ecotourism*, CABI Publishing, New York, NY, p.73–82, 123–129.

Weston J. (1997), *Planning and Environmental Impact Assessment in Practice*, Longman, Harlow, Essex, p.4–9.

White, L. (1967), "*The Historical Roots of Our Ecological Crisis*" Science, 155 March (10): 1203–1207.

Whitelaw, K. (1997), *ISO 14001 Environmental Systems Handbook*, Butterworth Heinemann, Boston, p.161, preface x.

Whitman. D. B. (2000). *Genetically modified foods: harmful or helpful?* Website: http://www.csa.com/hottopics/gmfood-overview.php

Wiebers, U. (1993), *Integrated Pest Management and Pesticide Regulation in Developing Asia*, Washington, DC: World Bank.

Wong, D. (2003). *Genetically modified food labelling*, Research and Library Services Division, Legislative Council, Hong Kong, 35pp. Website: http://www.legco.gov.hk/yr02-03/english/sec/library/0203rp05e.pdf

Wong, M.H., Cao, Z.H. and Luo, Y.M. (2003), *Chemosphere: Soil and Environmental Quality in Modern China*, Oxford: Pergamon.

Wood, C. (1995), *Environmental Impact Assessment: A Comparative Review*, Longman, Harlow, Essex, p.2, 17, 31–37.

World Commission on Economic Development (1987), *Our Common Future (the Brundtland Report)*, Oxford University Press, p.43–45, 49.

Zhang H and Blumwald E (2001). Transgenic salt tolerant tomato plants accumulate. salt in foliage but not in fruit. *Nature Biotechnology,* Vol 19; Aug 2001; pp. 765–768.

Zhou, J.M. (2001), *Sustainable Development in Asia, America and Europe with Global Applications: A New Approach to Land Ownership*, Cheltenham, UK and Nothampton, MA: Edward Elgar.

Contacts of CIES 裘槎環科所聯絡方法

Croucher Institute for Environmental Sciences, Hong Kong Baptist University
香港浸會大學 裘槎環科所

Telephone / 電話 : 3411 7746
Website / 網址 : www.cieshk.org
Email / 電郵 : cies@hkbu.edu.hk

A

Agenda 21 78
Agenda 21 for the Travel and Tourism Industry 94
Agriculture 29–32
Agriculture, Fisheries and Conservation Department (AFCD), HK Gov't 41
Animalia (kingdom) 7
Anthropocentric ethics 55

B

Basel Convention 171–173
Biodiversity – definition 14
Biodiversity – instrumental values 17
Biodiversity – loss 18
Black-faced spoonbill 44, 45

C

Carrying capacity 5
CERES principles 59
Chinese white dolphin 44, 45
Community – definition 4
Competition – definition 4
Conservation biology – definition 19
Convention on International Trade in Endangered Species of Flora and Fauna 18
Croucher Institute for Environmental Sciences xiii
Cycle – definition 5

D

DDT 109–111, 113–117
Deep ecology 57
Desertification 32
Dioxin 109–112, 114, 115
Dirty Dozen 109, 115

E

E-waste - definition 165
E-waste - impacts 167, 168
E-waste - recycling operations 166
Earth Citizen concept 79
Ecofeminism 58
Ecological footprint – definition 65
Ecological footprint – calculation 66
Ecological niche 5
Ecological pyramid 6
Ecology – definition 4
Ecosystem – definition 4
Ecosystem diversity 14
Ecotourism – definition 91
Ecotourism – impacts 95–97
Ecotourists 93
Energy consumption 31
Energy flow 6
Energy pyramid 16
Environment – definition 4
Environmental ethics 53
Environmental impact 180
Environmental Impact Assessment (EIA):
- definition 180, 181
- steps 182
- ordinance 183
Environmental Management System (EMS):
- 14001 model 194, 195
- 14001 format 196
- benefits 197
- definition 193
- pyramid 193
Environmental ordinances of Hong Kong 42
Environmental Protection Department (EPD), HK Gov't 41
Environment, Transport and Works Bureau, HK Gov't 41
Extinction – ecological 18
Extinction – mass 17
Exxon Valdez 34, 58, 59

F

Food chain 6
Forest fire 32
Fossil records 15
Frontier ethic 31

Fungi (kingdom) 7

G

Gaia hypothesis 56
Genetic diversity 14
Genetic engineering 123
Genetic erosion 18
Genetically modified (GM) food
- benefits 126–128
- disadvantages 128–131
- definition 123
- process 124, 125
- reducing risks 131, 132
GM cropping 124
Green turtle 46
Guiyu, China 169, 171

H

Habitat – definition 4
Habitat destruction 31
Harbour Area Treatment Scheme (HATS) 42
Holistic ethics 55
Hong Kong Baptist University xiii
Hong Kong Newt 46
Human population 29

I

Industrial revolution 29, 30
International Ecotourism Society 92
International Organization for Standardization (ISO) 192
Introduced species and diseases 33
ISO 14000 192, 193

K

Kingdoms of organisms 7

L

Land ethics 55
Landfilling 145
Lantau island development 185
Law – definition 3
Law of thermodynamics 3
Limiting factors 6
Long Valley 184

M

Monera (kingdom) 7

N

National Environmental Policy Act 181
National Tourism Administration of China 89
Non-anthropocentric ethics 55

O

Overexploitation of resources 33

P

Persistent toxic substances – definition 109
Pesticides 111
Plantae (kingdom) 7
Pollution 33
Polychlorinated biphenyl (PCB) 112, 114
Post-industrial accelerated change 30
Principle – definition 3
Productivity – definition 6
Protista (kingdom) 7

R

Regionally Based Assessment of Persistent Toxic Substances Project 115
Recycling of waste 146

S

Sewage composition 155
Sewage sludge 157, 158
Sewage treatment 156, 157

Sewerage Master Plans (SMPs) 159
Shallow ecology 56
Silent Spring 54
Sites of Special Scientific Interest (SSSI) 97
Social ecology 58
Species – definition 4
Species diversity 14
Species extinction 34
Stockholm Convention 115
SUSDEV 21 of Hong Kong 80
Sustainable development – definition 76
Symbiosis – definition 5

T

Theory – definition 3

U

United Nations Conference on Environment and Development 20, 77, 78
United Nations Environment Programme (UNEP) 109

W

Waste management and disposal 144
Waste Management Principle 149
Waste Reduction Framework Plan 42, 148
Waste types 143
Water cycle 5
Wildlife in Hong Kong 44
World Commission on Economic Development (Brutland Commission) 67, 77
World Commission on Economic Development 77
World Summit on Sustainable Development 78
World Tourism Organization 91

Ncte 記事

Note 記事

Principles of Bioeconomics

PRINCIPLES OF BIOECONOMICS

Edited by:

Venkatesh.C Pande

SOCIETY Publishing

www.societypublishing.com

Principles of Bioeconomics

Venkatesh.C Pande

Society Publishing
224 Shoreacres Road
Burlington, ON L7L 2H2
Canada
www.societypublishing.com
Email: orders@arclereducation.com

ISBN: 978-1-77469-483-1 (Hardcover)

ABOUT THE EDITOR

Mr. Venkatesh C. Pande (2022) is presently serving as Assistant Professor, Dept. of Agril. Economics, College of Agriculture, Naigaon Bz., Vasantrao Naik Marathwada Krishi Vidyapeeth, Parbhani. He obtained his B.Sc. (Agri.) Degree from College of Agriculture, Naigaon Bz., under Vasantrao Naik Marathwada Krishi Vidyapeeth, Parbhani in 2008. He completed successfully his Post Graduation and secured M.Sc. (Agri.) Degree with specialization in Agril. Economics & Statistics, from Govt. College of Agriculture, Latur under Vasantrao Naik Marathwada Krishi Vidyapeeth, Parbhani in the year 2010. He started his career in 2010 as Marketing Officer & Supervisor in an Agro Chemicals Manufacturing and Selling firm by serving this company for an successful period of 1.4 Years. After his successful work in Agro Marketing Industry, He started his Academic career as Assistant Professor in 2012. His field of specialization is Agril. Economics & Statistics. He published 01 Research Paper in Journal of MSAE- Maharashtra Society of Agril. Economics. He Actively Participated & Presented (Oral) 01 Research paper in National Conference of Maharashtra Society of of Agricultural Economics (MSAE) on the theme "Marketing & Pricing Policies of Agricultural Commodities, Marketing Reforms". He has delivered 01 Radio Talk which was a boon to guide Farmers of Marathwada region as well as Agriculture Graduates. He also has participated National Seminars, National as well as International Conferences & Webinars. He Actively Participated & Successfully completed a One Week Online Training Course on "Organic Farming" organized by Regional Centre of Organic Farming (RCOF), Nagpur, Organic Farming Research and Training Centre, VNMKV, Parbhani & Centre for Sustainable Agriculture, Hyderabad (Telangana State). He is having an wide experience of Teaching the subject of Agril. Economics to the under graduate students of Agriculture Faculty since last 10 years. Besides this, He has acted as Education & Admission Incharge at Agriculture College for 08 Years.

TABLE OF CONTENTS

List of Figures .... *xi*
List of Abbreviations .... *xv*
Glossary .... *xix*
Preface .... *xxv*

Chapter 1 Basic Concepts of Bioeconomy and Its Principles 1

1.1. Bioeconomy: Origin and Evolution 2
1.2. Bioeconomy Sectors 10
1.3. Knowledge Based Bioeconomy (KBBE) 11
1.4. Bioeconomy Strategies 13
1.5. Changing Perspectives on the Bioeconomy 15
1.6. Arising Criticism of the Concept 17
1.7. Bioeconomy and the Principles of Circular Economy 21
1.8. Bioeconomy as an Element of Great Societal Transformation 22
1.9. Strengthening the Demand for Bioeconomy Products 23
1.10. Principles for Circular Biomass Use 24
1.11. Conclusion 29
References 30

Chapter 2 Indicators to Monitor and Evaluate Bioeconomy Sustainability 31

2.1. Introduction 32
2.2. Sustainable Bioeconomy Principles, Criteria, and Impact Categories 34
2.3. Typology of Indicators 35
2.4. Combining Single Indicators to Report on Indicator Sets at Territorial and Product Level 38
2.5. Monitoring Approaches at Territorial Level 43
2.6. Good Practices as Indicators to Monitor and Evaluate Bioeconomy Sustainability 49

2.7. Stepwise Approach to Monitoring the Bioeconomy 53
2.8. Conclusion 59
References 60

Chapter 3 Challenges and Perspectives Towards a Sustainable Bioeconomy 61

3.1. Introduction 62
3.2. Fostering Sustainable Bioeconomics: The Role of Conscious Consumption 63
3.3. Biorefineries Supply to Sustainable Textiles 68
3.4. Wood-Based Solutions for Sustainable Built Environment 70
3.5. Environmental Sustainability Indicators for the Bioeconomy 72
3.6. Innovation and Sustainable Development: A Bioeconomic Perspective 74
3.7. Challenges in Sustainable Bioeconomy 76
3.8. Challenges and Gaps in Sustainability Assessment In Bioeconomy ... 78
3.9. Benefits and Impacts of the Bioeconomy 81
3.10. Constructing a Sustainable Bioeconomy: Multi-Scalar Perceptions of Sustainability 84
3.11. Ecological Limits to Sustainable Use of Wood Fuels 88
3.12. Conclusion 89
References 90

Chapter 4 Socio-Biology and Bioeconomics 91

4.1. Introduction 92
4.2. The Evolution of Socio-Biology 94
4.3. Socio-Biology and The Emotions 99
4.4. Socio-Biology Theory 101
4.5. Socio-Biology, Theory of Evolution And Bioeconomics 106
4.6. Conclusion 118
References 119

Chapter 5 Bioeconomy and Agricultural Production: Concepts and Evidence 121

5.1. Introduction 122
5.2. Significance of Agriculture for Bioeconomy 127
5.3. Challenges of the Agricultural Bioeconomy 130

5.4. Agricultural Production ... 132
5.5. Agricultural Production Economics ... 137
5.6. Primary Production Sustainability ... 138
5.7. Agricultural Sustainability ... 139
5.8. Agricultural Chains Sustainability Within the Concept of Circular Economy ... 141
5.9. The Challenges Toward the Adoption of a Circular Model in the Agri-Business ... 144
5.10. Evolutionary Scenarios for Agricultural Business Models ... 146
5.11. Unlocking the Potential of Agriculture with Evidence Based Production ... 147
5.12. Conclusion ... 149
References ... 151

Chapter 6 Bioeconomics of Fisheries Management ... 153
6.1. Introduction ... 154
6.2. Why is Fisheries Management and Regulation Needed? ... 155
6.3. The Social Trap and Free-Rider Behavior in Fisheries ... 159
6.4. Fundamentals of Fisheries Bioeconomics ... 160
6.5. The Basic Bioeconomic Model ... 162
6.6. Age Structured Bioeconomic Model ... 162
6.7. The Fisheries Management Process ... 163
6.8. Historical Perspective on the Development of the Paradigm ... 168
6.9. Economic Analysis of Fishery Regulation ... 170
6.10. Bioeconomics of Ecosystem Interdependencies ... 173
6.11. Spatial Management of Fisheries ... 175
6.12. Dealing With Risk and Uncertainty in Fisheries Management ... 177
6.13. Conclusion ... 181
References ... 182

Chapter 7 Bioeconomics of Aquaculture ... 183
7.1. Introduction ... 184
7.2. Bioeconomic Modeling and Salmon Aquaculture ... 186
7.3. Preliminary Analysis of the Culture Potential of the Freshwater Angelfish: ***Pterophyllum Scalare*... 192**
7.4. Farm Design ... 197

7.5. Marketing and Economic Considerations in the Production of ***P. Scalare*** ***203***

7.6. The Bioeconomics of Recirculating Aquaculture Systems 204

7.7. Bioeconomic Model 207

7.8. Technology and the Bioenergetic Model 210

7.9. Limitations and Future Research Directions 214

7.10. Conclusion 215

References 216

Chapter 8 Bioeconomic of Invasive Species 217

8.1. Introduction 218

8.2. Invasion Process and Feedbacks Between Biological and Economic Systems 230

8.3. Bioeconomic Impact of Existing Policy on Invasive Species 232

8.4. Integrating Economics and Biology for Invasive Species Management 234

8.5. What are the Disciplinary Impediments of Ecological Economic Modeling? 236

8.6. Trait-Based Risk Assessment for Invasive Species 238

8.7. Management of Invasive Species in the Great Lakes 240

8.8. Conclusion 247

References 248

Index 249

LIST OF FIGURES

Figure 1.1. Bioenergy system boundaries

Figure 1.2. Harnessing agricultural innovation to boost Africa's prosperity

Figure 1.3. Broadband applications direction microbiome-based precision treatment

Figure 1.4. Stakeholder dialog on bioeconomy (L-R): Johann Kottulinsky, Lenzing AG, Michael Proschek-Hauptmann, Schweighofer Group, Sten Nilsson, Forest Sector Insights AB, Markus Lassheikki The Central Union of Agricultural Producers and Forest Owners, Finland, JaanaBäck, University of Helsinki, Maartje Klapwijk, Swedish University of Agricultural Sciences, and Gerald Steindlegger, ISS integrated sustainability solutions

Figure 1.5. Bioeconomy challenges for the EU regions- in the context of the knowledge exchange platform (KEP) Belgium – Brussels – April 2016

Figure 1.6. An illustration of the circular economy concept: applying core engineering principles for the engineering lifecycle

Figure 1.7. Image showing economic development in Austria

Figure 1.8. Biomass storage silos and unloading point

Figure 2.1. Bio-waste and other waste farm produce used to make compost

Figure 2.2. Greens/EFA MEPs rallied outside European Parliament for CAP

Figure 2.3. A simple visualization of the four stages of product life cycle: introduction, growth, maturity, and decline

Figure 2.4. Life cycle assessment

Figure 2.5. Bio-based polymers used in healthcare industry

Figure 2.6. A three-tier agroforestry system (Jack fruit, papaya, brinjal) in Narsingdi village, Bangladesh

Figure 3.1. Sustainable development goals pyramid

Figure 3.2. Biological and technical nutrients in the cradle-to-cradle design framework

Figure 3.3. This was a picture taken during an Education for sustainable development ESD workshop in Kasese district Uganda

Figure 3.4. The pulp mill of MetsäFibre in Kemi, Finland

Figure 3.5. New IKEA building, near to Greenwich, Great Britain

Figure 3.6. Cross laminated timber is studied in a laboratory at Oregon State University

Figure 3.7. Laminated veneer lumber

Figure 3.8. Opening of the OECD global forum on development

Figure 3.9. 5th biennial high-level meeting of the development cooperation forum side event: Adaptation of the 2030 agenda for sustainable development

Figure 3.10. Types and generation of biofuels

Figure 3.11. Hydropower plant: Sir Adam Beck 1 & 2 generate 1.6GW of electricity for Ontario

Figure 4.1. Social behavior of lion cubs. Amboseli National Park, Kenya

Figure 4.2. A picture of Hymenoptera

Figure 4.3. The changeable hawk-eagle or crested hawk-eagle (*Nisaetus cirrhatus)*

Figure 4.4. Group of doves

Figure 4.5. A picture of peacock's colorful tail

Figure 4.6. Lamarck's theory of evolution compared to Darwinian evolution, Baldwin effect, and Waddington's genetic assimilation

Figure 4.7. Long waves of social evolution

Figure 5.1. A farmer fertilizing maize plants

Figure 5.2. Organic farming at the Batad rice terraces, Philippines

Figure 5.3. New crops-Chicago urban farm

Figure 5.4. A picture of home gardening

Figure 5.5. SICET, biomass power plant in Ospitale di Cadore, Italy

Figure 5.6. Measuring greenhouse gas emissions from smallholder systems

Figure 5.7. Application of agrochemicals including fertilizer to crop land and more nutrients entering water bodies

Figure 5.8. Conventional sprinkler irrigation at leafy greens in the Salinas valley of California operated by farmer

Figure 5.9. Grazing land at Bethlehem

Figure 5.10. A graphic describing the some of the areas of study that the triple bottom line framework

Figure 6.1. Women in fisheries: International maritime organization

Figure 6.2. Fisheries patrol vessel, Newlyn harbor, FPV Saint Piran is operated by the Cornwall Inshore Fisheries and Conservation Authority

Figure 6.3. Workers doing fishing job

Figure 6.4. Fisheries management meeting, Solomon Islands

Figure 6.5. Fishermen sorting fish for market

Figure 6.6. MPAs globally in 2020

Figure 6.7. Fish aggregating device (FAD) being deployed in Gwanatafu, a fishing community in North Malaita, Solomon Islands

Figure 7.1. Aquaculture fish farming in the fjords south of Castro, Chile

Figure 7.2. Freshwater angelfish – *Pterophyllum scalare*

Figure 7.3. Recirculating aquaculture system

Figure 7.4. Farmed Nile Tilapia in a fish market

Figure 8.1. The purple sea urchin (*Strongylocentrotus purpuratus*)

Figure 8.2. Sea lamprey (*Petromyzon marinus*)

Figure 8.3. House sparrow (*Passer domesticus*)

Figure 8.4. The green algae Caulerpa taxifolia – aquarium cultivar

Figure 8.5. Salvelinus namaycush fish

Figure 8.6. Gypsy moth (*Lymantria dispar*)

Figure 8.7. American chestnut (*Castanea dentate*)

Figure 8.8. Diagram showing the water pollution of the seas from untreated ballast water discharges

Figure 8.9. The Zebra mussel is an aquatic invasive species found at Diamond Lake in Umpqua National Forest in Oregon

Figure 8.10. A picture of *Eriocheir sinensis*

Figure 8.11. Upper Mississippi River National Wildlife and fish refuge

LIST OF ABBREVIATIONS

AAAA	Addis Ababa action agenda
BCI	bioeconomy contribution index
BERST	bioeconomy regional strategy toolkit
BIOEAST	based agriculture, aquaculture, and forestry in the bioeconomy
BMBF	Federal Ministry of Education and Science
BOD	biological oxygen demand
CAP	common agricultural policy
CBD	convention on biological diversity
CLT	cross-laminated timber
COPA	Committee of Professional Agricultural Organizations
CRM	customer relationship management
CSR	corporate social responsibility
CSSC	Chicago sanitary and ship canal
DMEY	dynamic maximum economic yield
DO	dissolved oxygen
EAF	employment application form
EAF	ecosystems approach to fisheries
EBP	evidence based production
EC	European Commission
EEZ	exclusive economic zone
EGSS	environmental goods and services
EU	European Union
FAO	Food and Agriculture Organization
FCR	feed conversion ratio
FFD	financing for development
FMIS	farm management information systems
GAM	generalized additive model
GHG	greenhouse gas
GPS	geographical positioning system
HYV	high yielding varieties

ICT	information and communication technology
ISBWG	International Sustainable Bioeconomy Working Group
JRC	Joint Research Center's
KBBE	knowledge based bioeconomy
KEP	knowledge exchange platform
KPIs	key performance indicators
LCA	life cycle assessment
LCC	life cycle costing
LCI	life cycle inventory
LCIA	life cycle impact analysis
LVL	laminated veneer lumber
M&E	monitoring and evaluation
MAGNET	modular applied general equilibrium tool
MEE	Ministry of Employment and Economy
MEY	maximum economic yield
MPAs	marine protected areas
NOBOB	no ballast on board
OECD	Organization for Economic Cooperation and Development
RE	renewable energy
SAT-BBE	Systems Analysis Tools System for the EU Bio-Based Economy Strategy
SCP	sustainable consumption and production
SDF	sustainable development finance
SDG	sustainable development goals
SERI	Sustainable Europe Research Institute
SLCA	social life cycle assessment
SRIA	strategic research and innovation agenda
SYMOBIO	systemic monitoring and modeling of the bioeconomy
TAC	total allowed catches
TAC	total allowable catch
TAN	total ammoniacal nitrogen
TFM	3-trifluoromethyl-4-nitrophenol
UAN	unionized ammonia nitrogen
UNEP	United Nations Environment Program

USDA	United States Department of Agriculture
VRA	variable rate application
WTO	World Trade Organization
WWF	world wide fund

GLOSSARY

A

Agro-Biotechnology – Agricultural biotechnology, also known as agri-tech, is an area of agricultural science involving the use of scientific tools and techniques, including genetic engineering, molecular markers, molecular diagnostics, vaccines, and tissue culture, to modify living organisms: plants, animals, and microorganisms.

Alleles – The term allele denotes the variant of a given gene. In genetics it is normal for genes to show deviations or diversity- all alleles together make up the set of genetic information that defines a gene. For example, the ABO blood grouping is controlled by the ABO gene, which has six common variants.

Altruistic – Altruism is the principle and moral practice of concern for happiness of other human beings or other animals, resulting in a quality of life both material and spiritual. It is a traditional virtue in many cultures and a core aspect of various religious and secular worldviews.

Anatomy – Anatomy is the branch of biology concerned with the study of the structure of organisms and their parts. Anatomy is a branch of natural science which deals with the structural organization of living things. It is an old science, having its beginnings in prehistoric times.

Audit – An audit is an "independent examination of financial information of any entity, whether profit oriented or not, irrespective of its size or legal form when such an examination is conducted with a view to express an opinion thereon."

B

Bioeconomy – Biobased economy, bioeconomy or biotechonomy is economic activity involving the use of biotechnology and biomass in the production of goods, services, or energy. The terms are widely used by regional development agencies, national, and international organizations and biotechnology companies

Bioeconomy- The term Bioeconomy generally refers to an economy using renewable natural resources to produce food, energy, products, and services. The important renewable natural resources include the biomass in forests, soil, fields, bodies of water and the sea and fresh water.

Bioenergy- Bioenergy is one of many diverse resources available to help meet our demand for energy. It is a form of renewable energy that is derived from recently living organic materials known as biomass, which can be used to produce transportation fuels, heat, electricity, and products.

Biofuels – Biofuel is a fuel that is produced over a short time span from biomass, rather than by the very slow natural processes involved in the formation of fossil fuels, such as oil. Since biomass can be used as a fuel directly, some people use the words biomass and biofuel interchangeably.

Biomass – Biomass is plant based material used as fuel to produce heat or electricity. Examples are wood and wood residues, energy crops, agricultural residues and waste from industry, farms, and households. Since biomass can be used as a fuel directly, some people use the words biomass and biofuel interchangeably.

Biotechnology- Biotechnology is "the integration of natural sciences and engineering sciences in order to achieve the application of organisms, cells, parts thereof and molecular analogs for products and services."

C

Caulerpa – Caulerpa is a genus of seaweeds in the family *Caulerpaceae*. They are unusual because they consist of only one cell with many nuclei, making them among the biggest single cells in the world. A species in the Mediterranean can have a stolon more than 3 meters long, with up to 200 fronds.

E

Evidence-Based Production – Evidence-based production (EBP) is set to have a major impact on the productivity of agricultural supply chains and is defined by Proagrica as "farming that embraces technology and utilizes data to inform production

F

Fisheries Bioeconomics – Bioeconomics is closely related to the early development of theories in fisheries economics, initially in the mid-1950s by Canadian economists Scott Gordon and Anthony Scott.

Fossil Fuels – A fossil fuel is a hydrocarbon-containing material formed naturally in the earth's crust from the remains of dead plants and animals that is extracted and burned as a fuel. The main fossil fuels are coal, crude oil and natural gas.

G

Green GDP – The green gross domestic product is an index of economic growth with the environmental consequences of that growth factored into a country's conventional GDP. Green GDP monetizes the loss of biodiversity and accounts for costs caused by climate change.

Greenhouse Gas – A greenhouse gas is a gas that absorbs and emits radiant energy within the thermal infrared range, causing the greenhouse effect. The primary greenhouse gases in Earth's atmosphere are water vapor, carbon dioxide, methane, nitrous oxide and ozone.

H

Herbivore – A herbivore is an animal anatomically and physiologically adapted to eating plant material, for example foliage or marine algae, for the main component of its diet. As a result of their plant diet, herbivorous animals typically have mouthparts adapted to rasping or grinding.

Hymenoptera – a large order of insects that includes the bees, wasps, ants, and sawflies. They have four transparent wings and the females typically have a sting.

Hypotheses – A hypothesis is an assumption that is made based on some evidence. This is the initial point of any investigation that translates the research questions into predictions. It includes components like variables, population, and the relation between the variables.

I

Industrialized – to make industrial industrialize an agricultural region.

Invasive Species – An invasive species is an introduced organism that becomes overpopulated and harms its new environment. Although most introduced species are neutral or beneficial with respect to other species, invasive species adversely affect habitats and bioregions, causing ecological, environmental, and/or economic damage.

K

Kin Selection – Kin selection is the evolutionary strategy that favors the reproductive success of an organism's relatives, even at a cost to the organism's own survival and reproduction. Kin altruism can look like altruistic behavior whose evolution is driven by kin selection.

M

Mechanization Agriculture – Mechanized agriculture is the process of using agricultural machinery to mechanize the work of agriculture, greatly increasing farm worker productivity. In modern times, powered machinery has replaced many farm jobs formerly carried out by manual labor or by working animals such as oxen, horses, and mules.

Metapopulations – A metapopulation consists of a group of spatially separated populations of the same species which interact at some level.

Monocropping – In agriculture, monocropping is the practice of growing a single crop year after year on the same land. Maize, soybeans, and wheat are three common crops often monocropped. Monocropping is also referred to as continuous cropping as in continuous corn.

O

Oceanographic – Oceanography covers a wide range of topics, including marine life and ecosystems, ocean circulation, plate tectonics and the geology of the seafloor and the chemical and physical properties of the ocean.

P

Paradigm – In science and philosophy, a paradigm is a distinct set of concepts or thought patterns, including theories, research methods, postulates, and standards for what constitute legitimate contributions to a field.

Petromyzon Marinus – The sea lamprey is a parasitic lamprey native to the Northern Hemisphere. It is sometimes referred to as the "vampire fish."

Photosynthesis – the process by which green plants and some other organisms use sunlight to synthesize nutrients from carbon dioxide and water. Photosynthesis in plants generally involves the green pigment chlorophyll and generates oxygen as a by-product.

Physiology – the branch of biology that deals with the normal functions of living organisms and their parts.

R

Racism – Racism is the belief that groups of humans possess different behavioral traits corresponding to inherited attributes and can be divided based on the superiority of one race over another.

Reciprocal Altruism – In evolutionary biology, reciprocal altruism is a behavior whereby an organism acts in a manner that temporarily reduces its fitness while increasing another organism's fitness with the expectation that the other organism will act in a similar manner at a later time.

Renewable Resources – A renewable resource, also known as a flow resource, is a natural resource which will replenish to replace the portion depleted by usage and consumption, either through natural reproduction or other recurring processes in a finite amount of time in a human time scale.

S

Salvelinus Namaycush – The lake trout is a freshwater char living mainly in lakes in northern North America. Other names for it include mackinaw, namaycush, lake char, touladi, tog, and gray trout. In Lake Superior, it can also be variously known as siscowet, paperbelly, and lean.

Sexism – Sexism is prejudice or discrimination based on one's sex or gender. Sexism can affect anyone but, it primarily affects women and girls. It has been linked to stereotypes and gender roles and may include the belief that one sex or gender is intrinsically superior to another.

Social Trap – In psychology, a social trap is a conflict of interest or perverse incentive where individuals or a group of people act to obtain short-term individual gains, which in the long run leads to a loss for the group as a whole.

Sociobiologists – an expert in or student of the development, structure, and functioning of human society.

Socio-Biology – Sociobiology is a field of biology that aims to examine and explain social behavior in terms of evolution. It draws from disciplines including psychology, ethology, anthropology, evolution, zoology, archaeology, and population genetics.

Strongylocentrotus Purpuratus – This sea urchin species is deep purple in color and lives in lower inter-tidal and nearshore sub-tidal communities. Its eggs are orange when secreted in water.

Sustainability – Sustainability is a societal goal with three dimensions: the environmental, economic, and social dimension. This concept can be used to guide decisions at the global, national, and at the individual level. A related concept is that of sustainable development. Both terms are often used synonymously.

U

Urbanization – Urbanization refers to the population shift from rural to urban areas, the corresponding decrease in the proportion of people living in rural areas, and the ways in which societies adapt to this change.

PREFACE

This book takes the readers through several aspects of Principles of Bioeconomics. This book gives a brief introduction to basic concepts of bioeconomy and its principles, socio-biology, and bioeconomics, challenges and perspectives towards a sustainable bioeconomy, bioeconomy and agricultural production, bioeconomics of fisheries management, invasive species, and aquaculture. The book briefly explains indicators to monitor and evaluate bioeconomy sustainability.

The first chapter stresses on the basic concepts of bioeconomy and its principles, origin and evolution of bioeconomy, bioeconomy sectors, knowledge-based bioeconomy, and bioeconomy strategies, so that the readers are clear about the philosophies behind that form the utmost basics in the field. This chapter will also emphasize on principles for circular biomass use and changing perspectives on the bioeconomy.

The second chapter takes the readers through the concepts of sustainable bioeconomy principles. This chapter will provide highlights on the various key aspects like the typology of indicators, monitoring approaches at the territorial level, stepwise approach to monitoring the bioeconomy, etc.

The third chapter explains the challenges and perspectives toward a sustainable bioeconomy. It also explains fostering sustainable bioeconomics, biorefineries supply to sustainable textiles, wood-based solutions for the sustainable built environment, bioeconomy and sustainable development goals, challenges in sustainable bioeconomy, and benefits and impacts of bioeconomy.

The fourth chapter introduces the readers to socio-biology and bioeconomics. This chapter also explains the evolution of socio-biology and emotions and met-socio-biology theory. The chapter also sheds light on socio-biology, the theory of evolution, and bioeconomics.

The fifth chapter throws light on bioeconomy and agricultural production. The chapter explains concepts of the significance of agriculture for the bioeconomy, challenges of the agricultural bioeconomy, agricultural production, primary production sustainability, and evolutionary scenarios for agricultural business models.

The sixth chapter takes the readers through the bioeconomics of fisheries management. The readers are then told about why is fisheries management and regulation needed, the fundamentals of fisheries bioeconomics, the fisheries management process, and spatial management of fisheries.

The seventh chapter explains the bioeconomics of aquaculture. It also explains bioeconomic modeling and salmon aquaculture, farm design, the bioeconomics of re-circulating aquaculture systems, and technology and bioenergetic model.

The last chapter of this book sheds light on the bioeconomics of invasive species, the bioeconomic impact of existing policy on invasive species, integration of economics and biology for invasive species management, trait-based risk assessment for invasive species, and management of invasive species in the Great lakes. This chapter also mentions the invasion process and feedback between biological and economic systems.

This book has been designed to suit the knowledge and pursuit of the researcher and scholars and to empower them with various aspects of principles of bioeconomics so that they are updated with the information. I hope that the readers find the book explanatory and insightful and that this book is referred by scholars across various fields.

CHAPTER 1

BASIC CONCEPTS OF BIOECONOMY AND ITS PRINCIPLES

CONTENTS

1.1. Bioeconomy: Origin and Evolution 2
1.2. Bioeconomy Sectors 10
1.3. Knowledge Based Bioeconomy (KBBE) 11
1.4. Bioeconomy Strategies 13
1.5. Changing Perspectives on the Bioeconomy 15
1.6. Arising Criticism of the Concept 17
1.7. Bioeconomy and the Principles of Circular Economy 21
1.8. Bioeconomy as an Element of Great Societal Transformation 22
1.9. Strengthening the Demand for Bioeconomy Products 23
1.10. Principles for Circular Biomass Use 24
1.11. Conclusion 29
References 30

In biotechnology and bioeconomy, goods, services, or energy are produced using biotechnology and biomass. The terms are commonly used by regional development agencies, national and international organizations, and biotechnology companies.

1.1. BIOECONOMY: ORIGIN AND EVOLUTION

Bioeconomy can be seen as knowledge based production and use of natural/biological resources, together with biological processes and laws that allow providing economic goods and services in an environmentally friendly way. According to the EBCD, the bioeconomy has a climate change mitigation potential of between 1 billion and 2.5 billion tons of CO_2 equivalent per year by 2030.

There are various ways to define the bioeconomy. Accordingly, the definition of bioeconomy in the Communication of the European Commission (EC) on 13 February 2012 of the European bioeconomy was adopted as the basis for the definition.

According to this definition, the bioeconomy involves the production of renewable resources of biological origin on land and in the sea and the use of these resources and waste streams to produce value added products such as food, feed, bioproducts, and bioenergy. Bioeconomy based on the use of renewable resources of biological origin is to gain a new character due to:

- Renewable resources;
- Resources with low greenhouse gas (GHG) emissions or neutral in this respect;
- Resources repeatedly used (cascade) in production processes; and
- Resources with high potential for beneficial properties concerning end products, such as lower or no toxicity, higher stability, higher durability and strength, limited water consumption, etc.

Agriculture, forestry, and fisheries sectors as well as all other related sectors of the economy (production of food, animal feed, wood, paper, biofuels, etc.) should be a part of bioeconomy. The new approach to this economy should strive to implement innovation (research and innovation at the interface of many different sectors and industries) in combination with the industrial application of biotechnology. The priority of the bioeconomy should be the economic growth achieved based on traditional and new (emerging) industries based on biological materials.

The increase will be realized through the development of new value chains based on biological resources to provide high value added products to the market. For the bioeconomy to develop, it requires research and innovations that go beyond particular sectors, a coherent policy and bioeconomy strategies defined at the level of national and regional governments as well as international and inter-sectoral co-operation.

The basis for the development of the bioeconomy should be intensified primary production. The goal and tool will also be the creation and development of new markets and the likely increase in the competitiveness of the entire economy.

The emergence of the post oil society is however subject to many conditions related to the current state of technology of converting fossil resources into widely used products as well as the continuous development of these technologies in terms of obtaining protection for growing social needs.

Processing of fossil resources is a huge area of the global economy whose transformation or extinction of specific sectors of this economy may be difficult or even impossible. Biomass resources, the foundation of the bioeconomy as well as the so-called "green economy" can be used as a substitute for fossil fuels not just for energy applications, but also for the production of chemicals and materials.

It includes the production and use of biological resources both within and between countries as well as economic activities related to the invention, development, production, and use of biological products and processes according to the European Bioeconomy Panel.

This includes the production of food and non-food crops and the technological processes that turn them into food, feed, bio-based products, agro-fuels, and bioenergy. More specifically, the bioeconomy encapsulates numerous sectors such as agriculture, forestry, fisheries, construction, food processing, pulp, and paper, biotechnology, environmental technology, industrial goods, textiles, chemicals, and pharmaceuticals and recycling and waste collection.

Concerns about resource sustainability are growing, particularly concerning food scarcity and food security, limited national (or global) capacity to produce goods, climate change and environmental degradation. The bioeconomy presents biotechnologies (processes that use and manipulate biological systems and organisms to develop new products) and biomass (material produced from vegetable or animal matter) as solutions for global resource shortages (Figure 1.1).

Several projects at the European Bioeconomy Panel aim to achieve this goal, such as processing plants that use hydrothermal carbonization to convert agricultural pulp waste. According to Enriquez and Martinez's 2002 Harvard Business School working paper, "Biotechonomy 1.0: A Rough Map of Biodata Flow," genetic material is flowing into and out of three major genetic databases: Gen-Bank, EMBL, and DDBJ.

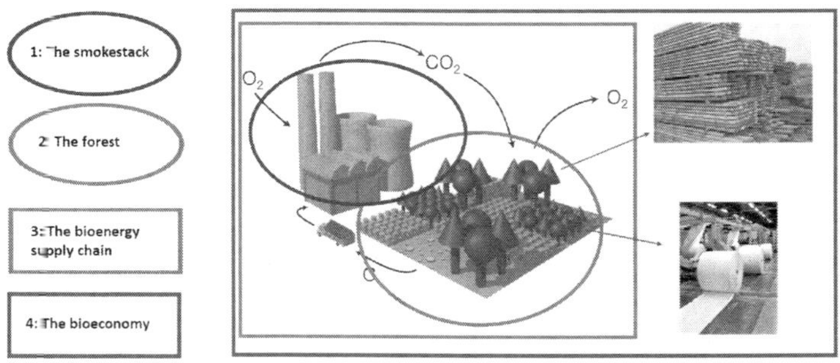

Figure 1.1. Bioenergy system boundaries.

Source: Image by Wikimedia commons.

The authors then hypothesized about the economic impact that such data flows might have on patent creation, the evolution of biotech start-ups and licensing fees. An adaptation of this paper was published in Wired magazine in 2003.

The term 'bioeconomy' became popular in the mid-2000s with its adoption by the European Union and Organization for Economic Co-operation and Development as a policy agenda and framework to promote the use of biotechnology to develop new products and markets and uses of biomass.

Since then, both the EU (2012) and OECD (2006) have created dedicated bioeconomy strategies as have an increasing number of countries around the world. Often these strategies conflate the bioeconomy with the term 'bio-based economy.' For example, since 2005 the Netherlands has sought to promote the creation of a biobased economy.

Pilot plants have been started i.e., in Lelystad (Zeafuels) and a centralized organization exists (Inter-departmental program biobased economy) with supporting research (Food & Biobased Research) being conducted. Other

European countries have also developed and implemented bioeconomy or bio-based economy policy strategies and frameworks. Mr. Obama announced in 2012 that he would encourage biological manufacturing methods with a National Bioeconomy Blueprint.

1.1.1. The First Use of Terms: "Bioeconomics" and "Bioeconomy"

The term "bioeconomics" can be traced back to Zeman in the late 1960s, who used it as a designation for an economic order that acknowledges the biological basis of almost all economic activities. As Bonaiuti (2014) further explained, Georgescu-Roegen "liked the term and from the early 1970s made it the banner summing up the most important conclusions he had come to in a lifetime of research."

An essential element in Georgescu-Roegen's use of the term bioeconomics was his concern that unlimited growth would not be compatible with the basic laws of nature (Bonaiuti, 2014). It should be noted that this usage of the term "bioeconomics" differs quite a bit from the earlier use of the term "bioeconomy," which referred to the application of biological knowledge for commercial or industrial purposes. One can consider this rather contrasting use of the two terms as an "irony of fate."

According to von Braun (2014), the term was first defined by the two geneticists Juan Enriquez Cabot and Rodrigo Martinez. A paper published by Enriquez in the Science magazine in 1998 (Enriquez, 1998) is also quoted as a source for this use of the term (Gottwald, 2016).

In this paper, which is entitled "Genomics and the World's Economy," Enriquez discusses that the application of the discoveries of genomics will lead to a restructuring in the role of companies and industries "in a way that will change the world's economy." He outlined "the creation of a new economic sector, the life sciences" in this paper (Enriquez, 1998).

It is important to note that even though this paper does not use the term "bioeconomy," the source represents one of the roots of the concept: advances in the biological sciences and biotechnology. These advancements have the potential to transform many industrial production processes.

The view that the "biological revolution" would eventually transform the industry was however not new at that time. "The industrial impact of the biological revolution" was already formulated in the early 1980s (Glick, 1982).

1.1.2. The Development of the Concept of the "Knowledge-Based Bioeconomy" in the European Union (EU)

Bioeconomy was first introduced by scientists who wanted to understand the industrial implications of advances in biology. However, it was a deliberate policy decision made by EC staff members to promote the concept that made the bioeconomy an important policy concept in Europe.

One of the key actors in this effort was Christian Patermann, the former Program Director of "Biotechnology, Agriculture, and Nutrition" in the Directorate General for Research, Science, and Education of the EC. According to his account, the term "bioeconomy" was used by a conference of Ministers of Environment.

Footnote Members of that conference had not further defined the term, but Patermann and his colleagues felt that it had the potential to be a policy concept that would allow the EU to respond to new opportunities. One opportunity was making economic use of the emerging new potential of using biotechnologies.

Another opportunity inherent in the concept of the bioeconomy is the replacement of fossil-based resources with bio-based resources, both for energy and for material use. In the early 2000s, decision-makers in the EU felt a strong incentive to find new concepts, because the need for increasing agricultural productivity to meet future needs for food and biomass was not very well recognized at the time. Funding for agricultural research, which is the key to increasing agricultural productivity had declined throughout the 1990s despite the emerging need to produce biomass for other uses than food (Geoghegan-Quinn, 2013).

During the EU's development of the bioeconomy concept, the label "knowledge-based" was added so that the bioeconomy became a knowledge-based economy. The label "knowledge-based" was in line with the EU innovation policy that prevailed at the time.

At a meeting in Lisbon in 2000, the European Council had committed to establishing the most competitive and dynamic knowledge based economy in the world (EU, 2000). A knowledge based economy is defined by the idea that economic growth can be attained through high technology industries that require investments in innovation and highly skilled workers.

The efforts of the EU to promote the concept of knowledge based bioeconomy (KBBE) proved remarkably successful. In 2005, the EC held a conference entitled "New Perspectives on the Knowledge Based Bioeconomy" (EC, 2005).

At this conference, JanezPotočnik, the European Commissioner for Science and Research gave a speech entitled "Transforming life sciences knowledge into new, sustainable, eco-efficient, and competitive products" (Potočnik, 2005). In the so called Cologne Paper of 2007, this title has been quoted as a definition of the KBBE (Figure 1.2).

Figure 1.2. Harnessing agricultural innovation to boost Africa's prosperity.

Source: Image by bioecenomy.co.za.

The Cologne Paper was based on a workshop held under the German Presidency of the Council of the European Union in 2007 in the city of Cologne. The workshop was attended by experts from research organizations and companies covering different fields, including crop production, biotechnology, bioenergy, and biomedicine (EU, 2007).

The Cologne Paper emphasized the two dimensions of the bioeconomy:

- On the one hand, the paper identified the role of biotechnology as an important pillar of Europe's economy by 2030 indispensable to sustainable economic growth, employment, energy supply and maintaining the standard of living (EU, 2007). One can label this dimension of the bioeconomy "the biotechnology innovation perspective."

- On the other hand, the Cologne Paper stressed the use of crops as "renewable industrial feedstock to produce biofuels, biopolymers, and chemicals" (EU, 2007). In addition to the then existing gasification technologies, the paper predicted that by 2020, lignocellulosic biomass will be converted by enzymatic hydrolysis to provide large feedstock supplies for bioprocesses and the production of transport fuels. One can label this dimension of the bioeconomy "the resource substitution perspective."

The changing emphasis of these two perspectives over time, the development of the concept of bioeconomy was accompanied by increased funding especially in the EU's Framework Programs for Research and Technological Development most notably in the current 8th Framework Program, which is entitled "Horizon 2020" (EC, 2013).

The development of the bioeconomy concept by the institutions of the EU was mirrored by efforts to establish this concept in the EU member states. Germany, for example, established a Bioeconomy Council at the federal level in 2010 under the leadership of the Federal Ministry of Education and Science (BMBF).

In 2010, a "National Research Strategy Bioeconomy 2030" was published (BMBF, 2010) and the federal government pledged to spend 2.4 billion euro for the bioeconomy research until 2016 (BMBF, 2014). In 2013, Germany published a "National Policy Strategy on Bioeconomy."

The policy had the subtitle "Renewable resources and biotechnological processes as a basis for food, industry, and energy," which reflects both the biotechnology innovation perspective and the resource substitution perspective (BMEL, 2013).

Other European countries also developed policies and strategies related to the bioeconomy. However, there was considerable variation regarding the extent to which these policies and strategies were specifically focused on the bioeconomy or rather on related aspects such as biotechnology or renewable energy (RE).

For example, by 2015, neither France nor Great Britain nor Italy had a strategy that specifically focused on the bioeconomy (BÖR, 2015a). Finland, in contrast, had already published a bioeconomy strategy in 2014. Austria and Norway, to mention two other examples, were in the process of preparing a dedicated bioeconomy strategy in 2015 (BÖR, 2015b).

1.1.3. The Rise of Bioeconomy as a Global Concept

The EU is not the only region of the world where the concept of the bioeconomy has been promoted since the early 2000s. The term bioeconomy was probably first used at a meeting of the American Association for the Advancement of Science in 1997.

In 2012, the Obama administration released an official strategy on the bioeconomy entitled the "National Bioeconomy Blueprint" (White House, 2012). This strategy defines the bioeconomy as follows:

- A bioeconomy is one based on the use of research and innovation in the biological sciences to create economic activity and public benefit. The U.S. bioeconomy is all around the US new drugs and diagnostics for improved human health, higher yielding food crops, emerging biofuels to reduce dependency on oil, and biobased chemical intermediates to name just a few (White House, 2012).

This definition also reflects the two perspectives of the bioeconomy, the biotechnology innovation perspective and the resource substitution perspective. In the first two decades of the 21st century, both industrialized and developing countries published bioeconomy related policies and strategies.

For example, Malaysia published a "Bioeconomy Transformation Program" in 2012 and South Africa released a bioeconomy strategy in 2013 (BÖR, 2015b). While, the number of countries that have dedicated bioeconomy policies is still limited. There are a large number of countries that have strategies related to biotechnology and/or renewable resources (BÖR, 2015b).

The first Global Bioeconomy Summit was held in Berlin in December 2015. The event was organized by the German Bioeconomy Council in collaboration with an international advisory committee. It brought together more than 700 bioeconomy experts from more than 80 countries (BÖR, 2015c).

The emergence of the bioeconomy as a global concept is not only reflected in the growing number of countries that have bioeconomy related strategies and policies, but also in the scientific literature. The number of publications listed in Scopus that refer to the bioeconomy has increased rapidly from 2005 onwards.

1.2. BIOECONOMY SECTORS

1.2.1. Forest Bioeconomy

Forest bioeconomy involves a wide range of different industries and production processes based on forests and their natural resources. Forest bioeconomy includes, for example, the processing of forest biomass to provide products relating to energy, chemistry or the food industry.

Thus, the forest bioeconomy covers a variety of different manufacturing processes that are based on wood material and the range of end products is wide. Besides different wood-based products, recreation, nature tourism and game are a crucial part of the forest bioeconomy.

Carbon sequestration and ecosystem services are also included in the concept of forest bioeconomy. Pulp, paper, packaging materials and sawn timber are the traditional products of the forest industry. Wood is also traditionally used in the furniture and construction industries (Figure 1.3).

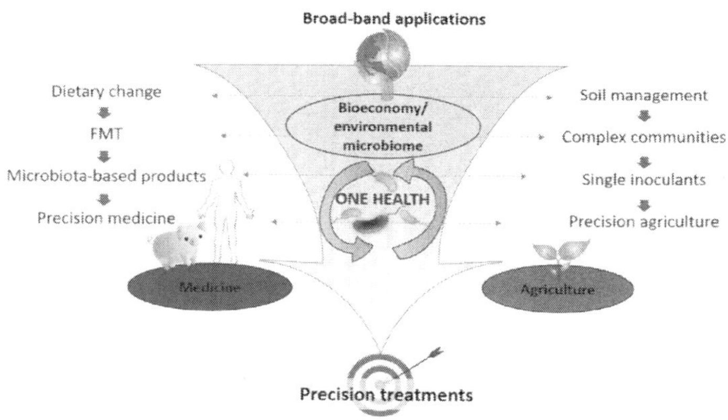

Figure 1.3. Broadband applications direction microbiome-based precision treatment.

Source: Image by Wikimedia commons.

But in addition to these, as a renewable natural resource, ingredients from wood can be valorized into innovative bioproducts alongside a range of conventional forest industry products.

Thus, traditional mill sites of large forest industry companies, for example in Finland, are in the process of becoming biorefineries. In different processes, forest biomass is used to produce textiles, chemicals, cosmetics, fuels, medicine, intelligent packaging, coatings, glues, plastics, food, and feed.

1.2.2. Blue Bioeconomy

Water related expertise areas as well as businesses based on the sustainable use of renewable aquatic resources, make up the blue bioeconomy. It covers the development and marketing of blue bioeconomy products and services.

In that respect, the key sectors include business activities based on water expertise and technology, water based tourism, making use of aquatic biomass and the value chain of fisheries. Furthermore, the immaterial value of aquatic natural resources is also very high.

Water areas also have other values beyond being platforms for economic activities. It provides human well-being, recreation, and health. According to the EU, the blue bioeconomy has a focus on aquatic or marine environments, especially on novel aquaculture applications, including non-food, food, and feed.

In the European Report on the Blue Growth Strategy – Towards more sustainable growth and jobs in the blue economy (2017), the blue bioeconomy is defined differently from the blue economy. The blue economy means the industries that are related to marine environment activities, e.g., shipbuilding, transport, coastal tourism, renewable energies (such as off-shore windmills), and living and non-living resources.

1.3. KNOWLEDGE BASED BIOECONOMY (KBBE)

The Knowledge-Based Bioeconomy (KBBE) is an important factor in understanding the current bioeconomy agenda. It is a specific approach to bioeconomy policy making that emerged from the EU's life sciences research agenda which has mainly focused on making agriculture more sustainable and efficient since the 1990s.

Since 2007, the European Commission has based its research priorities on the KBBE, which is a hybrid of the OECD's bioeconomy project and the EU's Knowledge Based Economy and links knowledge with technological innovation (Figure 1.4).

Figure 1.4. Stakeholder dialog on bioeconomy (L-R): Johann Kottulinsky, Lenzing AG, Michael Proschek-Hauptmann, Schweighofer Group, Sten Nilsson, Forest Sector Insights AB, Markus Lassheikki, The Central Union of Agricultural Producers and Forest Owners, Finland, JaanaBäck, University of Helsinki, Maartje Klapwijk, Swedish University of Agricultural Sciences, and Gerald Steindlegger, ISS integrated sustainability solutions.

Source: Image by flickr.

The KBBE can be understood as a new political economic strategy and plays a role in shaping policies, institutional practices and societal changes to create sustainable capital. Simply put the EU's KBBE agenda presents technological advancement as the equivalent of societal progress and improved life quality.

However, the KBBE does not address the long term consequences of constantly striving towards new more efficient technologies and the development of projects that promote the commodification of nature. According to the KBBE perspective, anything that can be re-grown is considered to have an infinite supply and technology that manipulates organisms should be utilized to create these renewable products.

In short, this requires the commodification of nature. The goal becomes sustainable capital, which drives a new trajectory of sustainable capitalism that is essentially no more than an expansion of the corporate driven market system. There are two analysis of the KBBE, each offering divergent priorities and models for the future of global production.

They promote different diagnoses and remedies for the instability of the current agricultural system and contrasting ideas about the future that can be shaped by mobilizing networks and resources and changing institutional practices. The dominant life sciences perspective argues that the inefficiency of farms, their inputs, processing methods and outputs is a major threat to society.

Plant scientists, multinational companies, some small and medium sized businesses and the Committee of Professional Agricultural Organizations (COPA) argue that Europe's agricultural industry is disadvantaged in terms of global competitiveness and technological progress.

Conversely, the alternative 'agro-ecology' perspective argues that agro-industrial monocultures force farmers to become dependent on external inputs, undermine their knowledge base and distance them from consumers. Proponents of this perspective include the organics industry and organic institutes (including the organic section of COPA) and environmental NGO's.

1.4. BIOECONOMY STRATEGIES

An increasing number of countries have adopted bioeconomy strategies or bioeconomy policies. In the following in recognition that the two terms are often used interchangeably, we refer to policy documents or strategy documents that have been officially released by national governments or parliaments as "bioeconomy strategies."

The rationale for government intervention in support of the bioeconomy, to better understand the bioeconomy strategies that governments have developed, it is useful to take the comparative advantage into account that a country has for developing different components of the bioeconomy.

The "diamond model" developed by Porter (1990) provides a conceptual framework, which can be used for determining the competitive advantage of a country's bioeconomy (Birner et al., 2014).

1.4.1. Basic Elements of a Bioeconomy Strategy

The four basic elements of the "diamond" model, which determine the competitive advantage of a country for developing its bioeconomy are:

- factor conditions;

- demand conditions;
- firm structure, strategy, and rivalry; and
- related and supporting industries.

Bioeconomy strategies typically aim to promote the bioeconomy by targeting several or all of these four groups of factors. The Global Competitiveness Report of the World Economic Forum (2016) provides a wide range of indicators related to these groups of factors for 138 countries.

Although these indicators are not bioeconomy specific, they are a useful source of information for countries to assess their general conditions for bioeconomy development.

1.4.2. Upgrading Factor Conditions for the Bioeconomy

Based on Porter (1990), one can distinguish five types of factor conditions that are relevant to the development of the bioeconomy:

- **Natural Conditions:** Land resources and agro-climatic conditions contribute significantly to a country's competitive advantage in biomass production. Countries with large land endowments, favorable agro-climatic conditions and low population density typically have a comparative advantage for emphasizing the resource substitution perspective of the bioeconomy as they can have the potential to produce biomass for bioenergy and bio-based materials (e.g., bioplastic) on a large scale and at comparatively low cost.

Brazil, which has a competitive advantage in producing sugarcanes is an example of this type of country. Countries that have access to marine resources may emphasize these resources in their bioeconomy related strategies. Norway is an example (BÖR, 2015b).

Countries with less favorable natural resource conditions and/or limited land resources will have to focus more on biotechnology innovation than on resource substitution to develop their bioeconomy.

- **Labor Resources:** The basic natural conditions cannot be influenced by government interventions however governments can have a substantial impact on the qualification of their labor force especially by investing in education and professional training.

The development of the bioeconomy requires specific skill sets and education programs need to be adjusted and developed to enable the labor force to gain those skills. As an example, the University of Hohenheim in Stuttgart, Germany introduced an interdisciplinary Master's program called "Bioeconomy" in 2014.

In Porter's framework such investments in education are referred to as "factor upgrading" which is an important strategy that countries can use to improve their competitive advantage for the development of their bioeconomy.

- **Knowledge Resources:** Investing in public bioeconomy research to foster innovation is one of the most important instruments that governments can use to develop their bioeconomics. The concept of the "KBBE" emphasizes this aspect.

Accordingly, investments in research and innovation are an important element of most bioeconomy related strategies (BÖR, 2015). Since research by the private sector also plays a key role in developing the bioeconomy, creating a conducive environment for research in the private sector is important as well.

- **Capital Resources:** The development of the bioeconomy relies on investments along the entire value chain for bioeconomy products including research, product development and marketing. It is therefore essential for the development of the bioeconomy to have access to capital specially venture capital for risky investments.
- **Infrastructure:** Governments can also help develop the bioeconomy by providing the necessary infrastructure including transport as well as information and communication technology (ICT). An important task is the identification of infrastructure needs that are particularly relevant for the bioeconomy strategy selected.

1.5. CHANGING PERSPECTIVES ON THE BIOECONOMY

The development of the concept of bioeconomy was characterized by two perspectives:

- the resource substitution perspective; and
- the biotechnology innovation perspective.

Over a time, the emphasis shifted between these two perspectives. Even though innovation in biotechnology was recognized as an opportunity for the bioeconomy from the very beginning, resource substitution perspectives gained a lot of momentum in the first decade of the 21st century.

Resources substitution is driven by the concept of "peak oil," which implies that oil extraction rates have reached a peak and will decline after that peak while oil prices will continuously increase (Bardi, 2009). Biomass for energy and materials becomes more competitive as oil prices rise.

The resource substitution perspective of the bioeconomy was supported by this line of reasoning. The resource substitution perspective of the bioeconomy. This diagram was developed by the German Bioeconomy Council in 2010 (BÖR, 2010).

Essential components of the bioeconomy are the production of biomass in various forms, its conditioning and conversion using different procedures and the production and marketing of food, feed, fiber, fuel, and fun. The term "fun" refers to products such as flowers.

The oil price crisis of 2007/2008 re-affirmed the perception of peak oil. The increasing use of food crops for biofuel contributed to the spike in food prices that were observed following the oil price crisis. This development was primarily promoted by high oil prices (Headey and Fan, 2008).

Biofuel policies, such as biofuel subsidies and mandates to add biofuel to commercial petrol, became subject to increasing criticism, as research established the impact that they can have on food prices (De Gorter et al., 2013).

The bioeconomy was affected by these developments in two significant ways: First, the possible tension between ensuring food availability and using biomass as a source of energy dominated public policy debates surrounding the bioeconomy. Additionally, the need to increase biomass production productivity has gained attention (Figure 1.5).

Figure 1.5. Bioeconomy challenges for the EU regions- in the context of the knowledge exchange platform (KEP) Belgium – Brussels – April 2016.

Source: Image by flickr.

In order to produce and use biomass in a manner that does not conflict with food needs this was done. Among these options are second generation bioenergy technologies and the use of waste products and by products as sources of energy. Both energy and food prices fell considerably after 2010 and they also became more volatile as compared to the 1990s (Kalkuhl et al., 2016).

The development of the oil price remains difficult to project (Baumeister and Kilian, 2016) but, because of the prevailing low oil prices, scarcity of oil was no longer a prominent argument for the resource substitution perspective.

Climate protection became the major argument for eliminating fossil based resources. While this argument was not new (e.g., WBGU, 2011) the Paris Agreement under the United Nations Framework Convention on Climate Change became a major rationale for resource substitution.

1.6. ARISING CRITICISM OF THE CONCEPT

The global rise of the concept of the bioeconomy has not been without its critics. One can distinguish two major types of criticism which one can label the "fundamental critique" and the "greenwashing critique." An example of the fundamental critique is the writings by Birch and co-authors (Birch, 2006; Birch et al., 2010).

They criticize the bioeconomy as the "neo liberalization of nature." The authors analyze the emerging discourse of the KBBE in the EU and criticize that the development of the concept has been dominated by what they refer to as a "neoliberal ideology."

Accordingly, the criticism of the bioeconomy concept is linked to a more general critique of a neoliberal regime in which market values are installed as the over-riding ethic in society and the market rule is imposed on all aspects of life (Birch, 2006).

Related to this type of criticism is the claim that the concept has been promoted to pursue the interest of big companies which are interested in commercializing innovations in the life sciences and in applying technologies that are contested in society such as genetic engineering and synthetic biology.

An example of this criticism is a paper by Gottwald and Budde that was published in 2015 on the occasion of the Global Bioeconomy Summit of 2015. These authors also argue that the bioeconomy would promote "land grabbing" and threaten world food security (Gottwald and Budde, 2015).

The second type of criticism is not fundamentally opposed to the concept of the bioeconomy but rather warns against the use of this concept for "greenwashing." An example of this type of criticism is a report by the World Wide Fund (WWF) for Nature published in 2009 (WWF 2009) which is entitled "Industrial biotechnology- More than green fuel in a dirty economy."

This report acknowledges the potential of the bioeconomy to make modern economic systems more environmentally sustainable but points out that the approaches that have been promoted under the label bioeconomy do not necessarily realize this potential.

The thrust of this criticism is to ensure that the label "bio" is not misused to portray an essentially non-sustainable economic system as environmentally friendly but to ensure that innovations in the life sciences are indeed used to ensure a transition toward a sustainable economic system.

It is possible that the increasing criticism against the bioeconomy contributed to the emergence of two prominent trends in the development of the concept. It would be beneficial to incorporate the concept of bioeconomy more explicitly into concepts such as sustainable development and the green economy.

Secondly, the bioeconomy is focusing more on consumers and society as a whole rather than the supply side i.e., companies and technology are moving to the market to commercialize them.

1.6.1. "Greening" the Bioeconomy

The early definitions of the bioeconomy quoted did not include explicit references to environmental goals. This is although ecological sustainability was implicitly assumed both in the context of biotechnology innovation and in the context of resource substitution.

As the bioeconomy concept was further developed in the second decade of the twenty first century, it was increasingly recognized that environmental goals need to be explicitly included in the concept as the use of biotechnological innovations and the use of bio-based resources are not automatically more environmentally friendly than alternative options.

The increasing criticism of the use of bioenergy which was associated with the food price crisis of 2008–2009 is a particularly pronounced example of this shift in emphasis. A green economy is characterized by a low carbon footprint, resource efficiency and social inclusion.

Investments in such economic activities, infrastructure and assets drive growth in employment and income in a green economy. Reducing carbon emissions and pollution, improving energy efficiency and preserving biodiversity are all possible with these activities.

Taxation and regulation reforms as well as targeted public spending are needed to enable and support these green investments. UN Environment promotes a development path that understands natural capital as a critical economic asset and a source of public benefits especially for poor people whose livelihoods depend on natural resources.

The notion of a green economy does not replace sustainable development but creates a new focus on the economy, investment, capital, and infrastructure, employment, and skills and positive social and environmental outcomes across Asia and the Pacific.

The Role of Green Economy, Sustainable Consumption and Production and Resource Efficiency for Sustainable Development: Sustainable Consumption and Production (SCP) aims to improve production processes and consumption practices to reduce resource consumption, waste generation and emissions across the full life cycle of processes and products while, Resource Efficiency refers to how resources are used to deliver value to society and aims to reduce the number of resources needed and emissions and waste generated per unit of product or service.

The Green Economy provides a macroeconomic approach to sustainable economic growth with a central focus on investments, employment, and skills.

1.6.2. The Bioeconomy as a Component of the Green Economy

At the Rio 20 Conference in Rio de Janeiro in 2002, the participants adopted a resolution entitled "The future we want" (UN, 2012). This resolution re-affirms the principle of sustainable development and it highlights the concept of the "green economy" as "one of the important tools available for achieving sustainable development" (UN, 2012).

The United Nations Environment Program (UNEP) defined a green economy:

- As one that results in improved human well-being and social equity, while significantly reducing environmental risks and ecological scarcities in its simplest expression, a green economy can be thought of as one which is low carbon, resource efficient and socially inclusive. (UNEP, 2011)
- In the academic literature, the concept of the green economy has a long history (Loiseau et al., 2016). The question arises as to how the concept of bioeconomy is linked to the concept of the green economy. Ultimately, this is a matter of definition. One option is to consider the bioeconomy as an integral component of the green economy. According to this view, one may consider RE sources that do not rely on biological resources, such as wind and solar energy as part of the green economy but not as part of the bioeconomy.
- In the UN resolution "The world we want," the international community also agreed on a process to establish sustainable development goals (SDGs) as a follow up to the Millennium Development Goals that were agreed upon in 2000 and covered the period until 2015 (UN, 2012). A set of 17 "SDGs" were adopted by the UN in 2015.

1.7. BIOECONOMY AND THE PRINCIPLES OF CIRCULAR ECONOMY

A concept related to the bioeconomy has gained prominence in recent years as well as the concept of the green economy and the concept of circular economy.

- **Global Bioeconomy Summit Communiqué:** The idea of a circular economy aligns well with the principles of a sustainable bioeconomy which would involve issues across sectors (nexus thinking) particularly policies aimed at optimizing bioeconomy value chains and reducing waste and losses (Bioeconomy Summit, 2015).

This concept of the circular economy was popularized in a classical textbook on environmental economics by David Pearce and Kerry Turner in 1989 (Pearce and Turner, 1989).

These authors trace it back to a landmark essay by Kenneth Boulding published in 1966 in which Boulding emphasized the need to manage the economy not as an open system but as a spaceship where man must find his place in a cyclical ecological system that is capable of continuous reproduction of material form (Boulding, 1966).

In recent reviews, the circular economy concept has been associated most with adopting closed loop manufacturing patterns in an economy to maximize resource efficiency in particular when it comes to urban and industrial wastes (Ghisellini et al., 2016). Consequently, the circular economy is a more narrowly defined concept than the green economy and bioeconomy.

To ensure that the bioeconomy is sustainable, a link between the bioeconomy and the principles of circular economy is vital. In addition, the focus on renewable resources and biotechnology innovations which are central components of the bioeconomy can play a vital role in implementing the principles of the circular economy.

The goal to link the bioeconomy with the principles of a circular economy has also led to the development of the concept of a "biomass based value web" (Virchow et al., 2016). This concept takes into account that the cascading use of biomass and the use of by-products from the processing of biomass lead to an interlinkage of different value chains.

These can be analyzed as a "value web," Scheiterle et al. (2017) present a case study of Brazil's sugarcane sector. Based on sugarcane biomass, the concept of a value chain is presented. Based on the diagram, it can be seen that waste products from sugarcane processing such as filter cake, vinasse, and bagasse are used for producing biogas or bioelectricity rather than being disposed of as waste.

In addition to being used for new types of bioeconomy products, such as flavors or pharmaceuticals, they can also serve to open up new branches of the biomass based value web (Figure 1.6).

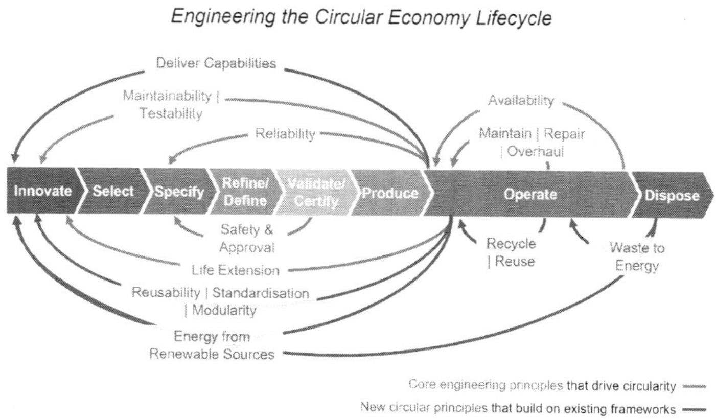

Figure 1.6. An illustration of the circular economy concept: applying core engineering principles for the engineering lifecycle.

Source: Image by Wikimedia commons.

1.8. BIOECONOMY AS AN ELEMENT OF GREAT SOCIETAL TRANSFORMATION

From the above definitions, it can be seen that the development of the bioeconomy concept was characterized by a focus on the role of biological resources and biotechnological processes in providing goods and services.

In recent years, more emphasis has been placed on the demand side of the bioeconomy and more generally, on the role of the bioeconomy in society. Taking the societal embeddedness of the bioeconomy a step further, one can also consider the bioeconomy as an element in a process of societal transformation.

This is ultimately required to transform the current economic system into one that is economically, environmentally, and socially sustainable. The recognition of the challenges involved in this transformation has led to the hypothesis that it will not be sufficient to create economic incentives and implement conducive environmental policies.

What is ultimately required is a profound societal transformation, which encompasses profound changes to infrastructures, production processes, regulation systems and lifestyles and extends to a new kind of interaction between politics, society, science, and the economy (WBGU, 2011).

In line with this thinking, the bioeconomy in a larger historical context. Bioeconomy is conceptualized from this perspective as a must have element in a new era that will ultimately replace industrialization. The industrial society followed the agricultural society which in turn followed the society of hunters and gatherers.

The industrial society was made possible by the technological revolution and agricultural revolution that preceded it. The agricultural society, in turn was made possible by the Neolithic Revolution. The agricultural society and the industrial society were associated with a substantial increase in energy and material use.

The lower parts of the transitions to the agricultural and to the industrial society were associated with a steep increase in world population, which slowed down only in the later phases of the industrial society.

Since the transitions to the agricultural and the industrial society were caused by so called revolutions, it appears justified to assume that the shift to the bioeconomy requires a similar large scale change. This line of thinking is reflected in the idea of a great societal transformation (WBGU, 2011).

1.9. STRENGTHENING THE DEMAND FOR BIOECONOMY PRODUCTS

The demand for bio-based products by consumers is one of the most important incentives for the bioeconomy to develop. The government can encourage this demand by urging bio-based labels and conducting information campaigns that facilitate consumer choice as well as by fostering social dialog.

Public procurement rules that encourage the public to purchase bio-based products can also be implemented by governments. The analysis of national economic strategies around the world conducted by the German Bioeconomy Council (BÖR, 2015a) showed that such demand side instruments play an important role in many bioeconomy strategies.

An interesting example of this approach is the Bio Preferred Program of the United States, Department of Agriculture (USDA). This program combines a voluntary labeling initiative for biobased products with mandatory purchasing requirements for federal agencies and their contractors which encompasses 97 product categories (Figure 1.7).

Figure 1.7. Image showing economic development in Austria.

Source: Image by Iberdrola.com.

1.10. PRINCIPLES FOR CIRCULAR BIOMASS USE

Acknowledging that a single definition for a circular bioeconomy might not exist, we argue that a circular bioeconomy ought to minimize the depletion of resources (e.g., phosphate rock, fossil fuels or soils) encourage regenerative practices (e.g., restoring fish stocks) prevent the loss of natural resources (e.g., carbon, nutrients, and water) and stimulate the reuse and recycling of inevitable by products, losses or wastes in a way that adds the highest possible value to the system.

A circular bioeconomy is inherently limited by the biosphere with its natural cycles and sinks. Here, we focus on agroecosystems of biomass production; encompassing land based agriculture, fisheries, and aquaculture as well as natural and managed ecosystems.

1.10.1. Safeguard

To produce biomass, aquatic, arable, grassland, and forest agroecosystems must be healthy. The utilization of natural resources by farming, fishing, and forestry must not exceed their regeneration and absorption capacity to ensure the current and future availability of those resources.

As such, the first principle calls for sustainable production practices that conserve and regenerate the health of agroecosystems and the resources they provide. Furthermore, the safeguard principle encourages the use of renewable resources (such as solar and wind energy) to minimize and ideally halt the consumption of finite ones (such as phosphate rock or fossil fuels).

In addition to reducing the throughout flow of natural resources, this principle also implies the continuous regeneration of resource quality. To sustain biomass harvesting from agroecosystems, for example, we need to invest in restoring soil carbon stocks.

In the absence of technologies and strategies that reduce GHG emissions or reduce the absorptive capacity of ecosystems (e.g., carbon capture and storage systems can act as artificial sinks) a circular bioeconomy means a greater reliance on the speed of natural cycles.

Not exceeding the absorptive capacity means that waste or losses are generated at a lower rate than the assimilation capacity of agroecosystems. For instance, GHG emissions should not exceed these systems sink capacity. Forests, for example, can act as a carbon sink as well as a valuable source for paper and pulp, textile or bioenergy industries.

The capacity of agroecosystems to act as sinks should be protected and enhanced acknowledging that agro-ecological conditions and sink capacity differ across geographical scales. Biodiversity plays an essential role in providing a variety of buffering capacities and contributing to present and future ecosystems resilience.

Protecting biodiversity requires conserving the natural ecosystems that are left (e.g., zero deforestation targets) and regenerating or restoring degraded ones (e.g., regenerating soil health, and encouraging biodiversity enhancing practices such as diverse crop rotations).

1.10.2. Avoid

In the second principle, bio-based products are used for non-essential purposes while they are avoided for essential ones. Preventing unnecessary consumption of non-essential items helps conserve natural resources especially since production impacts are unlikely to be fully offset by recovery and recycling.

Furthermore, emissions to air, water, and soil can be minimized if production processes are prevented further upstream. A question then arises, what are essential and non-essential products and what is waste? Our society wastes one-third of all food globally every year and discards a third of all fish caught globally because it does not belong to the target species.

Overconsumption is not included in these statistics. People in high and middle income countries consume more calories than they need leading to more deaths from obesity than underweight. With their low nutritional value and negative health effects, ultra-processed foods can also raise questions.

As the lifetime of products has greatly diminished, waste is generated at a rapid rate in sectors other than food, such as clothing and electronics. The rise of fast fashion has led to a garment's lifespan being reduced to around three years.

Determining which products are essential or non-essential and what is waste rather than a by-product entails challenging questions and requires the engagement of different sectors of society.

1.10.3. Prioritize

With growing biomass demands, natural resources need to be used effectively. The third principle concerns the priority of utilizing biomass. It argued that priority should start with basic human needs (food, pharmaceuticals, clothes) and sectors without sustainable alternatives (such as the chemical industry).

This implies, for instance, avoiding bioenergy use for light road vehicles since better alternatives exist (e.g., electrification). Bioenergy could be used in aviation, heavy transport and shipping until technologically and economically viable alternatives arise.

The use of biomass as a source of food is not currently efficient. A large percentage of the world's arable lands are used as feed for livestock, some of which are available for human consumption. Similarly, most fish that end up in fishmeal could be consumed by humans. Direct human consumption of such feeds would be more resource efficient.

However, livestock and farmed fish can contribute significantly to a sustainable nutrient supply, especially highly bio-available ones (such as protein and iron) or nutrients largely absent in plant sourced food (such as vitamin B12 and long-chain essential fatty acids).

Under the priority principle, livestock and fish would only consume leftovers from arable land, fisheries, and grass resources (recycle principle) that is, biomass that would be unfit for human consumption. Future biorefineries might also upgrade inedible biomass streams such as cultivated grass into proteins that humans can eat.

People can also consume a larger diversity of aquatic foods from aquatic production systems each in proportion to its natural production capacity (safeguard principle). Biomass is also required for essential non-food biomaterials such as clothing. Using the biomass cascading principle, it is possible to maximize resource availability and efficiency as well as the sustainability of resources.

In addition to preventing the need for additional biomass harvest, direct biomass to materials rather than energy uses may sequester carbon for longer periods. Here we argue for prioritizing non-food materials such as furniture and housing that also sequester carbon if used sustainably.

For cascades to work, incentivizing the use of biomass for creating materials instead of producing bioenergy will be needed across countries. Such cascades can be directed by social norms. We propose that biomass cascades are best informed by frameworks of human needs and resource use efficiency rather than the economics that is leading to current cascade frameworks such as the value pyramid.

The production of bioenergy, for instance, would only be desirable or effective for biomass streams that are not safe for recycling. This includes waste streams containing human and veterinary pharmaceuticals. Just as with the prioritize principle, reframing biomass cascades will entail rethinking which products are essential for human development (Figure 1.8).

Figure 1.8. Biomass storage silos and unloading point.

Source: Image by Flickr.

1.10.4. Recycle

Even when food and non-food bio-based products are produced and consumed in a waste free manner, there will still be by-products such as manure and slaughterhouse waste. By re-using by-products and nutrients from the bio-based system, we prioritize the well-being of humans and the planet.

Under the avoid principle, by-products that are edible to humans such as the middling from white and brown bread are avoided. Inevitable by-products, such as straw or animal/ human excreta, do contain valuable nutrients and carbon. Recycling them into agroecological systems can enrich the soil, fertilize crops, feed farm animals and produce biomaterials.

Recycling implies, for example, reconnecting arable and livestock farming at the farm and regional scales, but also building new connections between, for example, cities, and their hinterlands. It is important that by-products are safe to recycle and do not have harmful effects on humans, animals or the environment.

For example, while plant-based human in-edible food waste can be immediately fed to farm animals, animal contaminated food waste must be heat treated to deactivate potential diseases before being fed to pigs, poultry, fish, and insects. That said, recycling may entail a trade-off between food safety and additional energy and transport.

1.10.5. Entropy

In agroecosystems, energy is the driving force behind nutrient and carbon recycling. While biogeochemical, such as nutrients and carbon are cycled through ecosystems, energy flows and cascades from useful to less useful forms of energy as during each energy transformation process, entropy or disorder increases.

Increased circularity and recycling costs energy and a fully circular bioeconomy are difficult to achieve given the losses in each consecutive cycle. The fifth principal advocates minimizing energy use by working with nature, moving towards renewables and efficiently utilizing the rare metals on which current renewable technologies depend.

Following ecological engineering and design principles, a system can maximize the use of free-flowing energy from natural sources, particularly the Sun. Examples of energy systems that work with nature and thereby minimize overall energy use include silvo-pastoral farming systems or multispecies aquaculture, both illustrate how it is better to conserve energy in ecosystems by prioritizing material uses over energy use (via the principle of cascading) than to increase the use of biomass for bioenergy as well as the need to move away from fossil fuels the future circular bioeconomy will need to be based on RE.

Current technologies depend on finite metals and minerals to produce solar panels or wind turbines, underscoring the need to ensure that finite non-renewable materials (e.g., lithium and cobalt) can be recycled more easily in the future.

1.11. CONCLUSION

In the conclusion of the chapter, it discussed about the basic concept of bioeconomy and its significance. It also discussed about the origins and evolution of bioeconomy and bioeconomy sectors. In this chapter, the role of KBBE has also been discussed. It also discussed about the various bioeconomy strategies. Towards the end of the chapter, it discussed about the changing perspectives on the bioeconomy. This chapter provide highlights on the various principles of circular economy.

REFERENCES

1. Biernat, K., (2019). Introductory chapter: Objectives and scope of bioeconomy. *Elements of Bioeconomy*, [online] Available at: https://www.intechopen.com/chapters/68851 (accessed on 21 September 2022).
2. Birner, R., (2017). Bioeconomy concepts. *Bioeconomy*, [online] pp. 17–38. Available at: https://link.springer.com/chapter/10.1007/978-3-319-68152-8_3 (accessed on 21 September 2022).
3. Muscat, A., De Olde, E., Ripoll-Bosch, R., Van, Z. H., Metze, T., Termeer, C., Van, I. M., & De Boer, I., (2021). Principles, drivers and opportunities of a circular bioeconomy. *Nature Food, 2*(8), 561–566 [online]. Available at: https://www.researchgate.net/publication/353784692_Principles_drivers_and_opportunities_of_a_circular_bioeconomy (accessed on 21 September 2022).

CHAPTER 2

INDICATORS TO MONITOR AND EVALUATE BIOECONOMY SUSTAINABILITY

CONTENTS

2.1. Introduction 32

2.2. Sustainable Bioeconomy Principles, Criteria, and Impact Categories 34

2.3. Typology of Indicators 35

2.4. Combining Single Indicators to Report on Indicator Sets at Territorial and Product Level 38

2.5. Monitoring Approaches at Territorial Level 43

2.6. Good Practices as Indicators to Monitor and Evaluate Bioeconomy Sustainability 49

2.7. Stepwise Approach to Monitoring the Bioeconomy 53

2.8. Conclusion 59

References 60

The bioeconomy idea tries to provide a cogent response to a number of global issues, including food security, energy security, healthcare, and climate change. The bioeconomy which is fueled by fresh research and development in the biological and engineering science simultaneously viewed as an innovation engine.

Despite these lofty declarations, there does not appear to be a single consensus on what the term bioeconomy actually entails. What does it include and what are its limitations. These are crucial inquiries to make when talking about monitoring and evaluation (M&E) in relation to the effects of the growth of the bioeconomy since, in order to begin measuring, one must first determine where to begin and where to end the process.

2.1. INTRODUCTION

According to some sources, the bioeconomy represents a brand new course for economic growth and a substitute for the current (petrol-based) economy. The master narrative of the bioeconomy is viewed as a means of addressing and overcoming the shortcomings of the existing economic system.

In this discussion, biomass or biological resources are central. To make our daily life greener, fairer, and more inclusive, it is thought that the manufacture and use of goods like plastics, textiles, and chemicals should switch from the use of fossil fuel resources to the use of biological resources

In a nutshell, it is believed that this new development path will result in a paradigm that is more sustainable. It has open, vast bounds. In this story, a new economic paradigm that can be applied locally, nationally, regionally or globally is the bioeconomy.

The bioeconomy's contribution to economic growth and its effects (both good and bad) on the environment and society are the main topics of M&E.

Other sources take a production focused approach to the bioeconomy. The bioproduct not the economy as a whole is at the Centre of the conversation.

This school of thought considers the biological and engineering sciences (among others) to have the potential to replace specific fossil resource driven processes and products with the variety of biotechnologies and innovations provided by research and developments opposed to a new development path.

At the product level, the emergence of a bioeconomy represents prospects for streamlining the various phases of current resource flows by substituting sources of biological origin for materials and energy derived from fossil fuels. These are excellent resources for value addition, whether they are agricultural residues or crops from the primary production stage, bio-waste from processing, or food waste at the consumer stage (Figure 2.1).

Figure 2.1. Bio-waste and other waste farm produce used to make compost.

Source: Image by Flicker.

This output picture is far more constrained than the broader vision that portrays the bioeconomy as a new development route. The biomass flows from production and processing to use and re-use are where the limits begin and terminate.

In this perspective, the bioeconomy and its sustainability considerations are restricted to a few specific processes and products of both old and new value chains. For this reason, it's crucial that more effort be put out at the research, commercial sector and governance levels to close the enormous gap between these two bioeconomy visions.

When discussing monitoring and evaluation (M&E) of the bioeconomy, it is important to distinguish between these two opposing viewpoints on what the bioeconomy is and what it entails. However, both viewpoints call for monitoring of the economy in addition to other areas like social and environmental concerns.

To that purpose, a set of Aspirational Principles and Criteria for Sustainable Bioeconomy has been put forth by the International Sustainable Bioeconomy Working Group (ISBWG), which was developed within the framework of the FAO's project on sustainable bioeconomy guidelines.

Based on this reasoning, FAO has created two methods for M&E of a sustainable bioeconomy:

- According to one method, which assumes that the bioeconomy represents a new development route, territorial indicators are used to assess the impact of the bioeconomy at a national, regional, or sub-national level.

Indicators in this situation aim to quantify the economic and social impact of the bioeconomy. Four territorial levels are taken into account in this study: the global level, regional level (macro regions like the EU), national level and sub-national level. The outcomes of these four tiers are frequently displayed collectively under the heading territorial for simplicity's sake.

- Using the premise that the overall impact of the bioeconomy builds on the effect of substituting biological resources for fossil fuel resources in the various stages of the value chain and the diversification of existing products made from the same biomass, a second method of measuring the impact of the bioeconomy at the product level uses value chain and product level indicators.

Territorial or product/value chain methods might complement one another. In spite of the goal of monitoring the (sustainable) bioeconomy at both levels, currently used approaches for data collection and assessment are frequently insufficient to determine the bioeconomy's contribution to the overall economy.

2.2. SUSTAINABLE BIOECONOMY PRINCIPLES, CRITERIA, AND IMPACT CATEGORIES

The list of sustainable bioeconomy P&Cs that the ISBWG agreed upon in 2016 serves as the starting point for the analysis. One or more impact categories are developed from each criterion based on the agreed-upon P&Cs to make it easier to identify indicators. To ensure that all facts of the sustainability of the bioeconomy are covered, the indicators are grouped around the effect categories.

Although the criteria frequently cut over multiple dimensions, they were grouped as economic, social, and environmental challenges for the purpose of this study in order to express the harmony between these three dimensions.

Eight criteria in particular were obtained for each sustainability pillar. In practice, each criterion may have a combination of impacts on the economy, society, and environment supporting a holistic view of sustainability. The M&E framework for a sustainable bioeconomy will have a balanced collection of indicators if each criterion is linked to a specific sustainability pillar.

2.3. TYPOLOGY OF INDICATORS

Indicators give information that makes reality easier to understand. They provide information on trends and changes but do not always offer an explanation for them or suggest causal relationships. Quantitative or qualitative indicators are both acceptable (numbers such as units, prices, proportions, rates of change and ratios).

Both relative and absolute phrases can be used to convey quantitative indicators. Levels/stocks (x), changes over time (x) and/or performance measured against an objective are all examples of indicators. Based on pertinent knowledge of the system under consideration or a common understanding of the community that the system affects the reference levels or goal values of indicators must be established (Wu and Wu, 2012).

The preferred direction of change should still be stated even when reference levels or targets are difficult to identify. To get into the specifics of reference/target setting would go beyond the scope of this study, but it is important to recognize that indicators as is do not convey anything until they are compared to an appropriate reference value.

For instance, if the reference value is a target value, the indicator can tell you how far you are from your objective and if the reference value is a value from earlier in time it can tell you how things are changing.

Dummy indicators, which can only accept two values are used to depict a specific case. An indicator that gauges the implementation of a good practice, presence of an irrigation and water distribution system that maximizes crop yield are examples of a dummy indicator.

By giving the dummy 'yes' and the dummy 'no' a maximum value (such as 1) and a minimum value (such as 0), this sort of indication can be converted into quantitative indicators (e.g., 0).

Indicators can be performance indicators (also known as normative or progress indicators used to provide an assessment of progress towards specified objectives and targets) or descriptive indicators (also known as contextual or situational indicators used to characterize a condition or trend) (European Union, 2014).

Performance indicators depict progress or lack thereof against goals, targets or a planned end state. They enable us to state whether the current state is better or worse than it was in the past.

Without making reference to how the situation should be, descriptive indicators describe a situation or trend as it is. The reference framework in which an indication is being utilized must be made clear because the same data can be used as both a descriptive and a performance indicator depending on the context.

The information on greenhouse gas (GHG) emissions, for example can be a descriptive indication that describes the amount of CO_2 emissions in a certain location or it can be a performance indicator if it is connected to an established reduction objective.

Indicators can be classified as direct, indirect or proxy depending on how closely they connect to the object of the investigation. This distinction is crucial.

When the subject of the analysis is abstract and cannot be measured directly (such as gender equality, good governance or living conditions) or when the subject can only be measured using a complex effort that could not be carried out systematically or frequently enough, indirect or proxy indicators are useful (European Union, 2014).

Direct indicators, on the other hand, give information specifically on the topic of the analysis. An indicator in and of itself is neither direct nor indirect in this instance. The fundamental query being addressed determines whether it is one or the other (European Union, 2014).

For instance, household income (in $) in country x during the past 10 years is a direct indication if the inquiry is about changes in household income in the nation over that time, but it is an indirect or proxy indicator if the question is about household living circumstances.

The adoption of best practices and the evaluation of their effectiveness can serve as proxy indicators for monitoring and evaluating the sustainable bioeconomy, either in place of or in addition to utilizing direct indicators.

Indicators on good practice adoption and performance can be used at both territorial and product/value chain levels to acknowledge and quantify progress in a robust and economical manner whenever the adoption of other indicators is too resource and time intensive or because there is a shortage of data.

A quick evaluation of the quality of implementation can be added to quantifiable good practice indicators. For instance, in 2015, the EU Common Agriculture Policy (CAP) connected a significant portion of income support to farmers to environmental standards via the greening measures within Pillar 1 (Regulation EU).

Two years after its implementation, the payment system was reviewed by the EC in a paper published in 2017 for its environmental and climate friendly best practices (European Commission, 2017a). Along with its advantages, the EU process has several drawbacks. According to the study, there is room to increase the effectiveness of this kind of mechanism in the future (Figure 2.2).

Figure 2.2. Greens/EFA MEPs rallied outside European Parliament for CAP.

Source: Image by Flickr.

Designing good practices should take into account particular regional and local concerns Similar to how bioeconomy plans must be tailored to specific circumstances, sets of indicators for tracking the growth of a sustainable bioeconomy must also allow for some degree of flexibility (Stepping and Stoever, 2014).

To find good practices as chances for improvement that address context specific sustainability concerns, multi stakeholder methods might be used. The development of a sustainable bioeconomy can be monitored and evaluated with the use of good practices that are tailored to local circumstances.

2.4. COMBINING SINGLE INDICATORS TO REPORT ON INDICATOR SETS AT TERRITORIAL AND PRODUCT LEVEL

According to the above mentioned, it is impossible to comprehend complex phenomena like the sustainable bioeconomy with just one indication. We therefore require comprehensive sets of indicators that represent the various components of a sustainable bioeconomy in order to track it.

Indicators for the M&E of the bioeconomy are combined, displayed, and communicated in various ways. The life cycle assessment (LCA) is then presented as a helpful tool to evaluate the impact of a product life cycle using indicators (Figure 2.3).

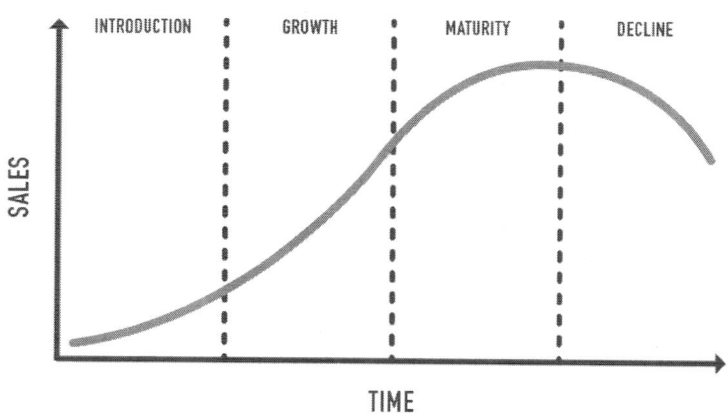

Figure 2.3. A simple visualization of the four stages of product life cycle: introduction, growth, maturity, and decline.

Source: Image by Wikimedia Commons.

The first section discusses the many approaches that may be used to group and show indicators both for at territorial and product levels.

An indicator set is a collection of indicators used collectively for a certain goal or project. It is difficult to comprehend a collection of numerous indicators especially when they have different change amplitudes and directions. Because of this, indexes are frequently created by mathematically averaging the indicators.

Aggregated indices have the benefit of representing a system's integrative properties and giving a clear, concise picture of the state or performance of a system (Wu and Wu, 2012). Normalization, weighting, and aggregation the three main components of index formation do not necessarily adhere to the tenets of sound science (Böhringer and Jochem, 2007).

The term "sustainability indicators" is frequently used to refer to both indicators and indices. Four of the various ways that indicators at the territorial and product level might be displayed and analyzed are outlined below (GGKP, 2016).

2.4.1. Dashboard of Indicators

A group of measures that represent data from multiple domains including combinations of environmental, economic, and social aspects.

2.4.1.1. Composite (or Aggregated) Indices

They have the benefit of giving a clear, concise picture of the state or performance of a system by combining various metrics into one by scoring and weighting the underlying data (Wu and Wu, 2012).

Normalization, weighting and aggregation the three key components of index formation are however highly value sensitive and don't necessarily adhere to standard scientific principles (Böhringer and Jochem, 2007).

2.4.2. Footprint Type Indicators

Footprint type indicators can be very helpful to explain results and can aggregate a variety of economic and environmental issues into a single indicator. They can be used to determine if present production/consumption patterns are sustainable or in accordance with the limits of the planet.

Examples of this type of indicator include the Sustainable Europe Research Institute's (SERI) Four Footprints (materials, carbon, water, and land) (SERI, 2013). The notions of carbon footprints and material footprints are very complex, however there are scientific rules and data that guarantee some degree of comparability.

On the other hand, there is still need for improvement in terms of accounting rules, data coverage, disaggregation, and quality for land and water footprints. Industrialized nations are increasingly using international trade to transfer environmental burdens to other regions with footprints being much larger than the corresponding territorial indicators (SERI, 2013).

This fact emphasizes how crucial it is to take into account footprint type indicators while analyzing territorial indicators and evaluating the sustainable bioeconomy in order to avoid drawing the wrong conclusions about policy. Indicators of the type of footprint could link the local and global aspects of the bioeconomy M&E for which there is currently a paucity of literature.

2.4.3. Adjusted Economic Measures

For instance, extended wealth, adjusted net savings, and green GDP by taking into account environmental or less frequently environmental and socially connected characteristics, they attempt to rectify standard economic factors.

The question of whether a sustainability assessment should be reductionist (i.e., break down a very complex natural and anthropogenic system into a few component pieces) or holistic (looking at systems as a whole) is up for dispute (Bond and Morrison-Saunders, 2011).

Reductionist methods are demonstrated through composite indices, which combine various components into a single value. The discussion is outside the purview of this study, but the four techniques key advantages and disadvantages are still relevant. It should be noted that the methods are not mutually exclusive because, for instance, dashboards might feature composite indications.

The life cycle assessment (LCA) is a potent tool to measure the influence of a product life cycle using indicators at the product level. Environmental and socio-economic impacts can be measured using indicators such as Life Cycle Costing (LCC), Social Life Cycle Assessment (S-LCA) and Environmental Life Cycle Assessment (LCA).

In order to examine potential environmental and socioeconomic effects from “cradle to grave” or depending on the boundary at hand, “cradle to cradle,” life cycle sustainability assessment frameworks are utilized.

In the event that attempts to reduce impacts in one process or life cycle stage unintentionally create (potentially larger) impacts in other processes or life cycle stages, they assist in identifying and preventing the shift of load between other processes or stages.

When fossil fuels are replaced with biofuels, for instance, GHG emissions from transportation are reduced, while emissions from harvesting and extracting the feedstock for biofuels are increased (Figure 2.4).

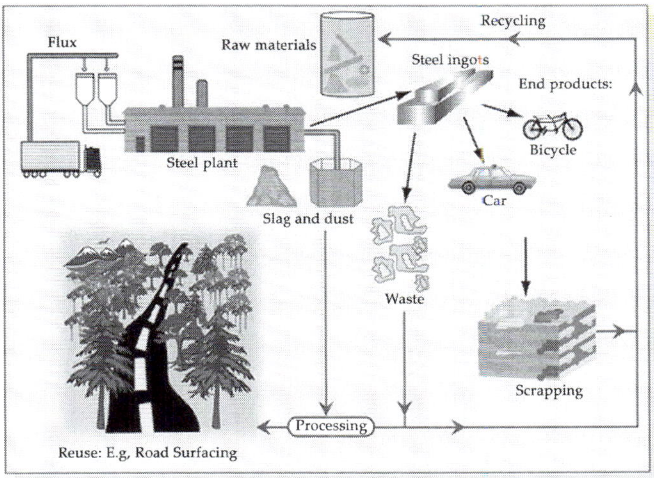

Figure 2.4. Life cycle assessment.

Source: Image by Wikimedia Commons.

These frameworks can be used to evaluate more complicated impacts resulting from the production and consumption of energy, transport, waste management systems and infrastructure even though they are typically employed to examine product systems. The function of the investigated entity serves as the central focus of the assessment in all applications which adopts a life cycle viewpoint.

The LCA framework's approach is divided into four iterative phases:

- The goal and scope definition phase outlines the objectives of the study, its purpose and the product, process or activity of interest. Notably, the functional unit is established, the system boundaries, including sub-units, inputs, and outputs, are recognized and the modeling strategies are provided during this phase.

- The product system (or systems) and its individual unit processes are described in the Life Cycle Inventory (LCI) phase. The basic flows between the product system and the environment, such as inputs (such as extracted raw materials, land used, labor) and outputs (such as emissions to air, water, and soil, food security, socio-economic advantages) are investigated and collated through data gathering for instance.

One functional unit, as specified in the Goal and Scope phase refers to the quantity of elementary fluxes exchanged by the product system and the environment.

- Based on the information acquired during the LCI process, the Life Cycle Impact Analysis (LCIA) phase evaluates the possible (environmental, social, and economic) effects of the product system.

This is accomplished by relating the LCI results to the impact categories and indicators as well as stakeholders. Selection of stakeholders, impact categories, indicators, and characterization models as well as the assignment of the LCI results to the various impact categories (classification) and computation of indicator results are all required components of LCIA (characterization). Optional components like normalization, grouping, and weighting can then come after this.

- To arrive at conclusions or recommendations, the Life Cycle Interpretation phase combines the results of the first two stages with the specified purpose and scope (Rios, Moore, and Jones, 2007). It is significant to remember that LCA offers an evaluation of potential impacts based on a selected functional unit.

Primary or secondary data may be used to complete the product life cycle impact evaluation. Primary data are company specific, site specific or supply chain specific (if there are numerous sites for the same product).

Meter readings, purchase records, utility bills, engineering models, direct monitoring, material/product balances, stoichiometry or other techniques for gathering data from certain processes in the company's value chain can all be used to get primary data.

Secondary data are those that come from sources such as a third party LCI database or other sources that are not directly gathered, measured, or estimated by the organization conducting the LCA study.

Secondary data can also be based on financial data, proxy data and other generic data and includes industry average data (such as from published production figures, government statistics and industry groups), literature studies, engineering studies and patents.

Frameworks for LCAs can be an effective tool for tracking and rating bioproducts. Once the indicators are chosen and suitable life cycle sustainability evaluation methodologies are applied (i.e., where inventorying is possible).

For instance, LCA frameworks allow comparison of the life cycle performances of sub-components of bioeconomy outputs at various levels of the value chain for the indicators (previously chosen, but not using the LCA methodology) at the product level (raw biomass production, bioproduct manufacturing end of life, etc.).

The life cycle sustainability assessment technique and literature would then provide a baseline for comparison between bioproducts and conventional goods and services after the indicators have been chosen. The evaluation of possible environmental and socioeconomic gains made through the shift to the bioeconomy is aided by this comparison.

2.5. MONITORING APPROACHES AT TERRITORIAL LEVEL

The primary techniques for territorial bioeconomy sustainability monitoring. Despite being limited to economic sustainability only a small number of the methods examined in this study are actually put into practice at the national level (e.g., in Argentina, Finland, and Malaysia).

Others were created as part of regional or international programs by international organizations, NGOs or academia but, they were never actually put into practice. Others create composite indices and footprint style indicators, while some use or recommend groupings of individual metrics.

2.5.1. Specific National Monitoring Approaches for Sustainable Bioeconomy

Environmental and social goals must be a part of any national bioeconomy policy that seeks to support sustainable development as well as any associated monitoring and measurement techniques. However, most nations now lack frameworks to track sustainable progress in achieving the goals established in the bioeconomy initiatives (FAO, 2016, 2018; German Bioeconomy Council, 2018).

The FAO report from 2016 on "How Sustainability is addressed in Official Bioeconomy Strategies at International, National, and Regional Levels," found that most bioeconomy strategies had weak M&E methods and many nations suggested that sustainability standards and guidelines should be developed and agreed upon on an international level (FAO, 2016).

Another FAO study, this one concentrating on Argentina, Australia, Germany, Malaysia, the Netherlands, South Africa and the United States of America examined national frameworks to determine the economic impact of the bioeconomy in 2018. (FAO, 2018).

This study also demonstrated that most nations do not currently have a system in place to track how well bioeconomy targets are being met. Most governments evaluate the contributions of the bioeconomy in terms of economic variables (such as value contributed and employment) while social and environmental considerations are typically only lightly considered.

Even at the national level, there are numerous research and initiatives aimed at creating thorough bioeconomy monitoring systems complete with social and environmental indicators. Some of these papers examine the analysis created over the previous few decades for sustainable biofuels, forests, biomass, and bioenergy.

For instance, to improve the results based management of national forest projects, many nations are implementing criteria and indicators for sustainable forest management (FAO, 2017a; Tegegne, Cramm, and Brusselen, 2018). An initial set of indicators for tracking the development of a low carbon, forest based bioeconomy is provided by the Canadian bioeconomy strategy (German Bioeconomy Council, 2018).

Several nations plan to include monitoring and measurement operations in their bioeconomy policy, according to the German Bioeconomy Council (German Bioeconomy Council, 2018).

According to the Council, an increasing number of nations (including Argentina, Australia, Brazil, Canada, China, France, Italy, Latvia, New Zealand, Spain, the UK, and the USA) are encouraging measuring activities to keep track of new (bio) technologies, biomass flows, bio-based goods and services, as well as their economic, ecological, and social effects (German Bioeconomy Council, 2018).

Additionally, a number of nations are putting forth proposals to expand monitoring and measuring efforts including the installation of integrated information and observatory systems, while yet other nations are in the process of creating thorough monitoring systems.

Argentina is one of these nations, which now tracks its bioeconomy exclusively in terms of the sectors' contributions to GDP, ignoring environmental and social factors (Bracco et al., 2018; Lechardoy, 2018). This study uses the Argentinian monitoring framework as an illustration of methods that only consider economic factors. Finland and Malaysia offer further instances.

A fascinating illustration of a composite index is the Malaysian Bioeconomy Contribution Index (BCI). The Malaysian bioeconomy's five economic indicators bioeconomy value added, biobased exports, bioeconomy investments, employment, and productivity performances are combined to create this index (Figure 2.5).

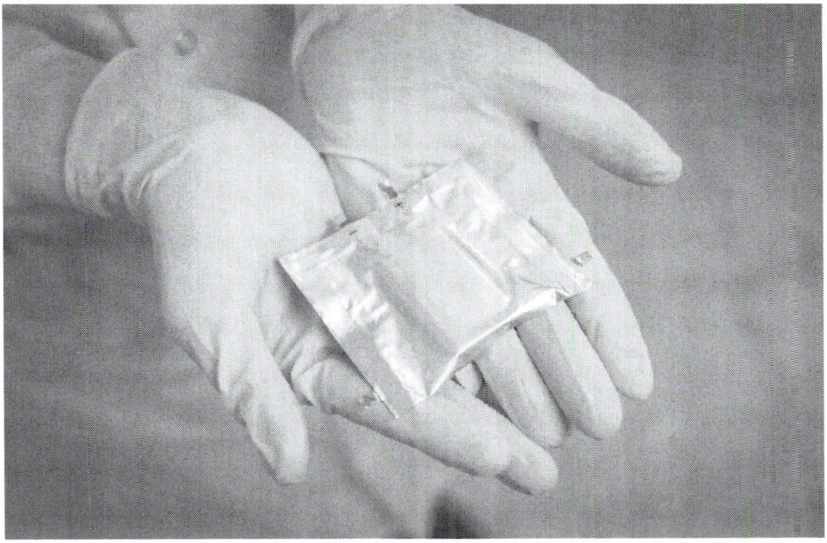

Figure 2.5. Bio-based polymers used in healthcare industry.

Source: Image by Data Bridge.

The BCI, which compares the country's economic performance in the bioeconomy to the baseline year 2005 at 100 points is produced. The BCI only assesses economic flows at the moment, but it might be improved to include environmental or socioeconomic performance.

For instance, it could take into consideration CO_2 emissions or the degree of regional biodiversity or it could include strategies for reducing poverty or income inequality in the bioeconomy sector (Al-Amin, 2015; Bracco et al., 2018).

The Ministry of Employment and the Economy (MEE) set up a project to write the Finnish Bioeconomy Strategy in collaboration with many other ministries (MEE, 2014).

Five economic indicators output, value added, investments, employment, and exports of bioeconomy goods have been used to track the Finnish bioeconomy sectors from 2010 to 2017 (biomass producing sectors, food, wood products, pulp, and paper, bioenergy, bio-construction sectors, treatment, and supply of water and bioeconomy services of nature tourism and recreation, recreational fishing and hunting).

Despite the fact that the Finnish bioeconomy policy is currently being modified, these Finnish indicators are nevertheless included in our study (EC JRC, 2018a).

Germany is now creating a thorough M&E strategy to track the bioeconomy. The monitoring of biomass flows, the establishment of economic key performance indicators (KPIs) and the Systemic Monitoring and Modeling of the Bioeconomy (SYMOBIO) are the three primary projects that make up the collaborative inter-ministerial effort.

These three projects findings cannot be presented in this study because they won't be finished until the following biennium. The Thünen Institute, based on the German sustainable development plan and related to Germany's strategic framework to achieve the SDG Agenda are the indicators that are the subject of this report's analysis.

Italy has created a set of sustainability indicators with quantifiable effects on economic growth, food security, the sustainability of natural resources, reliance on non-renewable resources and climate change.

A provisional set of EU KPIs created by the Bioeconomy Regional Strategy Toolkit (BERST), an EU-funded project to enable regional stakeholders in Europe by designing clever strategies to explore their bioeconomy potential is what the nation hopes to link bioeconomy implementation and monitoring too.

These indicators support the use of benchmarking analysis and make reference to national and Euro stat data. The "Systems Analysis Tools System for the EU Bio-Based Economy Strategy" (SAT-BBE) (Presidency of Council of Ministers, 2017), which addresses the sustainability aspects of the bioeconomy, is another document that Italy seeks to connect its indicator framework to.

2.5.2. Specific EU Monitoring Approaches for Sustainable Bioeconomy

The European Commission (EC) is funding a number of initiatives to track the growth of the bioeconomy in Europe, primarily through the Joint Research Center's (JRC) "Bioeconomy Knowledge Center" project.

The JRC evaluates a set of socioeconomic indicators (employee count, turnover, value added and labor productivity) for several bioeconomy sectors before estimating member state performance on a road to improved productivity (Ronzon, Camia, and Barek, 2018). This strategy is examined in this study because it offers a first step in tracking the bioeconomy in the European Union (EU).

The second feature of the EU bioeconomy is the biophysical dimension, which is examined in many studies that record biophysical indicators along the biomass supply chain such as resource cascading and recycling flows (this stream of work has not been reviewed since it goes beyond the boundaries of this study).

Thirdly, the LCA approach is used to evaluate any potential environmental effects connected to a system a process or a product throughout its life cycle.

Last but not least, the JRC is developing simulation models and forward looking studies to evaluate closed economic systems, address many goals in a unified framework, capture behavioral factors (consumer and producer decisions) and test various conceptual frameworks and policy options (EC JRC, 2018a).

Specifically, the bioeconomy has been assessed for policy coherence and SDGs using the economic modeling system Modular Applied General Equilibrium Tool (MAGNET) (Philippidis et al., 2018). This strategy is not examined in this study since it has no immediate bearing on the determination of sustainability metrics.

The EC is also trying to define a thorough and all-encompassing monitoring mechanism for the EU bioeconomy in the wake of the 2018 revision of the EU bioeconomy plan.

The Central and Eastern European Effort for Knowledge Based Agriculture, Aquaculture, and Forestry in the Bioeconomy (BIOEAST) which was founded by the Central and Eastern European countries to enforce cooperation, research, and policy making is another intriguing European initiative.

Through the BIOEAST Strategic Research and Innovation Agenda (SRIA), the BIOEAST countries want to influence the scope, emphasis areas and objectives of the future EU bioeconomy monitoring system although they do not directly contribute to its creation (EC JRC, 2018a).

A sustainable and reliable framework for monitoring the bioeconomy and its many effects on the EU and its Member States is what the Horizon 2020 research project Bio Monitor seeks to build. The experiment began in June 2018, however since the results are still pending, they were left out of this review.

2.5.3. Other Approaches Relevant for Bioeconomy Monitoring at Territorial Level

There are numerous monitoring strategies that have been designed especially for the bioeconomy and that address the sectors and problems that are pertinent to it. There are a number ways that are pertinent to the bioeconomy though discussing them all would go beyond the scope of the article.

First, since countries have already committed to the SDGs there are already SDG indicators in place and countries are taking steps to measure their implementation, the M&E framework for the bioeconomy could be as closely aligned with that of the SDGs as possible to avoid re-inventing the wheel and lower the risk of duplication.

We offer pertinent SDG indicators that might be used to assess the P&Cs of a sustainable bioeconomy in light of these factors (Alcolu and Bogdanski, forthcoming).

National accounts from certain nations also include data on social economic and environmental issues. A good example is the information being gathered by EUSTAT about the EU member states.

For example, the EU's environmental accounts report on two broad categories of activities and/or products: environmental protection (e.g., all activities related to preventing, reducing, and eliminating pollution and any other form of environmental degradation) and resource management (preserving and maintaining the stock of natural resources).

The environmental goods and services (EGSS) account offers data on the employment and gross value added associated with the manufacturing of environmental products (Eurostat, 2019a). Since 2015, Eurostat has also been putting in place a framework for tracking the development of a circular economy.

Additionally, this framework uses an online database and specialized website to provide ten indicators and 23 sub-indicators that capture the key components of a circular economy (production and consumption of goods, waste management, secondary raw materials, competitiveness, and innovation) (Eurostat, 2019b).

Along with the extensive number of indicators, this framework has given EU member states trend analysis and a substantial amount of data on each indicator.

2.6. GOOD PRACTICES AS INDICATORS TO MONITOR AND EVALUATE BIOECONOMY SUSTAINABILITY

At both the territorial and product/value chain levels, indicators on the acceptance and quality of the implementation of good practices can be used to acknowledge and quantify progress toward the sustainability of the bioeconomy in a fairly robust and economical manner.

The reporting of the adoption of best practices has been offered by indicator typologies as a potential replacement or addition to intricate quantitative measurement systems.

When measuring direct indicators would require an excessive amount of time and money or if data were unavailable, good practices may be utilized as a stand-in. Reporting on the application of good practices can also aid in quantifying milestones in relation to goals and targets, making it an effective system for recognizing progress that can be utilized for both governmental incentives and regulations as well as private project finance.

Frequently, certification programs also call for tracking the application of ethical standards (Bracco et al., forthcoming).

It should be emphasized that some of the indicators gathered from literature reviews and mentioned below are actually good practices, such as when they provide qualitative or fictitious data on the existence of a practice, policy, or strategy.

The presence of an irrigation and water distribution system that maximize crop productivity, the presence of a cluster organization that coordinates, manages, and facilitates the bio cluster, and SDG 13.2.1: Number of countries that have communicated the establishment or operationalization of an integrated policy, strategy or plan that improves their capacity to adapt to the negative effects of climate change and foster climate resiliency are a few examples of the indicators that have been gathered.

This is pertinent at the territorial level, particularly for the indicators linked to the criterion. Policies, rules, and institutional framework pertinent to bioeconomy sectors are suitably synchronized;

The proper mechanisms for inclusive consultation and the participation of all relevant societal sectors are based on open information exchange and policy consistency between the supply and demand of food and non-food items. Indicators for these criteria actually provide information on things like:

- number of nations ratifying, approving, and putting into practice through institutional, legal, and policy frameworks, tools, and monitoring frameworks;
- existence of an incubator or a cluster organization;
- institutionalized or non-institutionalized platforms;
- the quantity of participants from various stakeholder groups in consultations;
- free access to the information needed to support stakeholder viewpoints in a way that is timely, transparent, open, and approachable;
- number of nations with national action plans for sustainable consumption and production (SCP) or with SCP mainstreamed as a priority or an objective into national policy;
- progress made by nations in terms of how well they are implementing international agreements; and
- funding from both governmental and private sources ($).

Good practices can therefore be especially helpful to track and enhance inclusivity, as well as Harmonization and coherence in the development of the bioeconomy, for which precise quantitative measuring frameworks and indicators are still lacking.

However, when implementing precise technical measurements on a broad scale is challenging or impossible, good practices can also be employed in their place.

A "proxy" indicator, such as the adoption of land protection practices, can be used to determine whether "biodiversity conservation is ensured," such as the "pace of biodiversity loss."

"Protected areas and land with significant biodiversity values and biodiversity conservation and management" and "SDG 15.1.2 Proportion of important sites for terrestrial and freshwater biodiversity that are covered by protected areas, by ecosystem type" are two examples of indicators for this practice that were retrieved from the reviewed literature.

A concrete illustration of a direct or indirect indication that might be used to monitor soil degradation. Conservation agriculture is a viable strategy for combating soil deterioration in the agricultural sector when it comes to the biomass generation stage.

Thus, "number of hectares under conservation agriculture" or "number of farmers having adopted conservation agriculture" can be proxy indicators for soil deterioration. Good practices can also be implemented at the federal level as policies or plans.

Policies for soil conservation are established, implemented, and controlled, for instance at the territorial level, potential proxies for indicators include "the amount of money invested each year to carry out the action plan outlined in the soil conservation policy" and "the number of hectares/percentage of land under soil protection measures."

Adopting good practices, however, does not ensure that the expected effects will materialize. Various drawbacks and trade-offs that may result from using agricultural practices in a causal loop in a particular situation (TEEB, 2018b).

If "agroforestry," for instance, is chosen as a good practice, an indicator to measure the extent of its implementation could be the "number of hectares of mono-cropping replaced by agroforestry" or the "number of farmers practicing agroforestry," supplemented by a quick evaluation of the quality of implementation.

In order to assess progress, it is crucial to keep track of both the quality and rate of adoption of good practices. This is especially true in years when accurate statistics are not yet available. In actuality, given the expenses associated with data retrieval and measuring specific indicators, these may only be measured every five or ten years, with good practice monitoring taking place in between.

Indicators can be used to examine whether excellent practices truly result in change or to monitor and assess the performance of their implementation. Adopting good practices can directly assist in addressing identified sustainability challenges, both at the product and territorial level, beyond simply using them for reporting purposes.

Groups of stakeholders and experts should identify and prioritize good practices as improvement opportunities related to sustainability concerns in the bioeconomy, such as hotspots in bio-based value chains and sectors. This is necessary for a sustainable bioeconomy M&E (Figure 2.6).

Figure 2.6. A three-tier agroforestry system (Jack fruit, papaya, brinjal) in Narsingdi village, Bangladesh.

Source: Image by Flickr.

The TSC method, a private sector project that exemplifies how good practices, so-called improvement opportunities, can address sustainability issues of main concern (hotspots) within a particular value chain, is a good example at the product level.

2.7. STEPWISE APPROACH TO MONITORING THE BIOECONOMY

A methodical approach with the goal of keeping an eye on the bioeconomy's and bioproducts long-term viability. This approach's design is based on the findings of the analysis.

Following the "cockpit approach," governments or bioproduct producers and manufacturers are given access to a lengthy range of scientifically sound indicators from which to select a small number of indicators that best suit their needs and conditions. This strategy, which reflects a balanced weighting exercise, is crucial for communication, measurement viability, and policymaking.

By allocating weight to the indicators, the assessment is calibrated in accordance with their applicability and importance for certain policies, plans, programs or initiatives in a particular nation or region at a specific time (Villeneuve et al., 2017).

The M&E framework must be adjusted in this step to account for the circumstances and context in which it will be used. Is this indicator the "SMARTest" (most Certain, Measurable, Achievable, Relevant, and Time bound) measure for measuring and evaluating the criterion in a specific country/region at a specific time? This question must be answered for each indicator.

The chosen metrics might be modified at the territorial level to reflect the national priorities for the bioeconomy. They might already have been gathered by the nations (for instance, to report on SDG indicators), and they ought to be presented in a form that is understandable to all stakeholders.

The chosen indicators at the product/value-chain level can be modified for each bioproduct based on the pertinent value chain and its hotspots. If the bioproduct is approved or branded, for example, the data for these indications may already be available. Again, it's crucial to make sure that all users and buyers can easily understand the findings that are displayed.

Indicators of best practices can be used in conjunction with the stepwise approach to supplement the measurement of impacts. The development of a framework for monitoring the bioeconomy may be driven by a variety of reasons.

This strategy is primarily aimed at technical staff members of Ministries or public organizations who are responsible for creating an M&E system for a sustainable bioeconomy on a national level. This document targets the private sector players at the product and value chain levels who desire or are required to report on the sustainability of their bioproducts and value chains.

The preceding step-by-step process is further explained in the sections that follow. the procedures to be followed in order to choose a customized "cockpit" of:

- country-relevant criteria and indicators that deal with sustainability issues along their value chains (i.e., hotspots) and that adhere to sustainability goals for market-uptake.
- Bio products relevant criteria and indicators that deal with sustainability issues along their value chains (i.e., hotspots).

Below are descriptions of each approach's steps at both the territorial and product/value-chain levels.

2.7.1. Step 1: Stakeholder Engagement

A collaborative process must be used to choose the indicators for the "cockpit screen." The selection of pertinent priorities, hotspots, and indicators depends heavily on the involvement and participation of all important bioeconomy stakeholders and specialists.

To guide the process from the definition of bioeconomy priorities to the construction of an M&E framework, a multi-stakeholder platform comprised of representatives from the public and private sectors as well as the civil society can be developed.

However, if there are numerous stakeholders with opposing viewpoints, participatory procedures may be difficult. As a result, the stakeholders may occasionally be given a higher priority or included in separate rounds.

The EC, for instance, iterates between idea formulations, proposals to high priority stakeholders, reformulations of ideas, returns to high priority stakeholders with updated proposals and gradually expanding to all stakeholders, once a formulation is agreed upon by core stakeholders.

2.7.2. Step 2: Choice of System Boundaries

- **Territorial Level:** A country should first decide whether it wants to assess its performance at the national or sub-national level. The BERST approach is an example of a regional (sub-national) level framework in contrast to the majority of indicators found in the literature that indicate national performance.

Additionally, IINAS (2015) offers local, sub-national, and national criteria and indicators for biomass sustainability. A nation may use the metrics provided in these methodologies to analyze sub-national performance.

- **Product/Value-Chain Level:** According to their particular sector, relevant stakeholders must specify the parts and boundaries of their value chain. Sector specific indicators from the literature review are presented. When adopting a market and/or sectoral sustainability policy, policy makers and/or actors from public institutions may also be interested in pursuing this strategy. The private sector can monitor and evaluate specific items or entire sectors.

2.7.3. Step 3: Identification of Sustainability Issues and Relevant Indicators

- **Territorial Level:** The study links indicators to standards from the sustainable bioeconomy P&Cs adopted by the ISBWG. There are 24 criteria, including ones for social, environmental, and economic factors. Therefore, the nation can address the three facets of sustainability by choosing one or two indicators for each criterion.

For every P&C of the sustainable bioeconomy, the nation can choose the "best" indication. The "best" indicator for assessing and evaluating the criterion in the nation should be the "SMARTest": the most Specific; Measurable; Achievable; Relevant and Time-bound indicator. A country may choose two or more choices if there are more than ten indicators mentioned for each criterion; otherwise, it may choose just one indicator for each criterion.

The nation may suggest an indicator for the criteria for which none is already available. In order to cap the total number of indicators in the framework for this exercise, a new indicator should only be introduced if an existing indicator is deleted.

At this point, the nation may add and select various indicators using a participative method and professional judgment, so long as they adhere to the effect categories of the sustainable bioeconomy P&Cs.

This process will result in the identification of a group of around 24 balanced indicators that verify compliance with the P&Cs and span economic, environmental, and social aspects. Frequently, an SDG indicator can be used to quantify the criterion.

Countries may use SDG indicators since they are already required to report on their SDGs (many SDG indicators are currently gathered by UNSTATS), but they should be aware of the attribution problem.

- **Product/Value-Chain Level:** The TSC technique is used to identify hotspots and the sustainability challenges they raise at the product and value chain levels. For each product category, this method pinpoints hotspots which are "activities in the life cycle that produce one or more social or environmental consequences" (TSC, 2015).

As it ensures adequate examination of the most crucial concerns with fewer resources and time, this concentration on hotspots is highly cost-effective. For each product/value chain, the hotspots and associated sustainability challenges should be identified through study and expert engagement using a multi-stakeholder participatory method.

2.7.4. Step 4: Choice of Indicators That Reflect Bioeconomy Strategy Objectives or Hotspots

- **Product/Value-Chain Level:** In order to complete this stage, stakeholders can choose the SMARTest indicators from the extensive list supplied to track and assess the pertinent sustainability concerns related to the pinpointed hotspots.

Depending on the regional context and unique requirements, a small number of indicators such as 1 or 2 will be chosen for each hotspot in order to make the measurement viable. The stakeholders may suggest new indicators if there isn't one for a given criterion or if they find superior ones.

Unless the sustainability challenges connected to the identified hotspot address a generic component of sustainability, the indicators chosen need not be balanced across the three sustainability pillars (i.e., environmental, social, and economic).

SCL indicators address a number of criteria that stakeholders should prioritize if they intend to adopt or have already adopted a SCL (and have the data for that indicator) (in order to avoid duplication of measurement efforts).

2.7.5. Step 5: Discussion and Selection of Reference Values for Each Indicator

Once the indicators have been chosen, they can be used to assess the effectiveness and effects of bioproducts and the bioeconomy. Each indicator must have a benchmark, target, or reference value for this activity in order to evaluate its change or trend.

Reference values might be historical data or time series, or they can be reference values that are supposedly based on normative policy or science as is frequently the case in sustainability studies. Positive indicators are those that demonstrate a large amount of or moderate progress toward a goal.

Negative indicators are those that suggest either a lack of progress toward or a complete departure from an aim.

Local stakeholders should choose the direction of indicators while bearing in mind the ecological boundaries of the globe (O'Neill et al., 2018). Positive social indicator patterns, however, are preferred. Economic indicators are frequently more important because their improvement may be linked to excessive resource exploitation or may occur at the expense of social advancements.

A growth in GDP (at the territorial level) or company revenue (at the product/value chain level), for instance, could go hand in hand with a rise in inequality or be connected to the overuse of natural resources.

2.7.6. Step 6: Definition of Data Collection Methodology and Assessment of Data Availability

This process is crucial for gathering the information required for the indicators. Stakeholders can choose to use pre-existing datasets and databases or conduct field analysis to gather data. The latter, however, could be dangerous because it is difficult to locate data that is only attributable to the bio-based value chains of a bioproduct and to the bio-based activities and sectors at a national level.

If there are data gaps at the product/value chain level, the country can disaggregate the data at the territorial level and the other way around if there are data gaps at the product/value chain level.

Stakeholders may think about repeating stages 3 through 5 at the country level and steps 4 and 5 at the product/value chain level as needed after the assessment of data availability has been completed and data gaps related the chosen indicators have been discovered. Stakeholders have the option to modify the list of indicators by repeating these processes according to the accessibility and availability of data.

2.7.7. Step 7 (Optional): Selection of Good Practices to Address Sustainability Issues

Through a participatory stakeholder debate and expert opinion, this step identifies best practices or improvement possibilities to address sustainability challenges and hotspots at the product/value chain level and at the territorial level.

Additionally, it establishes routine evaluation of advancements made following the adoption of best practices. Additionally, good practices can be used as proxy indicators and to support more complicated indicators saving time and money in M&E efforts.

2.7.8. Step 8: Assessment of Progress towards Bioeconomy Strategy Objectives and Sustainability Goals

- **Territorial Level:** This step offers data and information at the territorial level on the bioeconomy's progress toward the strategy's objectives and sustainability goals. The performance of the bioeconomy can be assessed after a benchmark condition has been created and the ideal trend for each indicator has been determined.

2.7.9. Step 9: Effective Communication of the Results

The findings ought to be communicated clearly yet plainly. This is necessary to convey the findings to decision-makers and enable informed decision making as well as to the general public to raise consumer awareness and promote the use of bioproducts. Spider diagrams and/or interactive visuals, for instance, could be a useful technique to demonstrate progress in the examined indicators. It could be easier to communicate the findings if indicators were grouped under the three sustainability pillars of economic, social, and environmental indicators.

How to handle the trade-offs and synergies of the growth of the bioeconomy because many areas of sustainability are frequently interconnected. For instance, inclusivity, such as the types of jobs produced, fair treatment of employees, favorable working conditions and climate change mitigation could suffer in the name of economic progress.

2.8. CONCLUSION

In the conclusion of the chapter, it discussed about the various indicators to monitor and evaluate bioeconomy sustainability. It also discussed about the different principles of sustainable bioeconomy, its criteria and impact categories. In this chapter, several typology of indicators have also been discussed in the chapter. Towards the end of the chapter, it provides highlights on stepwise approach to monitor the bioeconomy.

REFERENCES

1. Befort, N., (2020). Going beyond definitions to understand tensions within the bioeconomy: The contribution of sociotechnical regimes to contested fields. *Technological Forecasting and Social Change, 153*, 119923. [online] Available at: https://linkinghub.elsevier.com/retrieve/pii/S0040162519307929 (accessed on 21 September 2022).
2. Bracco, S., Tani, A., Çalıcıoğlu, Ö., San, J. M., & Bogdanski, A., (2019). *Indicators to Monitor and Evaluate the Sustainability of Bioeconomy* (p. 3). [e-Book] Rome: Food and agriculture organization of the United Nations. Available at: https://www.fao.org/3/ca6048en/CA6048EN.pdf (accessed on 21 September 2022).
3. Jander, W., Wydra, S., Wackerbauer, J., Grundmann, P., & Piotrowski, S., (2020). *Monitoring Bioeconomy Transitions with Economic–Environmental and Innovation Indicators: Addressing Data Gaps in the Short Term* (p. 5). [e-Book]. Available at: https://www.mdpi.com/2071-1050/12/11/4683/pdf (accessed on 21 September 2022).
4. Wohlfahrt, J., Ferchaud, F., Gabrielle, B., Godard, C., Kurek, B., Loyce, C., & Therond, O., (2019). Characteristics of bioeconomy systems and sustainability issues at the territorial scale. A review. *Journal of Cleaner Production, 232*, 898–909 [online]. Available at: https://www.sciencedirect.com/science/article/abs/pii/S0959652619319274?via%3Dihub (accessed on 21 September 2022).
5. Zeug, W., Bezama, A., & Thrän, D., (2021). A framework for implementing holistic and integrated life cycle sustainability assessment of regional bioeconomy. *The International Journal of Life Cycle Assessment, 26*(10), 1998–2023 [online]. Available at: https://link.springer.com/article/10.1007/s11367-021-01983-1 (accessed on 21 September 2022).

CHAPTER 3

CHALLENGES AND PERSPECTIVES TOWARDS A SUSTAINABLE BIOECONOMY

CONTENTS

3.1. Introduction 62

3.2. Fostering Sustainable Bioeconomics: The Role of Conscious Consumption 63

3.3. Biorefineries Supply to Sustainable Textiles 68

3.4. Wood-Based Solutions for Sustainable Built Environment 70

3.5. Environmental Sustainability Indicators for the Bioeconomy 72

3.6. Innovation and Sustainable Development: A Bioeconomic Perspective 74

3.7. Challenges in Sustainable Bioeconomy... 76

3.8. Challenges and Gaps in Sustainability Assessment In Bioeconomy ... 78

3.9. Benefits and Impacts of the Bioeconomy 81

3.10. Constructing a Sustainable Bioeconomy: Multi-Scalar Perceptions of Sustainability 84

3.11. Ecological Limits to Sustainable Use of Wood Fuels 88

3.12. Conclusion 89

References 90

The term "bioeconomic" refers to all economic sectors engaged in the production, processing, and use of biological resources (plants, animals, and microorganisms) for the production of food and nourishment, the assistance of biofuels as an asset and also the production of bio-based chemicals and products as well as bioenergy.

3.1. INTRODUCTION

As a knowledge based bioeconomic sustainable use of limited resources should be accomplished through the incorporated knowledge of the application of biological resources and processes. The economic, ecological, and social dimensions of sustainability are therefore considered for instance in terms of protecting against the advancement and adaptation to climate change as well as demographic shifts.

3.1.1. Sustainable Bioeconomy for the 21st Century

Since the idea of the knowledge based bioeconomy (KBBE) was first released in Europe in 2005, the bioeconomy view has made its way into the strategy development of top countries across the world. By 2020, 54 nations had already initiated bioeconomy-related policy initiatives.

Current schemes are progressively associated with United Nations' 2015 Sustainable Development Goals (SDGs). The ideas and remedies of a feasible bioeconomy donate to thirteen of the seventeen SDGs.

Both the EU Member States as well as the European Commission (EC) have made significant pledges to a sustainable bioeconomy in Europe. It is now an important element of European funding for research and plays a vital role in various EU techniques and guidance notes.

This also plays a major role in the advancement of a circular economy. The "European Green Deal" embraced by the new European Council in December 2019 intends to build a number of bioeconomy regulations.

Germany is a global leader in the development of bioeconomy policy and research techniques. In 2009, the federal government has formed Bioeconomy Council, which has since provided policy advice. The federal government's high-tech approach has long addressed different aspects of prospective bioeconomy studies.

Following the publication of the National Research Strategy Bioeconomy 2030 as well as the National Policy Strategy Bioeconomy, the federal government released the National Bioeconomy Strategy in January 2020, which consolidated the previous work. Bioeconomy had been chosen as the subject of the Science Year 2020/21 in order to encourage open conversation and also to identify and entail all social circles (Figure 3.1).

Figure 3.1. Sustainable development goals pyramid.

Source: Image by Wikimedia Commons.

3.2. FOSTERING SUSTAINABLE BIOECONOMICS: THE ROLE OF CONSCIOUS CONSUMPTION

In the United States, sustainability is usually addressed in silos. Cradle-to-cradle production and regulation to reduce greenhouse gas (GHG) emissions are proposed as remedies for demonstrable deterioration, although little attention is given to how a society can allow sustainability as a cultural practice.

Furthermore, the involvement of the individual economic representative as a customer, shareholder, and government participant appears to be underappreciated (Figure 3.2).

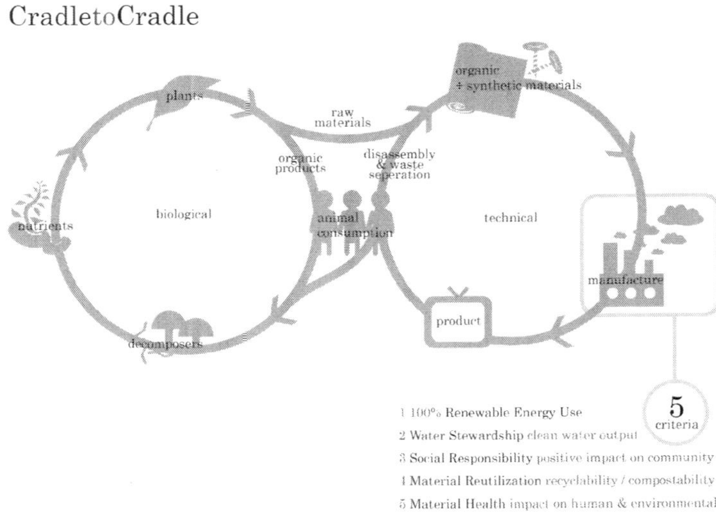

Figure 3.2. Biological and technical nutrients in the cradle-to-cradle design framework.

Source: Image by Wikimedia Commons.

To a great extent, the population majority delegates the rights vested in the three roles to a minority primarily through indifferent conveyance built on honesty, leaving results affecting society as a whole dependent on the rewards of a few that might or might not be allied to public benefit.

Given the facts of marketed demand facilitated by a consumer based economy, allowing conscientious usage at the individual scale is perhaps the most important, powerful as well as traction inducing vehicle for establishing sustainability.

Debatably, the culvert for conscientious usage would then be education, which would include not only the meaning of sustainability, but also the justification for sustainability, the patience required for the implementation, and acceptance of sustainability as a social norm of behavior.

The foundation for conscious consumption, on the other hand, is identified in understanding the basis of current consumption decisions and finally the virtues that form the behaviors that lead to noticeable economic results.

3.2.1. Significance of Educating for Sustainability

In the United States, consumer spending accounts for more than 65% of GDP, which has been the international measurement for economic growth ever since the 1940s.

Considering this connection and the correlating emphasis on GDP growth as a proxy for advancement, consumption decisions could have a substantial cascading effect across an economic system and also on the bounded global resource base.

Take into account the recycling of milk boxes. Wax-lined, printed paper milk crates have indeed been developed to carry and preserve milk from the point of production to the final consumption.

Even so, the carton's elements were not designed with waste disposal in mind; instead, the box was designed with increased distribution and marketing in mind. As a matter of fact, the milk cartons non-biodegradable as well as re-usable structure, which is primarily connected to the central grounds of its formation, it serves a utilization intent without regard to the environmental impact or possible future human and animal health.

larger consumption scale offers a straight-forward point of view for evaluating the core beliefs recorded in consumption behavior. From this vantage point, manufacturing for consumption can be described as a myopic activity that prioritizes the immediate satisfaction of a need or desire over the long-term influence or wider impact of the satiation.

> The morals ingrained and conveyed within buyers and sellers decide how a requirement or want is met. To a degree there is no debate on both the values and behavioral factors presumed and reflected in demand and supply, the morals and behavior patterns become inherent to the economic system.

As a consequence, clear, and specific awareness of current behavioral presumptions, such as consumers' unlimited wants, producers profit maximization motivating factors, and the subtle resource shortages caused by externalized costs, holds the great possibility to alter active and ingrained behavior.

A knowledge of economics that is particularly geared toward facilitating the development of rational economic operative behavior could indeed better understand the importance of consumption habits as it refers to sustainable growth in which the definition of sustainable development is consistent with the economic perspective of resource utilization delimited by intergenerational equity, which according to definition recognizes the requirements of the present and future relates to existing available resources (Figure 3.3).

Figure 3.3. This was a picture taken during an Education for sustainable development ESD workshop in Kasese district Uganda.

Source: Image by Wikimedia Commons.

The knowledge and understanding of resource scarcity in conjunction with space-time allotment, in turn, nurtures the planning and application of unconscious and conscious sustainability reassurance that are required components in driving a culture of sustainability.

3.2.2. Conscious Consumption and the Sustainability

Generally, the clear and unambiguous discussion of the engrained presumptions directing the decision behavior makers is not part of the economic education process.

As a result, the economic debate doesn't really encourage or place the evaluation of possible solutions to the extent that personal economic officials, production companies or customers of goods and services are constrained by reasoning which does not include addressing the effect of externalized or non-quantifiable costs.

The economic discussion implicitly and endogenously sustains and maintains a consumption-to-production circular flow, concentrating on the satisfaction of consumption and profit-taking from production while totally ignoring external costs and comprehensive dynamics.

Coming back to the milk carton example from the introduction, the economic discussion will be restricted to the utility gained from consuming the milk and the corresponding profit maximization of the production company.

Waste will be viewed as an externality instead of an endogenous element of the decision-making procedure. Furthermore, expenses are built into the product using effective market presumptions.

In general, a customer expects the sales price to reflect the total cost of the product, whereas producers see costs of production as being linked to market-priced inputs rather than environmental effects all through or as part of the product's life cycle.

In current practice, economic efficiency is measured by the use of resources to maximize production and consumption, rather than by the moral desirability of the physical processes and social institutions used to accomplish this goal.

An economic evaluation's factors are restricted to tangible quantitative costs, and costs are ignored in which either a market or regulatory oversight has not supplied a financial explanation. From this vantage point, the effect of consumption decisions on the environment, economic disparity, or the extinction of other species is unimportant.

The market system disenfranchises the customer from the wellbeing of those affected by his or her consumption and helps to promote the impression that cost on its own is reflective of a good's actual cost. The prospect that consumption must be lowered since it is not good for the heart and does not make people happy has had no part in the economic value system.

The basic belief is that customers are compelled to desire more. As a result, economic modeling tends to assume that current usage reductions are only discussed through the lens of future demand increases. The fact that the assumption of unquenchable lust can be taught and reinforced through a market model is not even acknowledged in the economy.

3.3. BIOREFINERIES SUPPLY TO SUSTAINABLE TEXTILES

MetsäFibre Ltd.'s latest bioproduct mill in änekoski is the largest forest industry asset in Finland, with an overall investment of approximately EUR 1.2 billion. It is encouraging companies of all sizes to join its "ecosystem" with each specializing in a distinct bioproduct or service at the same mill site. MetsäFibre Ltd.'s primary product will continue to be softwood pulp, but it will also generate heat and electricity.

The goal is to use all byproducts from the bioproduct mill as much as possible in the ecosystem, such as tall oil, turpentine, cosmetics various other eco-friendly chemicals and bio-composites.

Textiles are one of the potential product development lines and the business has declared a partnership with the Itochu Corporation, a trading firm that owns 24.9% of MetsäFibre Ltd. and serves as the company's pulp marketing agent (Figure 3.4).

Figure 3.4. The pulp mill of MetsäFibre in Kemi, Finland.

Source: Image by Wikimedia Commons.

The company has taken part in technical education introducing better uses for pulp, along with other huge forest industry sectors in Finland, such as the FuBio Future Bio-refinery programs in the Finnish Bioeconomy Cluster FIBIC and the Acel–Advanced Cellulose to Novel Products Program.

The Ion-cell technique has been introduced in several project phases, each organized by Aalto University and involving several universities,

R&D partners and industrial partners from the forest industry, the chemicals industry and the textile industry.

The Ioncell technique eliminates carbon disulfide, a toxic chemical used in the production of viscose with an ionized solvent. Other teams are developing similar technologies in collaboration with the same R&D partners as well as in parallel.

A number of technology start-ups (e.g., Spinnova Ltd., The Infinite Fiber Company Ltd.) and new biomill investment plans (e.g., KaiCellFibres Ltd.) are attempting to advance their technology to a pilot plant or even industrial scale production through international collaboration.

The Ion-cell technique is still being developed for solvent recovery during the procedure and yet MetsäFibre Ltd. has already presented its first pilot clothing to generate public and potential investor understanding of testing and scale-up production.

Innovative applications of advanced cellulosic materials are being developed at Aalto University through the collaborative efforts of two schools: chemical technology and arts, design and architecture. Industrial partners include clothing retailer H&M and furniture retailer IKEA (Figure 3.5).

Figure 3.5. New IKEA building, near to Greenwich, Great Britain.

Source: Image by Geograph.

The Ion-cell approach was first tested using dissolving pulp from Stora Enso Ltd.'s Enocell mill, and a prototype outfit was created for the Finnish design firm Marimekko. Stora Enso Ltd. currently sells dissolving pulp to China as a raw material for the manufacture of viscose fibers and thus for conventional fabric production methods.

3.4. WOOD-BASED SOLUTIONS FOR SUSTAINABLE BUILT ENVIRONMENT

In Finland, the use of wood in multistory building projects has been backed by government strategic programs, such as the development of regulations and standards (Ministry of Employment and the Economy 2015; Ministry of Economic Affairs and Employment 2017).

As a whole, the use of wood for multistory construction in Finland has risen in the last five years as evidenced by the number of construction projects and companies that produce engineered wood products like cross-laminated timber (CLT) and laminated veneer lumber (LVL), as well as prefabricated wood components and modules (Figure 3.6).

Figure 3.6. Cross laminated timber is studied in a laboratory at Oregon State University.

Source: Image by Flicker,

Many projects, which include technological development, pilot, and demonstration locations have produced solutions. The Wood City is being used as an example; the project began in 2011 in collaboration between the construction company SRV Ltd. and Stora Enso Ltd., the supplier of the massive wood construction method though construction did not begin until the end of 2016.

The reasons for the project's postponement are beyond the scope of this paper. However, the time frame illustrates the complexity of such projects.

Large wood product companies, like Stora Enso Ltd., perform in global markets and have established assistance to support product use such as building information prototype tools for architects and engineers, as well as region-specific requirements like acoustics and fire safety in major market sectors (Figure 3.7).

Figure 3.7. Laminated veneer lumber.

Source: Image by Wikimedia Commons.

Metsä Wood Ltd.'s Timber Academy, a virtual education tool for professionals and Stora Enso Ltd.'s open building system presented in 2016are two more examples. As a result, companies provide potential customers with information and tools to help them apply and use their goods.

Pilot projects are showcases that provide references for wood-based solutions; the use of wood competes with the more established solutions of concrete and steel construction, and the construction industry has been slow to embrace new technologies.

The Wood City area in Helsinki consists of business and hotel space as well as two residential buildings for ATT Ltd, a Helsinki-based housing developer. The Wood City partners describe numerous benefits depending on the supply of the examined materials.

The Wood City quarter is promoted with a central position, marine surroundings, quarter design (adding a sense of community and local facilities) and benefits from the wood material for human welfare. Wooden structures are associated with the concept of more effective performance for possible office tenants and with the concept of acoustic and esthetic attractiveness for hotel guests.

Moreover, the wood material's value propositions emphasize its ecological, sustainable, recyclable, regenerative and cost-competitive (quick building time) construction. Its contribution to emissions reduction is also highlighted.

Accoya is an exterior cladding material made in the Netherlands from Pinus Radiata, a wood species grown in places like New Zealand. Its advantages are also highlighted based on material qualities; technology based on wood acetylation makes tropical hardwoods a durable and dimensionally stable material, allowing for reduced maintenance expenses.

According to SRV Ltd., both wood and concrete are suitable materials when it comes to carbon footprints, but wood gives more adaptability to user demands; the spaces are easier to adjust afterwards. In partnership with a service design consultancy the Wood City space and service idea were created.

The goal was to provide a user experience for shared areas that would benefit the tenants companies. The proposal was included in the Helsinki Design Week program along with other service design projects implemented in the area.

3.5. ENVIRONMENTAL SUSTAINABILITY INDICATORS FOR THE BIOECONOMY

Bio-based products are manufactured from reusable forestry and agricultural feedstock such as biological waste streams and marine biomass. Conventional methods of biomass include food, feed, lumber, pulp & paper and electricity.

Presently, biomass may be used to substitute the use of fossil resources in a variety of different applications and solutions, including the manufacture of fuels, chemicals, advanced materials, polymers, composite fiber sand medicines. Basically, the utilization of renewable fuels has been advocated as an alternative to finite fossil fuels.

Furthermore, the bioeconomy is regarded as a means of reducing environmental consequences such as carbon intensity and global warming. As a result, various political actions to support the bioeconomy have been implemented, and the number of national bioeconomy strategies has grown.

The European Union (EU) has announced and endorsed a bioeconomy plan for Europe, and the Organization for Economic Cooperation and Development (OECD) has produced a policy agenda to promote and stimulate the transition to a bioeconomy (Figure 3.8).

Figure 3.8. Opening of the OECD global forum on development.

Source: Image by Flicker.

One crucial component of these plans and agendas is sustainability. There are also other sectoral accreditation schemes available for bioeconomy items such as biofuels.

Although the bioeconomy can be viewed as a solution to global environmental concerns and is a clear strategic objective it may also have negative consequences.

According to the World Economic Forum's Global Risk Report, the most serious known global environmental hazards include failures in climate change mitigation and adaptation, biodiversity loss and ecosystem collapse, plus worldwide food and water disasters.

The expanding bioeconomy business may be one option for global warming mitigation, but if not handled effectively, detrimental consequences for biodiversity or ecosystems might arise across the supply chain. Moreover, biomass production necessitates the use of land, water and fertilizers, creating a situation in which soil degradation and negative consequences for biodiversity and water are possible.

In addition, a greater volume of biomass is often designed to supply the same functional unit (e.g., energy content) as fossil fuels, implying that biomass requires more transportation and processing capability. As a result, GHG emissions from the whole value chain may be considerable in biobased value chains as well.

In order to address all these aspects and provide an overview on the environmental sustainability indicators of relevance for the bioeconomy and concretize these with an example, this study summarizes the main environmental concerns related to biobased value chains and presents an indicator framework which addresses these concerns in a holistic manner.

3.6. INNOVATION AND SUSTAINABLE DEVELOPMENT: A BIOECONOMIC PERSPECTIVE

Sustainability is a connection or balancing act between numerous continuously changing aspects (social, environmental, and economic realities and restrictions).

Sustaining a dynamic equilibrium between a growing population as well as its requirements, he changing capabilities of the physical environment to dissolve the waste materials of human activity; the changing options managed to open up by new knowledge and technological changes as well as the values, aspirations, and organizations that channel people's activity as a result, views of a sustainable future must naturally evolve in reaction to changes in any aspect of this dynamic interaction.

The bioeconomy or the rising cross-cutting economic sector that generates, transforms and consumes bio-based materials and products is at the heart of governments long-term economic goals.

The term "bioeconomy" refers to a technological transformation namely biotechnology that employs bio-resources. Competition in such a system would largely depend on bio-based goods and process technology developments.

3.6.1. Bioeconomy and Sustainable Development Goals

Developing the SDGs is associated with attaining a sustainable bioeconomy. Approximately half of the 17 SDGs are tied directly to the bioeconomy. SDG demands sustainable agriculture, particularly in the battle against hunger.

In the most recent version of the yearly UN hunger survey (The State of Food Insecurity in the World 2015-SOFI), nearly one in every nine people was diagnosed with chronic malnutrition. To satisfy the food demands of the world's expected population of about 9 billion people by the middle of this century, the bioeconomy must create a stable food supply system that meets the fundamental needs of the worldwide population.

To reduce severe poverty by 2030, as envisioned by SDG, a thorough understanding of rural poverty and the paths out of it is essential. In this light, the connection between agriculture and environmental issues must be appreciated.

Unsustainable agriculture methods are threatening the majority of the planet's limits today. The SDGs and the bioeconomy intersect in the field of biofuels. Considering all of the food system's problems, any large-scale deployment of biofuels might severely compete with food or other ecological services.

However, there may be non-competing uses for the bioeconomy, like SDG on accessibility to renewable energy (RE) sources that the bioeconomy might possibly research and participate in. Another area where bioeconomy could help is public health by providing important treatments for illnesses as well as healthy ecosystem services.

3.6.2. Technology and Bioeconomy: A Systems Perspective

Solutions to sustainable issues and difficulties would necessitate the use of scientific understanding and relevant technology. In this context, a much more completely integrated S&T system with a transdisciplinary emphasis will be preferable to the typical single-discipline focus of S&T.

To face the difficulties in reaching the complicated and closely interconnected SDGs, a broad mix of disciplines and skills must be considered in order to build a region-specific bioeconomic agenda that takes into account the local context, microclimate and micro-ecological circumstances. New technologies, like biotechnology, provide a plethora of opportunities for addressing development concerns such as access to safe drinking water, energy availability, adequate health care, food, and nutrition security, etc.

Distributed technology can support some development difficulties, such as providing safe water and clean electricity. The complicated and multidisciplinary character of emerging technologies, on the other hand would necessitate a significant reorientation and restructuring of R & D procedures.

This has significant ramifications for developing nations that are tied to a path of knowledge production, development and commercialization within an institutional framework that may become more irrelevant in terms of evolving technology requirements.

The necessity for institutional adjustments in creating S & T capabilities will be significant in this respect. In the preceding context, institutions are groups of shared behaviors, routines, practices, norms, or regulations that govern the connections and interactions of individuals or groups.

Because institutions influence interaction patterns, learning behaviors and information sharing to a considerable extent, their role in developing capacities is critical.

Despite the prevalence of technology, developing nations have significant obstacles to accessing safe drinking water, electricity and sustainable agricultural production systems. This suggests that the answer to these development difficulties is not exclusively dependent on technology but also to a considerable measure on the governance structure.

Elements that influence the link between knowledge, dissemination and results in agriculture, water and energy mostly center around resources, infrastructure, institutional strength and policy quality. Even if information is disseminated and services are accessible, the human factor including cultural and individual behavioral patterns plays a role.

3.7. CHALLENGES IN SUSTAINABLE BIOECONOMY

3.7.1. Lack of Standardization

In the existing bio-economies, there seems to be an obvious absence of standardization. That has a significant influence on the practicality of the solutions. Enhanced viability could be gained by economies of scale by modularizing some sections of the value chain, both in terms of technical and business idea modularization. This will also improve the dependability of the installations functioning.

3.7.2. Inward-Oriented Business Models

Most of those actors participating in the bioeconomy industry lacked an outward-oriented business plan, which is required for bio-economies to prosper. An outward-oriented business plan implies a business model that considers not just a specific firm and its revenue logic but also the other actors in the network.

3.7.3. Challenges Related to Authorities

The government can handle various issues with bio-economic solutions that the government can handle directly or indirectly. These are their names:

- Strict guidelines for materials that may be used in biofuel production.
- Stringent rules for the resources that can be utilized in biofuel production.
- Restrictions on the use of biofuel production by-products (e.g., for animal feed or fertilizing).
- The need for biofuel quotas at the national level.
- Tax breaks and other perks
- Making decisions based on life-cycle philosophy.
- Legislative changes and the ensuing uncertainties.
- A lack of permissions and regulations, for example, for new forms of manufacturing within the framework of the bioeconomy.

3.7.4. Challenges Related to Knowledge and Information Flows

Biomass producers who may be utilized as a raw material for biofuel production typically originate from nonenergy related companies.

Companies frequently lack the essential information on the nature of the biomass flow they have, such as peak loads, volumes, seasonal fluctuations in the flow, biomass quality and content and so on, because they do not require this data for their primary business or even trash handling.

This states that in order to be incorporated into a bioeconomic solution, biomass providers would need to spend money on new procedures and resources to assist in monitoring the necessary data. Moving forward, they may need to increase the quality and other biomass flow properties in order to integrate into or enhance the overall solution's functionality.

3.7.5. Related to Investments

Some bio-economic approaches seem to be appealing from a sustainability standpoint, but they do not generate quick and considerable revenues. As a result, investments in these kinds of solutions have a variety of characteristics:

- They protect the business unless environmental laws are tightened in the coming years to move towards sustainability.
- However, it is difficult to predict changes in energy, environmental and other related legislation and policy, which may also end up making certain components of the solution extremely difficult or impractical.
- The revenue generated by running bio-economic systems may be minimal but consistent. As a result, it is less enticing for large corporations or financial institutions to invest in and devote significant resources to such solutions.

According to the firms examined, in order to raise finance, an operational or production company with a variety of companies and institutions as shareholders should be formed. This would also solve the benefit and risk sharing issues as well as the basic business difficulties mentioned further.

3.8. CHALLENGES AND GAPS IN SUSTAINABILITY ASSESSMENT IN BIOECONOMY

Environmental sustainability evaluation of biomaterial value chain should include a suitable range of metrics to acquire a full and complete picture of the biomaterial value chain and include all important bioeconomy features and environmental consequences.

Moreover, systems for assessing sustainability should consider bio-product contributions to global environmental concerns. The outcomes of the sustainability assessment should also be simple to comprehend, meaningful, and indicate possible areas for improvement or "hot spots" in the value chain.

Furthermore, sustainability evaluation should be capable of providing related information for decision making. Several techniques and ongoing activities are attempting to determine the best answer for assessing environmental sustainability in the bioeconomy.

Numerous contemporary techniques use an LCA-approach that spans the whole value chain, but no unified or standardized methodology exists as of yet, despite the fact that many projects strive to standardize or harmonies a methodology.

Additionally, inconsistencies in scope, allocation methodology and effect evaluation approaches result in dissimilar outcomes. Cristobal et al., 2016 found that methodological standardization and more extensive LCA of bioeconomy value chains are urgently needed.

3.8.1. Biodiversity Finance is Part of the Broader Sustainable Development Finance (SDF) Challenge

Most of the other 2.7 billion people that live on less than $2 per day rely on biodiversity and healthy ecosystems directly. "The potentially catastrophic changes in biodiversity will also have severe ramifications for those living below the poverty line that depend on biodiversity for a disproportionately large percentage of their livelihood."

In fact, in Africa, wood fuel accounts for up to 70% of total energy use. Rural poverty and sustainable development alternatives are severely restricted by global yearly net forest loss on the order of 3.3 million hectares between 2010 and 2015. Despite these important contributions to sustainable development, biodiversity is chronically neglected.

The SDGs are linked to the Financing for Development (FFD) mechanism, which seeks financial resources to execute the 2030 Sustainable Development Agenda. The Addis Ababa Action Agenda (AAAA) serves as a road map for funding the SDGs. It highlights the importance of biodiversity and ecological protection as well as the abolition of illegal trading in species and natural goods (Figure 3.9).

Figure 3.9. 5th biennial high-level meeting of the development cooperation forum side event: Adaptation of the 2030 agenda for sustainable development.

Source: Image by Wikimedia Commons.

As per the Intergovernmental Group of Experts on Sustainable Development Finance (SDF), achieving the SDGs in developing countries will cost US $3.3–4.5 trillion per year, or nearly 5% of the world's current GDP. With around US $1.4 trillion in present commitments, an annual investment gap of US $1.9–3.1 trillion may be calculated with biodiversity requiring roughly 10% of the whole.

The SDGs are linked to a method called Financing for Development (FFD), which pursues financial resources to implement the 2030 SDGs. In the context of a total stock of global financial assets valued at more than US $200 trillion, the potential for narrowing this financing gap is within reach.

Despite the fact that there is no scarcity of cash around the globe, most governments do not spend enough on biodiversity or sustainable development. A significant shift towards new investment and fiscal policy paradigms that effectively include the economic and financial advantages of biodiversity and sustainable development is necessary.

Biodiversity finance is the activity of acquiring and managing cash as well as employing financial incentives to assist long-term biodiversity management. There is a great deal of interest and a great deal of want for biodiversity financing instruments to be used for corporate and public investments in biodiversity and ecosystem services.

Biodiversity financing and investments in biodiversity focused economic development policies integrate biodiversity conservation and ecosystem service management with long-term economic development strategy and objectives.

Finance and economics provide essential aspects for making a compelling commercial case for biodiversity initiatives. The number of accessible financial solutions is growing and the means in which resources are mobilized and used are becoming more diverse.

Approaches to blended financing that profit from cooperation across public, charitable and private players have grown prevalent. Green finance markets are growing in value, thanks in part to the creation of green bonds and increasingly inventive forms of venture capitalism and impact investment.

3.9. BENEFITS AND IMPACTS OF THE BIOECONOMY

The creation of a bioeconomy has the potential to impact a variety of crucial environmental, societal and economic issues while also creating new possibilities. The bioeconomy would have an impact on major sectors in the form of new development opportunities, although some hurdles must yet be solved in order to capitalize on this potential.

The bioeconomy has a significant impact on the energy sector in that the requirement to transition from fossil fuels to a more sustainable choice may be satisfied in part by creating RE from biomass.

The promotion of bioenergy has the potential to reduce toxic gases and the loss of non-renewable resources, thus minimizing industry's negative environmental consequences. At the same time the nation's relative independence from coal and oil has a positive impact on the security of supply.

The better combustion properties of biofuels over some fossil fuels contribute to their lower environmental impact. But, more importantly, biofuels generated in a well-managed bioeconomy have a substantially reduced life cycle effect. When a biofuel is made from local garbage and consumed locally, emissions in the raw material manufacturing, transportation and distribution stages are decreased.

Agriculture may resurrect as an enterprise inside a bioeconomy. Agriculture's demise has been a major issue in a lot of countries, including Finland. The growing value of bio-resources has the potential to boost agricultural expansion into a viable enterprise.

It would imply a more efficient use of the country's natural resources and rural development. The development of rural regions is especially vital for Finland whose territory is mostly made up of remote regions and woods.

Agriculture's environmental effects might be lessened if it becomes a component of the bioeconomy. The sector would become more sustainable if local organic fertilizers were used in place of synthetic ones.

Such organic fertilizers, which are a by-product of biofuel manufacturing offer several advantages over synthetic fertilizers. In relation to the environment, the use of organic fertilizers not only gives nutrients to the soil but also ensures their retention in the soil and the restoration of the humus layer.

Because synthetic fertilizers are unable to provide these properties, there is nutrient run-off and a thin humus layer. The Nordic nations are particularly affected as is the eutrophication of the Baltic Sea.

A further possible advantage of the bioeconomic is the proper utilization of ley agriculture. Lands that have been kept vacant for an extended length of time could be utilized to produce energy crops like clover which do not require soil fertilization.

This would enhance the efficiency of natural resource usage while minimizing environmental and biodiversity damage. A further area impacted by the bioeconomy is waste management. The industry has progressed from being primarily concerned with garbage collection to producing biomass.

This provides not only financial potential but also major environmental benefits: a considerable quantity of bio-waste is refined into a range of high-value products rather than just being left to decay. Thus, nitrogen run-off from garbage is partially addressed.

Another way to solve this issue within the context of the bioeconomy is to use common reed for biofuel production as a way to reduce eutrophication in the Baltic Sea. The bioeconomy is, however, a social development opportunity.

Sustainable regional solutions ability to create jobs is sometimes undervalued. As previously said, a healthy bioeconomy enables rural economic growth. The issue of unemployment is becoming increasingly difficult for industrialized countries as many bulk occupations are being outsourced to other countries in pursuit of a cheaper labor supply.

This has been a problem for a number of significant businesses in Finland such as shipbuilding and communication technology. In general, the solution to this dilemma is to focus on high-value services and knowledge exports rather than physical items.

Similarly, during its growth, a bioeconomy would generate a large number of employment solutions and the knowledge gained would become a significant export commodity.

The local character of bio-economic solutions opens up the possibility of generating local goods such as fuels, food and materials, the safety and quality of which benefits Finnish society. The economic implication is that the bioeconomy promotes local producers.

On a national level, the economic situation may be stabilized and improved for a variety of reasons, including: energy efficiency throughout the country is a good opportunity to save money; and the export of ideas and expert knowledge in emerging bioeconomic solutions could become a new high-value product, continuing to improve the GDP and the nation's reputation in the world.

The bioeconomy will entail new goods, but more importantly, new services and skills that could be sold to the corporate sector. Businesses that participate in the bioeconomy benefit from company growth.

Because of the relevance of bio-economic solutions to the country's economy and general sustainability, further funding might be sought. Furthermore, resource efficiency, which is a need for the bioeconomy, is a direct chance to lower manufacturing costs for each individual firm or commercial unit.

3.10. CONSTRUCTING A SUSTAINABLE BIOECONOMY: MULTI-SCALAR PERCEPTIONS OF SUSTAINABILITY

3.10.1. Introduction: Global Transitions to a Sustainable Bioeconomy

Bioenergy, which refers to the use of renewable biomass to generate energy for electricity, heat and transportation fuel, has the potential to solve concerns about the impact of fossil fuel production and GHG emissions on global warming and world energy safety.

Considering these advantages, most governments at both the national and sub-national levels have enacted laws to encourage the use and production of bioenergy. Because some nations are unable to meet aggressive bioenergy objectives established by domestic government programs, international trade markets are emerging.

Furthermore, growing nations cheap biomass production might supply the worldwide bioenergy based transportation fuel market. Because bioenergy is encouraged to minimize reliance on unsustainable fossil fuels, sustainability is fundamental to its usage and supply. As a result, to optimize advantages, all bioenergy generation should be sustainable. Therefore, the generation of bioenergy may not always be sustainable.

According to studies, implementing forest-based sustainability certification programs is tough, especially in developing countries, due to factors such as a lack of demand for certified products, a disparity between existing and more sustainable management styles, poor policy compliance by corporate entities and a total absence of execution functionality.

Nevertheless, worldwide bioenergy commerce has generated a demand for sustainability certification agencies that analyze the sustainability of bioenergy systems in order to prevent unsustainable production technologies from being implemented, particularly in bioenergy exporting poor nations.

At times, bioenergy policy papers include directions to guarantee that sustainability figures prominently in policy objectives such as avoiding the use of food crops for biodiesel, generating rural jobs and reducing GHG emissions. Separately, several government bodies give feedstock harvesting procedure instructions.

There are various worldwide third-party certification organizations that check the sustainability of bioenergy production techniques. A few of these sustainability certification systems are targeted at certain countries or products.

To evaluate the sustainability of a bioenergy project, many certification processes incorporate environmental, economic and (to a lesser degree) social indices.

In this section, the authors have used a multi-scalar approach to investigate two prominent research studies of bioenergy development in an attempt to comprehend how this adds to the sustainability of the developing bioeconomy while taking into account the difficulties of developing a stable bioeconomy and assessing how different factors are taken into account in national policies and international sustainability certification programs.

In particular, researchers wonder what it means for such biofuel projects to be sustainable from many perspectives operating at various social scales from local community members through state and national policy implementation to sustainability criteria in international certification programs. Researchers first investigate how local people view the social sustainability consequences of regional bioenergy projects and then further analyze the roles of bioenergy policy in the creation and success of bioenergy transition initiatives.

Finally, we investigate how international bioenergy certification systems affect our examples with a particular focus on the social sustainability of bioenergy projects. Looking at these examples from a multi-scalar perspective can teach a lot. Analyzing both regulations and local community people's perspectives of what sustainability entails in the context of bioenergy development initiatives.

3.10.2. Understanding Sustainability: A Multi-Scale Examination

Complex systems are required for the creation of new biofuels. The carbon neutrality of bioenergy is determined by the type of feedstock utilized, how it is cultivated, the land-use changes connected with feedstock cultivation, and the techniques of bioenergy conversion and production.

There are serious issues involved with fuel sources that have conflicting uses, like food crops; sustainability necessitates a comprehensive, multi-scalar analysis. Because bioenergy initiatives are rooted in local communities, social sustainability should be considered in addition to ecological sustainability (Figure 3.10).

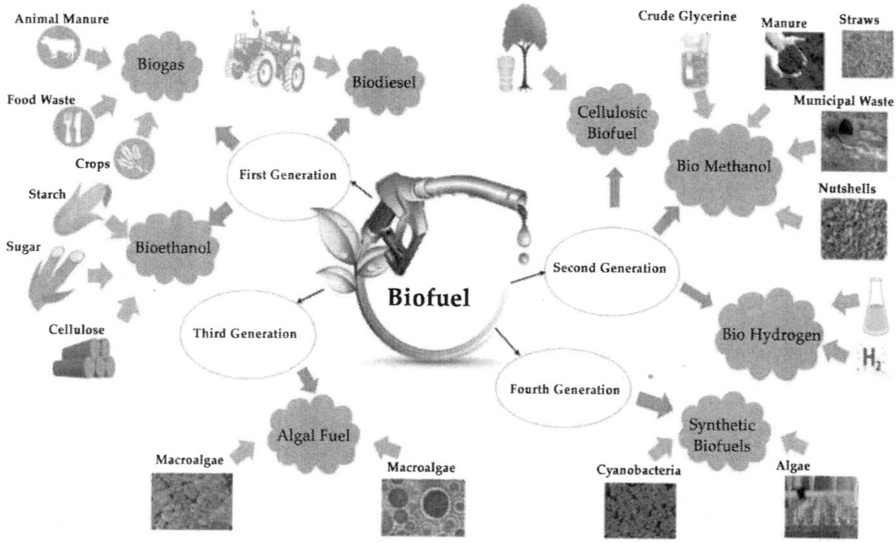

Figure 3.10. Types and generation of biofuels.

Source: Image by Wikimedia Commons.

Negative externalities of the projects may seep into local communities and have an influence on community-level socioeconomic situations. Furthermore, social ramifications are not always obvious and a better knowledge of sustainability from many social dimensions within the same bioenergy development case studies can assist in discovering the intricacies.

One method of analyzing a bioeconomy's social sustainability is to analyze how local residents view the socioeconomic consequences of bioenergy in their communities. Any new energy project in a community creates new material realities in the areas in which it is located, and bioenergy operations are no exception.

Aside from the biomass processing center, a new bioenergy project develops or improves current methods of biomass extraction, storage and transportation. All of these initiatives that help a bioeconomy expose individuals in adjacent areas to new hazards and possibilities.

However, how individuals perceive local bioeconomy activities as threats or possibilities is determined by their worldview. While based on a constructivist perspective because it demands considering constructions of reality seriously, paying close attention to disparities in public beliefs and expectations might help to know the true triumphs and shortcomings in planning for a socially sustainable bioeconomy.

Furthermore, public interest in or resistance to new technological initiatives in a community can either facilitate or deter the spread of new technologies. Understanding how individuals view bioenergy projects may therefore assist in preventing conflict and guaranteeing smooth day-to-day functioning of the projects as well as other supply chain management phases crucial to the projects.

Furthermore, advancing towards sustainability needs public support since sustainability concepts frequently contradict current financial systems of production and consumption, necessitating mass support for sustainability programs to thrive in the long term.

Understanding how consumers view programs aims to allow transformations to sustainable bioenergy systems may thus guide future policy processes supporting sustainability.

Government policies have a significant impact on the transition to new RE systems and can ultimately influence the extent to which bioenergy production is sustainable. Many nations have policies in place to encourage national bioenergy development.

However, not every country with a bioenergy policy has had comparable success in building a viable bioenergy economy. Because building a bioeconomy necessitates the engagement of a large number of market investors, suppliers and developers who are subject to various risks, government support for the sector is crucial to promoting their participation.

Local, regional, and national laws can also compel compliance with various sustainability initiatives, therefore aiding in the establishment of a sustainable bioeconomy.

International commerce in bioenergy markets, ranging from forest-based bioenergy products to oil seeds, is expanding. Largescale international commerce, on the other hand, can only arise from largescale bioenergy production, which can have an impact on the social, economic, cultural and ecological sustainability of the producing areas.

As a result, multiple accreditation bodies have formed, each with their own set of metrics for quantifying the sustainability implications of bioenergy projects, Programs for certifying sustainability.

3.11. ECOLOGICAL LIMITS TO SUSTAINABLE USE OF WOOD FUELS

The decreasing oil reserves, imperatives on accessible wellsprings of wind and hydro energy, and CO_2 outflows of carbon let out of fossil energy transporters, adding to the climb in worldwide temperature, have provoked corporate players to propel a bioeconomy plan, advancing the quest for sustainable substitutes for diminishing materials and energy assets to help further monetary development and movement. In any case, the ecological and social costs imposed by growing financial disparity, which have been identified as the two primary drivers of the mismatch between monetary and personal satisfaction, highlight a flaw in the ebb and flow framework. A new industrial revolution is thought to fix old and current issues by exploring the potential of changed economic relationships to materials and finite resources raised by an innovated economic system based on principles of protecting natural capital, optimizing yields from consumed resources, and eliminating externalities caused by pollution to foster effectiveness, whereas maximizing the productivity of assets or resources is viewed as a value driver. However, the criticisms leveled at the bioeconomy agenda for being little more than an extension of the corporate-driven market system are not resolved in the circular economy vision, which is clearly focused on benefitting from creative technology advances. Without addressing the fundamental economic questions of the purpose of production and the distribution of wealth produced, not to mention the limits to growth (Figure 3.11).

Figure 3.11. Hydropower plant: Sir Adam Beck 1 & 2 generate 1.6GW of electricity for Ontario.

Source: Image by Wikimedia Commons.

Within the EU research framework for a knowledge based bioeconomy, two competing visions of the economic status of a bio-based future knowledge based society have emerged, one of which is focused on nature's adjustment through genetic engineering and the other on agroecological engineering to design agricultural systems sustained by minor inputs regulating and increasing productivity and crop protection.

Wood and other photosynthetic products are accepted as renewable and CO_2-neutral sources of energy, despite significant concerns, visible conflicts and proof of rising CO_2 levels in the atmosphere since wood fuel overtook coal in thermoelectric energy plants.

Despite previous arguments and general information that CO_2 is emitted when burning any carbon containing fuel, CO_2 emissions from burning wood and other biomass are still ignored and not taken into consideration.

Because carbon accounts for around 50% of wood and other plant biomass, each kilogram of wood burned releases 0.5 kg of carbon that is not instantly stored from the environment into fresh biofuels by photosynthesis.

For this reason, extensive manufacturing usage of bio-energy has been chastised for disregarding carbon footprints by fuel burning and the point at which a renewable resource becomes non-renewable, for which purpose it is advocated that it be omitted from the concept of RE.

Certain writers propose compensating for the carbon generated when burning wood using "carbon debt," "payback time" and "carbon balance."

3.12. CONCLUSION

As a knowledge based bioeconomics, sustainable use of limited resources should be accomplished through incorporating knowledge of the application of biological resources and processes.

The term "bioeconomics" refers to all economic sectors engaged in the production, processing and use of biological resources (plants, animals, microorganisms) for the production of food and nourishment, the assistance of biofuels as an asset and also the production of bio-based chemicals and products as well as bioenergy.

REFERENCES

1. Anand, M., (2016). *Innovation and Sustainable Development: A Bioeconomic Perspective.* [online] Sustainabledevelopment.un.org. Available at: https://sustainabledevelopment.un.org/content/documents/982044_Anand_Innovation%20and%20Sustainable%20Development_A%20Bioeconomic%20Perspective.pdf (accessed on 21 September 2022).
2. Biosc.de. (n.d). *Sustainable Bioeconomy*. [online] Available at: https://www.biosc.de/sustainable_bioeconomy_en (accessed on 21 September 2022).
3. Filho, W., (2018). Towards a sustainable bioeconomy: Principles, challenges and perspectives. *World Sustainability Series*, [online] Available at: https://link.springer.com/book/10.1007/978-3-319-73028-8 (accessed on 21 September 2022).
4. Gustafsson, M., Stoor, R., & Tsvetkova, A., (n.d). *Sustainable Bioeconomy: Potential, Challenges and Opportunities in Finland.* [online] ResearchGate. Available at: https://www.researchgate.net/project/TEKES-RECO-Designing-sustainable-ecosystems (accessed on 21 September 2022).
5. Leal, F. W., Pociovălişteanu, D., De Brito, P., & Borges De, L. I., (2018). Erratum to: Towards a sustainable bioeconomy: Principles, challenges and perspectives. *World Sustainability Series*, [online] Available at: https://link.springer.com/chapter/10.1007/978-3-319-73028-8_30 (accessed on 21 September 2022).
6. Lopes, M., (2015). Engineering biological systems toward a sustainable bioeconomy. *Journal of Industrial Microbiology and Biotechnology, 42*(6), 813–838 [online]. Available at: https://academic.oup.com/jimb/article/42/6/813/5995412?login=false (accessed on 21 September 2022).

CHAPTER 4

SOCIO-BIOLOGY AND BIOECONOMICS

CONTENTS

4.1. Introduction 92

4.2. The Evolution of Socio-Biology 94

4.3. Socio-Biology and The Emotions 99

4.4. Socio-Biology Theory 101

4.5. Socio-Biology, Theory of Evolution And Biceconomics 106

4.6. Conclusion 118

References 119

Socio-biology is the methodical investigation of the biological underpinnings of social behavior. The American biologist Edward O. Wilson popularized the term socio-biology in his book Socio-biology: The New Synthesis (1975).

By examining animal (and human) social behavior in the context of natural selection and other biological processes, socio-biology aims to comprehend and explain it. One of its key ideas is that animals will act in ways that increase their chances of passing along copies of their genes to following generations.

4.1. INTRODUCTION

This idea holds that genes (and their transmission through successful reproduction) are the primary motivators in animals struggle for survival. Given that behavioral patterns are partially inherited, it may be claimed that the evolutionary process of natural selection promotes the behavioral and physical characteristics that boost an individual's chances of success.

Numerous new perspectives on animal social behavior have been gained because to socio-biology. Since such behaviors typically favor closely related individuals whose genes resemble those of the altruistic individual, it explains how what appears to be altruistic behavior in some animal species is actually genetically selfish.

This understanding contributes to the understanding of why soldier ants give their life to protect their colony or why worker honeybees in a hive forego reproduction in favor of aiding the queen in her reproduction.

The varied methods that the sexes must employ in order to pass on their genes to future generations can often be explained by socio-biology as the cause of the disparities between male and female behavior in some animal species.

However, socio-biology is more contentious when it tries to explain various human social behaviors in terms of their worth as reproductive strategies for adaptation. According to one argument, many of these behaviors are more credibly seen as cultural artifacts or as evolutionary by-products without any overt adaptive function.

Wilson in particular has come under fire for allegedly attaching adaptive value to a variety of pervasive but morally repugnant behaviors (such as sexism and racism) in order to justify or explain them as inevitable or natural (Figure 4.1).

Figure 4.1. Social behavior of lion cubs. Amboseli National Park, Kenya.

Source: Image by Flickr.

Defenders of socio-biology respond that socio-biology does not imply strict biological determinism and that at least some aspects of human behavior must be biologically influenced (because competition with other species would select for this trait).

They also argue that evolutionary explanations of human behavior are not inherently flawed but should be evaluated similarly to other scientific hypotheses.

Toward the end of the 20th century, a brand new integrative discipline called "socio-biology" was born. It changed the field of research into animal social behavior into one that adheres to Neo-Darwinian theory, which sees evolution as a shift in gene frequencies in populations.

Because of this, it was feasible to create mathematical theories that could be verified by actual research, transforming conventional natural history into a more accurate science.

However, "socio-biology" did not take hold as the preferred term among working biologists. Many researchers engaged in what would have been considered legitimate sociobiological research consciously avoided the term, referring to themselves instead as behavioral ecologists or functional ethologists.

This was largely due to the controversy surrounding Harvard zoologist Edward O. Wilson's (1975) book Socio-biology: The New Synthesis, which gave socio-biology political overtones.

Particularly sensitive individuals worked in what might be referred to as "human socio-biology," preferring to be referred to as biosocial anthropologists, human ethologists, human behavioral ecologists, or Darwinian anthropologists. Many sociobiological studies adopted the new moniker "evolutionary psychology" starting in the 1990s.

As it turned out, actual "sociobiologists" did not adopt Wilson's specific, vaguely conceived sociobiological research program that he described in his book (although they did not call themselves that). Instead, The Selfish Gene by Richard Dawkins offered a relatively modest but cohesive group of theoretical findings that made up the actual sociobiological research paradigm.

4.2. THE EVOLUTION OF SOCIO-BIOLOGY

When one organism sacrifices itself for the benefit of the group, as is the case with ants, bees and wasps (Hymenoptera), Darwin is particularly captivated. Such behavior seems at first glance to be at variance with the kinds of egotistical actions that would result in personal triumph in the struggle for existence.

The domestic world had demonstrated how one can select vicariously, as it were, for traits in animals that will not reproduce, so Darwin was able to understand how the sterility of a worker ant, for example, might be transmitted through fertile nest members, but he was unable to understand how sterility itself would come into being.

Darwin was certain that all selection must be made for the individual, not the group, and that sociality—in particular, worker sterility posed a significant problem. Darwin came to the conclusion that a colony (of ants, for example) might be thought of as a form of superorganism on which selection can act generally, but he never fully addressed the issue of sociality.

The study of the evolution of social behavior trailed behind Darwin for a variety of reasons. First off, the emergence of the social sciences and their focus on behavior deterred biologists from tackling the issue.

Social scientists frequently conducted studies using rats and mice, made generalizations and then came to the conclusion that interspecies differences were unimportant. In addition, social scientists frequently worked in artificial settings, which made them less interested in natural behavior and less able to spot it when it did (Figure 4.2).

Figure 4.2. A picture of Hymenoptera.

Source: Image by Wikimedia Commons.

Second, during the first half of the 20th century, the Third Reich's racial theories persuaded many people that studying social behavior from a biological perspective would result in assertions about the innate tendencies of people with the ensuing devaluation of those outside one's own group.

were required in the social sciences so that evolutionists could dissect nature and extract its hidden Even while some argued that such worries shouldn't cast a shadow over all biological studies of behavior, the damage had already been done and persisted for many years after World War II.

Most crucially, no one truly understood how to develop Darwin's theory in a way that allowed social scientists to analyze social behavior while adhering to the laws of natural selection. New methods that avoided group selection truths.

- **Kin Selection:** There were advancements in the 1960s. In order to examine animal social behavior while maintaining the individualist or selfish aspect of selection, a number of models were developed.

The idea of kin selection, developed by the English biologist William Hamilton, is noteworthy because it demonstrated that close relatives have a biological incentive to assist one another because doing so indirectly promotes the survival of their own genetic inheritance or genes.

Hamilton used the ants, bees and wasps as examples of this way of thinking, pointing out that these creatures had an odd breeding system where only females have dads (males being born from unfertilized eggs).

Sisters are therefore more closely linked to one another than is typical. Sisters and mothers are half related as are daughters in the typical situation (e.g., humans). In the hymenopteran scenario, a female receives the same genetic input from her sister's father and then receives 50% of that same genetic input from her mother.

Sisters are therefore 75% linked to one another but moms and daughters are only 50% related. Raising fertile sisters rather than fertile daughters is in the worker's reproductive best interests; this activity is made easier rather than more difficult by the worker's own infertility.

The colony does not need to be viewed as a single entity because this is the most integrated and harmonious social setting in which one may observe individual interests being acted out.

- **Reciprocal Altruism**: Other models were developed, one of which Darwin himself sensed but did not completely express. According to the idea of reciprocal altruism, an organism has a right to receive assistance when required.

Even non-relatives or in the most extreme case, different species can benefit from reciprocal altruism. Although certain fish are significant predators, they allow other fish that swim right into their mouths and remove pathogenic fungi and germs from their gums.

Because the larger fish does not ingest the smaller fish in its mouth, the cleaners enjoy a decent meal while the predators exercise oral cleanliness.

- **Evolutionary Equilibrium**: In situations where players employ different strategies to succeed in light of the fact that other players (in biological terms, other members of the species) are also seeking to succeed, evolutionists have turned to game theory.

British biologist Richard Dawkins demonstrated how certain evolutionary situations achieve equilibrium or reveal what he called "Evolutionary Stable Strategies" when no one member of the group can achieve more than limited benefits due to the conflicting interests of the group in The Selfish Gene (1976), a provocative popularization of this theory.

Consider a group that consists of two different types of people to use one of Dawkins examples. Some of the group's members are hawks, who are combative and willing to fight in any scenario where there is a chance of conflict. Other members of the group are doves, who always flee whenever a battle appears imminent (Figure 4.3).

Figure 4.3. The changeable hawk-eagle or crested hawk-eagle (Nisaetus cirrhatus).

Source: Image by Wikimedia Commons.

One may predict that hawks would take the lead and that natural selection would result in a population devoid of doves. However, this is false. Every time a hawk comes into contact with another hawk, a conflict ensues, which result in one could bird being hurt or killed.

However, because they fly, doves are never hurt. So, on average, being a hawk comes with a price. Doves, however, are unable to rule either since they typically incur expenses. Hawks always prevail when hawks and doves square off.

As a result, neither the hawks nor the doves are able to increase their representation at the expense of the other, and the group of birds finds itself in an unpleasant but balanced middle position.

Edward O. Wilson.: The study of the evolution of social behavior, which is today known as sociobiology moved forward quickly and enthusiastically and soon it was prepared to join paleontology and the other fields in the Darwinian family and take its rightful place. But a dispute lingered (Figure 4.4).

Figure 4.4. Group of doves.

Source: Image by Wikimedia Commons.

In The Descent of Man (1874), Darwin fulfilled his desire to apply his theories to people as did other Darwinian scientists who followed him, most notably Harvard entomologist Edward O. Wilson.

Wilson argued that nearly every aspect of human life and nature is a function of biology or more precisely, the genes as shaped by natural selection, in a seminal overview of the field, Sociobiology: The New Synthesis (1975) and later in a work addressing the human species on Human Nature (1978).

Natural selection at work on the units of heredity has led to changes in sexual preferences, family structures, religion, conflict, language and many other things.

Even homosexuality may have a biological basis because gay and lesbian family members frequently provide care for their immediate family members, just like sterile mammals do at their nests. In addition, Wilson contended that while humans might be able to alter some aspects of biology, in many ways, people are stuck with who they are. Plans for transformation that are too idealistic will not work.

4.2.1. Early Objections to Socio-Biology

As would be expected, the emerging discipline of socio-biology was met with numerous criticisms. Because they believed that biologists were encroaching on their territory, social scientists became tense.

They feared sociology would disappear and be replaced by socio-biology (social-group divide), rather than seeing biology as a supplement or help to social science. Feminists detested what they saw as a direct assault on their philosophy, which maintained that sexual preferences and family arrangements were entirely cultural rather than biological inventions.

Socio-biology was seen as a justification for the status quo that oppresses women and children and Darwin was portrayed as the prototypical Victorian male chauvinist. Marxists believed that a biological approach was a mockery of the truth because it claimed that natural selection and evolution had achieved what was actually a function and outcome of economic deprivation.

This view was shared by some renowned biologists. Human socio-biology was the worst possible example of a reductionist approach to knowledge, according to their ideological forefather Friedrich Engels.

4.3. SOCIO-BIOLOGY AND THE EMOTIONS

In the 1970s and 1980s, socio-biology offered a fresh perspective on the development of emotion. Additionally, it shifted the focus of the research from the fundamental emotions to the moral and quasimoral emotions involved in interpersonal relationships.

Trust, loyalty, guilt and shame are only a few examples of the emotions that play a clear role in regulating the competitive social interactions that were the main topic of the majority of human socio-biology study.

Many sociobiologists briefly said that moral sentiments must have developed as psychological processes to carry out evolutionary stable social interaction methods (Weinrich, 1980).

According to Robert A. Frank (1988), the moral emotions developed as a response to "commitment issues." When the winning tactic in an evolutionary interaction entail making a legally binding but conditional promise to do something that would go against one's own interests if the condition were ever fulfilled, a commitment problem occurs.

In order for such a commitment to be believed, a unique mechanism that would lead the organism to act against its own interests is required.

According to Frank, feelings of fury and vengeance developed so that creatures might participate in credible deterrence by threatening to engage in self-destructive aggression in order to dissuade a more powerful aggressor.

In contrast, feelings like love and guilt evolved to enable creatures to practice reciprocal altruism in circumstances when there is no chance of reprisal if one partner does not reciprocate.

Ethology was condemned by sociobiologists as being nothing more than descriptive natural history since it lacked a theoretical framework to predict how humans will act (Barash, 1979; Barkow, 1979).

Comparatively speaking, socio-biology appeared to produce firm predictions that ran counter to some elements of the affect program hypothesis of fundamental emotions. It makes no sense from a sociobiological standpoint for creatures to exhibit involuntary expressive behavior that freely divulges details about the animal's motivational state to anyone who is interested in learning about it.

When evolutionary game theory is applied to emotional conduct, it is predicted that this behavior will be created to subvert other species expectations rather than to openly display emotional feelings. Some have used this theoretical defense to oppose the affect program theory of fundamental emotions.

According to Alan Fridlund (1994), emotional conduct should be viewed as a paralanguage of social signals whose production is dependent on an organism's social setting at least as much as it is on its emotional state.

The generation of the basic emotions has been shown to be influenced by "audience effects," according to Fridlund and other paralanguage theorists, who claim that this is incompatible with the affect program hypothesis.

Russell and Fernández-Dols (1997) also attempted to demonstrate that people tend to attribute emotions to others more based on context than on behavior and that the traditional facial manifestations of emotion do not always accurately reflect the underlying emotional state.

However, whether or not this information goes against the influence program theory's predictions is unknown. In many creatures, including domestic hens, signaling behavior demonstrates audience effects.

This activity is likely part of a highly stereotyped action sequence that is managed by a relatively basic cognitive mechanism.

According to Mark Hauser, Fridlund's views are relevant to issues regarding the biological basis of emotional conduct, whereas the affect program model is focused on the mechanisms that give rise to this behavior (Hauser, 1996).

However, there are other instances where Fridlund does seem to be talking about the underlying emotional processes themselves rather than just their biological purpose. According to Fridlund, a specific emotion defined behaviorally need not be the same as any specific motivational state (feeling) of the organism creating it, as emotions are signals produced due to the impact they will have on other organisms.

4.4. SOCIO-BIOLOGY THEORY

The scientific study of how natural selection changes the biological underpinnings of all social behavior is known as sociobiology (Wilson, 1975). The biological pressure to transmit genetic inheritance as broadly as possible is one biological imperative that can be used to explain patterns of human social behavior.

Comparing mental mechanisms to genes as the evolutionary determinant of adaptiveness, sociobiology sets itself apart from evolutionary psychology. According to sociobiology, social behaviors as a whole are products of Darwinian evolution and both human and animal social behavior is influenced by how genes and culture have co-evolved (Lumsden, 2011).

An important theory of theoretical and empirical methods to the study of animal social behavior from an evolutionary perspective emerged in the early 1970s. Similar to how they studied physical features like skin tone and eye color, scientists wanted to comprehend social behavior.

Others, however, believed that applying an evolutionary approach to human behavior was both politically incorrect and faulty scientifically (Segerstrale, 2015). Sociobiologists believe that the testing of hypotheses can shed light on social phenomena such as mating preferences, offspring sex ratios, warning calls, how parents treat their young, gregariousness, territorial defense and more.

The assumption that the social characteristics under consideration had been shaped by a history of Darwinian selection that promoted the traits that would have made the animals ancestors the fittest to reproduce could be used to explain this uptake in specific traits. This is because sociobiologists are also selectionists.

Scientists started to think of organisms as evolved reproductive strategists as a result of William's work, which means that the traits that organisms possess can be interpreted as strategies for reproductive competitiveness against others of their own kind. Plants and non-conscious species like animals both have reproductive strategies that are made up of specific reproduction methods.

4.4.1. Inclusive Fitness

William Hamilton's concept of inclusive fitness or kin selection was one that had a considerable influence on the early history of socio-biology (1964). No matter whether these genes are in the organism's direct descendants or other relatives, Hamilton hypothesized that selection will favor any phenotype or outward attribute, that appears to be reflective of the organisms own genes.

For instance, sterile worker ants brood care and colony upkeep may be chosen for favor if these behaviors encourage the reproduction of a queen who is closely connected to the workers. Hamilton claims that selection increases the number of alleles in all of the organism's relatives, not just his direct offspring.

Organisms are hence nepotistic (Alexander, 1979). Several human behaviors, including sex roles, violence, generosity and even moral and religious beliefs, according to Wilson, may have a biological basis (Wilson, 1975).

Wilson used comparisons to the behavior of other primates and cited earlier studies on a few traits from twin studies and human behavioral genetics to back up his claim.

This was unmistakably evidence in favor of the critical school's biologically deterministic theory of human nature, which claimed that since social inequity was in our genes, social changes would be useless (Segerstrale, 2015).

The criticism of socio-biology rapidly reached the level of a letter of condemnation signed by several academics from the Boston region and the avoidance of researchers using this word to describe their study.

Robert Trivers developed this idea further in his articles on parental investment, conflict between parents and their children and reciprocal altruism (1972).

4.4.2. The Problem of Altruism

The so-called "problem of altruism" is one of Socio-biology's main areas of interest. The topic of how prosocial behavior, such as altruism may have developed as a restraint in aggressive conflict or to promote active cooperation is raised by the problem of altruism.

Sociobiologists describe altruism as behavior that, on average, results in the actor's reproductive success declining while the reproductive success of another person directly increasing (Wright, 2015).

This definition disregards the desire to assist others, in contrast to the notion of altruism found in daily language. Darwin's theory of evolution initially makes it seem as though altruistic conduct will always be punished by selection; nonetheless, many animals engage in altruistic behaviors, such as caring for others young or performing a risky or aggressive action to warn others of danger.

Hamilton's inclusive fitness theory is the most widely referenced solution to the problem of altruism (Wright, 2015).

According to Hamilton's inclusive fitness hypothesis, if the beneficiary of an act of kindness is related to the actor and if the relative's gain outweighs the actor's sacrifice, the likelihood that the actor's genes will be handed on is increased, and altruism may become more prevalent.

In particular, Hamilton proposed the formula, commonly referred to as "Hamilton's rule," that an altruistic tendency will increase in prevalence under selection if $r * b > c$, where 'c' is the cost to expected direct fitness that the actor must incur, 'b' is the benefit to direct fitness by the beneficiary of the act and 'r' is the coefficient of relatedness between the actor and beneficiary; the extent to which the actor and beneficiary have the same genes (Wright, 2015).

A significant portion of sociobiological research examines whether altruistic behavior can be explained by Hamilton's rule.

Animals typically react differently depending on how closely related to their neighbors or other interact ants they are. For instance, depending on how closely related the caller is to the others nearby, colonial ground squirrels and prairie dogs choose whether to shout out warnings about predators.

Similar to humans, adult animals typically exclusively take care of their own young and those that assist in taking care of the young of others frequently judge the young based on how closely related they are to themselves (Daly and Wilson, 1988; Griffin and West, 2003). Other studies have concentrated on how animals can identify relatives when they exhibit nepotistic behavior (Krupp et al., 2011).

Some instances of this discrimination are based on genotypical traits like odor, whereas in others, it is just a pattern of behavior that continues until they are more inclined to walk away from kin.

When a bird flies away from its colony after reaching adulthood, for instance, they may cooperate up to that time because they are significantly less likely to run into others they are closely related to (Wright, 2015).

The issue of altruism that is unrelated to relatedness has also been addressed by theoretical sociobiologists. These theorists frequently employ game theory simulations. According to these ideas, animals are always exchanging favors for ones that have been promised in the future.

But despite efforts to undermine it, Hamilton's thesis is still crucial to explanations of social evolution (Gardner et al., 2011; Queller, 2011; Wright, 2015).

4.4.3. Sexual Selection and Sexual Conflict

Sociobiological considerations about sexual selection center on issues such how and why males and females differ in species where there are several separate sexes and why organisms reproduce with two parents rather than just one.

In his 1859 theory of evolution, Charles Darwin (1871) postulated the existence of a direct evolutionary process known as sexual selection. Sexual selection explains why some members of a species have more access to mates and are more likely to reproduce with partners than other members of the same species.

Darwin compared this to natural selection, which explains why certain animals may mate more easily than others because of their success in navigating challenges like getting food and avoiding predators.

The key difference between sexual selection and natural selection is that sexual selection can promote the evolution of characteristics that would disadvantage animals under natural selection (by, for example, raising their danger of predation), such as a peacock's colorful tail (Wright, 2015) (Figure 4.5).

Figure 4.5. A picture of peacock's colorful tail.

Source: Image by Wikimedia Commons.

The concept of sexual selection received little attention in academia before socio-biology. Williams (1966) and Trivers (1972) proposed that the degree to which sexual selection occurs differently in males and females, producing sex differences in anatomy, physiology, psychology, and behavior was determined by the degree to which one sex's reproductive efforts were a limiting resource. This revived the idea as a focus of socio-biology.

The sex that puts in less work competes for the chance to mate with the sex that puts in more effort during reproduction, such as raising children (Wright, 2015).

4.5. SOCIO-BIOLOGY, THEORY OF EVOLUTION AND BIOECONOMICS

A unified theory of social interactions between humans and other animals is formed by sociobiology. The maximization of genetic fitness by the individual is its guiding concept, uniting human and animal civilizations. A person's core "teleology" in society is the maximization of genetic fitness.

Sociobiology applies Darwinian evolutionary theory to the field of social and economic sciences as an application of the Neo-Darwinian Synthesis of evolution theory. Since individuals and species compete for the same limited resources in nature to maximize their inclusive fitness and the survival of their progeny, sociobiology becomes bioeconomic or the study of the economy of nature (Figure 4.6).

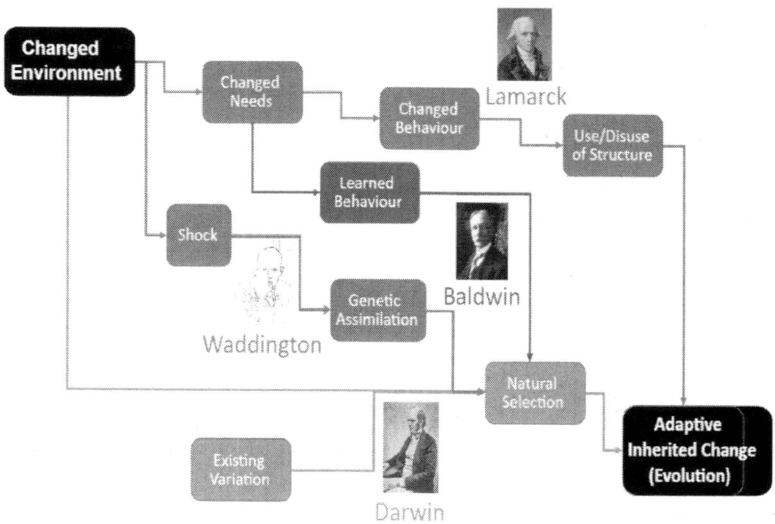

Figure 4.6. Lamarck's theory of evolution compared to Darwinian evolution, Baldwin effect, and Waddington's genetic assimilation.

Source: Image by Wikimedia Commons.

The key difference between sexual selection and natural selection is that sexual selection can promote the evolution of characteristics that would disadvantage animals under natural selection (by, for example, raising their danger of predation), such as a peacock's colorful tail (Wright, 2015) (Figure 4.5).

Figure 4.5. A picture of peacock's colorful tail.

Source: Image by Wikimedia Commons.

The concept of sexual selection received little attention in academia before socio-biology. Williams (1966) and Trivers (1972) proposed that the degree to which sexual selection occurs differently in males and females, producing sex differences in anatomy, physiology, psychology, and behavior was determined by the degree to which one sex's reproductive efforts were a limiting resource. This revived the idea as a focus of socio-biology.

The sex that puts in less work competes for the chance to mate with the sex that puts in more effort during reproduction, such as raising children (Wright, 2015).

4.5. SOCIO-BIOLOGY, THEORY OF EVOLUTION AND BIOECONOMICS

A unified theory of social interactions between humans and other animals is formed by sociobiology. The maximization of genetic fitness by the individual is its guiding concept, uniting human and animal civilizations. A person's core "teleology" in society is the maximization of genetic fitness.

Sociobiology applies Darwinian evolutionary theory to the field of social and economic sciences as an application of the Neo-Darwinian Synthesis of evolution theory. Since individuals and species compete for the same limited resources in nature to maximize their inclusive fitness and the survival of their progeny, sociobiology becomes bioeconomic or the study of the economy of nature (Figure 4.6).

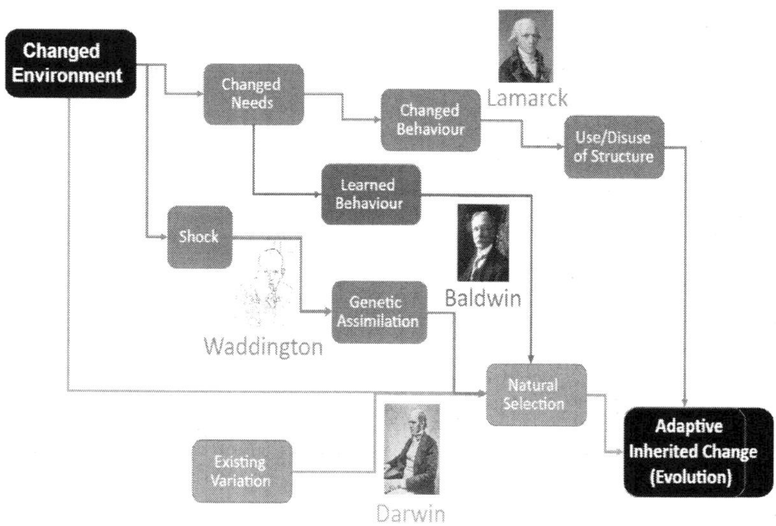

Figure 4.6. Lamarck's theory of evolution compared to Darwinian evolution, Baldwin effect, and Waddington's genetic assimilation.

Source: Image by Wikimedia Commons.

The key to maximizing inclusive fitness is to utilize the resources of nature as economically as possible. In order to analyze the economy of nature, the life sciences are driven by their own research to turn to economic theory. To live is to utilize one's resources Wisely as well as those of one's surroundings. Using resources inefficiently means that fewer living forms will arise.

The Neo-Darwinian Synthesis and evolutionism, on the one hand, and the older natural right tradition, on the other, both acknowledge the economics of nature and the significance of economic principles in the natural and in the social evolution as well as in their coevolution.

The differences in human culture that can be extended and adjusted but only with bio-economic costs are constrained by the laws of nature and its economics. Sociobiology and bio economics can help us understand these costs and encourage us to act rationally in light of nature's limitations.

It is not always wrong for biological theory and the life sciences to be applied to human society and the economy. Instead, it offers significant new insights into the characteristics and limitations of societal organization in humans.

The life sciences, social theory, and philosophy are all interested in sociobiology as a general area of study. However, sociobiology is employed in two different ways: as a field for studying human and animal societies, and as a specific theory in this subject that extends Darwinian evolution to a unified metaphysics and a comprehensive account of being and the cosmos.

E.O. Wilson's latter method transforms sociobiology into an ontology and metaphysics of all existence. As an ontological or metaphysical theory, sociobiology or evolutionism must therefore defend its claim to metaphysics using metaphysical or philosophical justifications.

However, the "metaphysical" sociobiologists reject this work as philosophy. Authors like Dawkins do not address it and E.O. Wilson only addresses it in broad strokes (Figure 4.7).

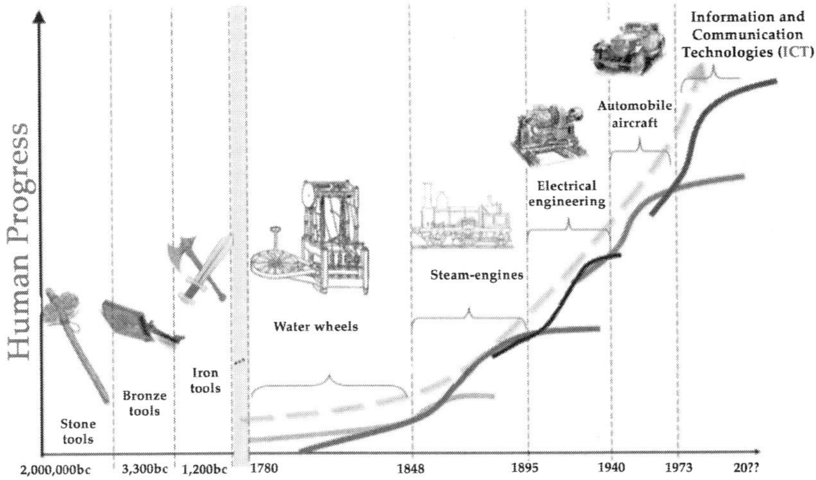

Figure 4.7. Long waves of social evolution.

Source: Image by Wikimedia Commons.

The group CNITAS has already published three volumes of collections of essays in the German language that analyze and critique the ontological and metaphysical claims of evolutionism and sociobiology.

The current volume aims to advance this discussion further and take it to new areas of research. The current volume examines particular instances of conflict within the Darwinian approach in addition to continuing the philosophical discussion on the Neo-Darwinian Synthesis. It talks about the theoretical crisis points and Darwinism's gaps. It investigates whether the theory of Darwinism, which holds that natural selection promotes divergence and greater complexity of forms, can be reconciled with the second rule of thermodynamics.

It looks into how genetic, somatic, and cultural-social systems of heredity interact as well as how these systems might be conceptualized in the course of biological and cultural evolution.

The boundaries between biological and hermeneutic interpretations of the genetic and biological code, as well as those between the evolutionist narratives and the narratives of mythology and philosophy are redrawn as part of an analysis of the interaction of genetic and non-genetic factors in evolution.

The discussion of the points of crisis of Darwinism in this volume is developed in a conversation between biologists and social scientists as well as in a discourse about new research directions in biology in contrast to the previous three CIVITAS volumes, which focused on the debate between philosophers and biologists.

The synthesis of evolutionary biology and evolutionary economics, of biological and economic theory, is the other novel contribution of this volume to the discussion on a unified theory of the life sciences and the social sciences.

An integrated bioeconomic analysis of symbiosis, mutualism, rivalry, and cooperation as well as the circumstances under which they emerge and become evolutionary stable strategies begins with an examination of the economy of nature.

What function do concepts like symbiosis, mutualism, and cooperation have in human society and economics compared to those of animal communities? Are there similarities between animal and human society in terms of competition and cooperation? Finally, the connection between natural selection and market selection is investigated.

The integration of biological and economic analysis in bio economics is important for understanding ecological equilibrium, environmental issues, and theories of economic growth. What are the developmental and biological limits on natural selection that affect economic development?

The growing area of bioeconomics includes the ecological-economic discussion of sustainable development and the issue of whether the economy should be viewed in mechanistic or organismic terms.

4.5.1. Sociobiology: A Definition and Some Criticisms

In addition to the ground-breaking works by Hamilton (1964), Trivers (1972) and Alexander (1974), socio-biology is typically and primarily linked with Richard Dawkins the Selfish Gene (1976) and Edward O. Wilson's Socio-biology: The New Synthesis (1975).

Wilson and Dawkins, who hold a position of great authority in the biology community, present a thorough synthesis of numerous earlier publications, including their own and specifically discuss the boundaries of what constitutes a sociobiological study in their view.

The main argument is rather straightforward, despite the complexity of a scientific argument that requires hundreds of pages to create and expose.

In fact, it doesn't appear like this is just rhetorical to say it's simple. It exemplifies the reductionist endeavor that at least in Wilson's and Dawkins conceptions is at work in socio-biology and describes the course that science must take.

As Wilson (1999 [1998]) argues, portraying reductionism as "an obsessional condition, decreasing towards a fatal stage one writer recently named "reductive megalomania" is "an actionable misdiagnosis."

The 'cutting edge of science' and "its" "main and fundamental activity" is, in fact, reductionism (Wilson 1999 [1998]). It provides explanations that are as concise as feasible.

Here, reductionism suggests that theories for sociobiology rely on a combination of natural selection, which is thought to be the exclusive, cause of modification (Ghiselin, emphasis added) and a certain form of genetics (Ghiselin).

As a matter of fact, sociobiologists suggested that "to explain social behavior, not by transferring the unit and beneficiary of natural selection upward to the group, and certainly not to the species, but downward to the genes" (Depew & Weber).

Thus, sociobiology is a part of the "English wing" of Neo-Darwinism, which is more severe than its American equivalent and "has become gene focused and reductionist, attributing causal effectiveness to "selfish genes," coming out of the work of Ronald Fisher" (Depew & Weber).

It is crucial to define the various strategies that have emerged within such a broad framework. According to Koslowski, "veterinary socio-biology" is "a part of biology, [... which] deals above all with the phenomena, functions, mechanisms, and the genesis of social behavior among animals and attempts to explain this behavior by genetic-evolutionary and eco logical evolutionary theories.

Koslowski describes socio-biology as having its roots in the application of its theory to social animals. Assuming that all living things obey the imperative of reproduction and of enhancing inclusive genetic fitness, veterinary socio-biology has since expanded into the study of human socio-biology.

The difference between the empirical and radical types of human socio-biology needs to be stressed once more.

The difference is that proponents of the first approach "concede a distinction between animal and human socio-biology and assume a plurality of fields of phenomena," whereas Wilson and Dawkins and other proponents of the latter approach accept "that the same laws apply in human society as in the animal world."

According to this theory, the imperatives of reproduction and maximizing genetic fitness for all members of a society are followed by all human societies.

Thus, teleological notions of the rational pursuit of objectives and of maximization are introduced into biology by radical human sociobiologists. Because of this, socio-biology, and bio economics have been placed on an equal basis.

More specifically, bioeconomic is generally understood as a subfield of socio-biology that makes use of economic techniques to study living things. Such a classification ignores the significant differences between the two methodologies.

Even while Ghiselin's 1974 book The Economy of Nature and the Evolution of Sex "has been looked upon as an early contribution to socio-biology [it] truly was about what is going to be called "bioeconomics,"" according to Ghiselin (1989).

The distinctions between radical socio-biology and bio economics are of some relevance since they shed light on the main factor that makes radical socio-biology contentious. In essence, according to Ghiselin, sociobiologists claims regarding genes (including their nature, existence and function) are not supported by science but rather are merely the result of metaphysical views.

Only metaphysics is left after the "false impression" (Ghiselin) that socio-biology is "good biology" has been dispelled.

The idea that genes are the fundamental reality and that organisms and species are in some way epiphenomenal, is a kind of ontological reductionism that is supported by a small majority of biologists, primarily individuals who are interested in specific facets of social behavior and who sometimes call themselves socio-biologists. This position is best summed up by the following quotation.

Because the ideology, theology, and metaphysical kitsch is rejected with which it has been tainted, a lot of people who have made significant contributions to the scientific study of the phenomenon in question prefer not to embrace that title (Ghiselin).

The metaphysical aspect of sociobiology's (pseudo)scientific discourse is what makes it so problematically extreme.

Therefore, radical sociobiology "explicitly declares their ontological basic tenets and their claim to complete explanation and they expressly champion the monistic ontology of a single process of evolution" in contrast to bio economics and similar approaches to human behavior, which "restrict themselves to a scientific approach" (Koslowski).

Koslowski draws the following conclusion: "The philosophical difficulty of sociobiology rests not at the level of the simultaneous analysis of animal and human civilizations, but at the point where socio- biology makes claims to a holistic, monistic-metaphysical theory of totality.

Socio-biology demonstrates the non-scientific nature of a philosophy or worldview, not as a theory pertaining to ecological populations but rather as a theory of the complete reality, as a totalistic theory.

The difference between totalism and totalitarianism today is not as great. According to Huttermann's reasoning, sociobiology is a totalistic theory that might give rise to totalitarian ideologies: "Socio-biology, in my opinion, as it is put out by Dawkins and all these individuals is endangering not only our religion but also our culture and our civilization.

It is important to remember that the two totalitarian ideologies of this century, communism, and fascism, were both affected by nineteenth-century ideas of Darwinism in how they saw the world.

Let's note that Huttermann's judgment unquestionably applies to Social Darwinism, a theory that has in fact been used to support totalitarian ideologies, but probably not to Darwinism because, as many have emphasized, one can only find support for Social Darwinism by stretching Darwin's theory beyond its 'original' limits.

Even though the connection between Darwinism and authoritarianism is not as obvious as it first appears, there is undoubtedly a connection between the theories and notions that a science produces and the perception that a society has of itself.

Because "the way a society defines itself simultaneously constitutes an aspect of this society's reality," Oyama claims that "every theory carves the universe in particular ways." Koslowski summarizes the issue's significance by saying that: "If human society interprets itself according to the model of sociobiology and makes the reproductive imperative the ultimate purpose of its social teleology, this does not remain without consequences for the consciousness of its members and the character of society."

The world view that sociobiology holds is best exemplified by Wilson's (1999 [1998]) definition of ought: "ought to be just shorthand for one form of factual assertion. Ought to be the output of a material process."

The so-called naturalistic fallacy can therefore be disregarded as being "itself a fallacy" because there is no gap between what is and what ought to be (Wilson 1999 [1998]). The justification put out is that since everything is "the physical products of the brain and culture," nothing can be different from what it is. Wilson (1999 [1998]) asks, "If ought to be not is, what is?"

Therefore, biology renders the naturalistic fallacy outmoded, and it is difficult to imagine that Kant, Moore, and Rawls would have reasoned as they did if they had access to modern biology and experimental psychology.

This viewpoint is consistent with a Panglossian understanding of evolution. Progress is not only implied by evolution, but it is also a necessary good because what occurs must take place.

Ghiselin thus describes the situation as "not merely the best of all possible worlds, with everything in it a necessary evil, but at every given time, it has to be bettered than any other instant."

In other words, this account of the evolution of living things emphasizes progress without reference to any historical setting and only with regard to the biological underpinnings of life, a point stated by Koslowski.

Naturally, challenging the biological analysis that underpins sociobiology results in rejecting related metaphysical implications about the naturalistic fallacy. Ghiselin contends that, even if progress exists, it cannot be equated with optimization (Ghiselin, 1999).

4.5.2. Beyond Radical Socio-Biology?

Wilson's socio-biology has an interesting counterargument in the form of bio economics. Numerous points need to be emphasized. First, as we've shown, socio-biology's solely genetic principles are not shared by bio economics.

Since genes are resources rather than causative mechanisms, they are seen as having a significant impact on how organisms grow (Ghiselin). But genes "like many other resources decay and become obsolete [and] need to be updated and replaced," according to one researcher (Ghiselin).

Here, sexual reproduction exerts its greatest influence. Ghiselin underlined that there wouldn't be any species if it weren't for sex. The exchange of genetic resources is made possible by sexual reproduction.

Such a viewpoint implies, and this is the second crucial point, that natural selection no longer occurs at the gene level but rather at higher levels such as family or organism: "Organisms are the units that have reproductive success or failure, and as a result, may reasonably be said to be selected."

The organism's component elements do not stand alone; rather, they are multiplied proportionately to the whole (Ghiselin).

As opposed to socio-biology, which brings the analysis downward, bio economics moves it upward, taking into account higher-level 'individuals' like tribes or even species as competitive units. Because of this, and this is the third important factor, one must pay attention to what occurs within these higher-level units.

In particular, it's possible that, contrary to what Darwin himself said, morality, and cooperation increase rather than decrease a group's competitive advantage over other groups with less-developed morals. Therefore, it implies that egoism and competition are not the main factors influencing evolution.

4.5.3. Socio-Biology, Bioeconomics: Methodological Consequences

A methodology that holds that "Biology is the key to human nature" (Wilson 1977) inevitably has methodological repercussions for the social sciences, particularly economic theory. Wilson (1977) is apparent from this angle: "biology can change the social sciences" because "social scientists cannot afford to disregard its evolving principles" (Wilson).

Revolution is in fact the appropriate word because Wilson's approach implies a "hierarchical ordering of knowledge" in addition to the fact that social sciences and biology are sufficiently related to permit thinking in terms of a single, unified science (Wilson 1999 [1998]).

In other words, according to Wilson, "the evolutionary theoretical science of biology has become, according to the social sciences, the science underlying them and offers them with interpretive models and essential ideas of social behavior drawn from genetics and behavior research."

This supports the idea of what Wilson will later refer to as the "consilience" of the major disciplines of knowledge, using a term William Whewell first used in 1840. Although some bio economists claim that biology is the "next economics frontier" (Landa & Ghiselin 1999), let's be clear that this does not imply the kind of robust consilience that Wilson's socio-biology needs.

Because economics is the "next frontier of biology," one should instead use the term "weak" consilience (Landa & Ghiselin 1999). Bio economics does not imply a hierarchical arrangement of information.

Let's examine how economists have responded to socio-biology in more detail. It is generally acknowledged that orthodox economists have readily accepted socio-biology or bio economics. Two different forms of reasoning support this.

First, it is claimed that the standard economic model and the standard theory of evolution have methodological goals and aspirations that are so similar that they are practically twin sciences. Saunders develops this position.

With the model of Newtonian mechanics very much in mind, if only implicitly, both biologists and economists have sought over-arching theories that promise to explain a wide range of complex phenomena, including life, evolution, and the organization and functioning of human societies.

'Saunders' analysis is predicated on the assumption that both economic analysis and evolution theory, in their conventional forms Economists refer to market forces in the same way that sociobiologists refer to "natural selection" as an exclusive mechanism.

Thus, ordinary equilibrium economics and evolution theory share the same upbeat faith in outside forces in both situations, with the implication being to "lower our personal responsibility for the state of the world and for our own activities" (Saunders).

In other words, regardless of what people do, what occurs is necessary and unavoidable. As Witt contends, there is unquestionably a significant difference between nature and culture, between evolution and the market, because people can act to change how the market functions.

However, these activities don't actually alter the course of the process, according to the Panglossian viewpoint of conventional socio-biology.

Undoubtedly, the people are in the finest state conceivable, or at the very least well on our way there, according to Saunders, who writes: "If system seeks to progress to the optimal state."

Therefore, "Evolution theory and equilibrium economics give a means of explaining the status quo as the inevitable product of the workings of an Invisible Hand or, which is somewhat similar, the Blind Watchmaker" (Sanders, ibid.).

Furthermore, there is no particular difficulty in translating this methodological conviction into its ideological equivalent. This is the second sort of argument used to support the tight relationship between economic and evolutionary theory.

The basic orthodox forms of economic theory and evolution theory both offer justifications for liberalism, laisse-faire, and capitalism.

It is made even simpler by the fact that Darwinism and the free-market ideology are formed of the same ideological fabric, as He argues in response to a claim made by Hodgson & Ormerod. As a result of their scientific and ideological similarity, biology, and economic theory, in their conventional forms are mutually attractive.

There are three facets to this argument. First, the idea that it is unavoidably necessary to make economics grow into a better heterodox model, especially with the help of heterodox biology, results from the belief that "economics is in crisis," just as biology.

The objective is to steer clear of the flaws in both conventional neo-classical economics and neo-Darwinian biology (that is socio-biology). From this vantage point, He suggests applying the developmentalist strategy previously stated to assess an economic system and, more specifically, to look into what constitutes a "healthy economic system" or a "sustainable economy."

A healthy economic system, like an organism, depends on the organization of the mobilization and circulation of a balanced flow of energy and resources throughout the system, the author claims, basing her case on the comparison between an economic system and an organism.

In order to ensure that resources may be mobilized, disseminated, and redistributed at will as needed, it enables her to focus on the significance of reciprocity, trust, or goodwill in relationships.

Second, claiming that biology and economics have similar problems implies that orthodox theorists have nearly universally favored the other discipline.

In actuality, this is untrue. Standard economic theory, in Wilson's opinion, presents some challenges because it "has not acquired or not even attempted serious consilience with the natural sciences" (Wilson, 1999 [1998]; emphasis added), which is evident in the fact that economists continue to downplay the biological and psychological underpinnings of human nature.

Wilson essentially builds on Gary Becker's theoretical works when he claims that recent changes in economic thought indicate economists propensity to close themselves off from other (particularly natural) sciences rather than attempting to open out.

When Becker responded to Wilson's Sociobiology thesis in 1976, he had already taken this stance. As a traditional economist who is confident in the power and merits of his field, Becker does not see the benefit of using evidence from other fields to support or even enhance economic theory.

Even sociobiology found it difficult to explain certain behaviors, like altruism, but the standard ecological model is capable of doing so. Gordon Tullock is a significant contributor to the discussions regarding the connections between biology and economics.

The first bio economist to attempt to reconcile economics and biology was Tullock (1970, 1971a, b). Despite this, he continues to be as cautious in his efforts to establish a new research program as he is suspicious of the expected outcomes.

Thus, Tullock (1979) believes that "there is little more direct benefit which any student of human society can obtain from sociobiology," despite the fact that he admits the close relationship between economics and biology. He came to the conclusion that studying animal communities would not likely teach them anything about human society (Tullock 1979).

Other traditional economists, like Hirshleifer. consider biology to be useful since it offers dependable scientific theories for the justification of human conduct. In reality, removing the barriers that separate the sciences of man and more fundamental studies of life is necessary in order to employ economic instruments to address a wide range of behaviors.

As a result, "the different social disciplines devoted to the study of man, economics among them, constitute simply a branch of the all-encompassing area of sociobiology" (Hirshleifer 1977, 1978).

Hirshleifer, a bio economist rather than a sociobiologist, does not intend to organize the various sciences in a hierarchical manner. The author asserts, however, that "the fundamental organizing principles of the leading analytical frameworks used in economics and sociobiology aue strikingly parallel" (Hirshleifer 1977; emphasis in the original).

As a result, orthodox economists may or may not be interested in biology and may or may not view this field as having any bearing on economics. Socio-biology and traditional economic theory do not directly relate to one another.

The properties of the various varieties of socio-biology are clarified by socio-biology and bio economics. The examination of Wilson's socio-biology's metaphysical underpinnings is particularly significant. It is beneficial to comprehend how several methodologies can be suggested as alternatives to radical socio-biology.

It is interesting to note from the perspective of the interplay between economics and biology that, beyond strictly ideological grounds, the acceptance or rejection of the hand extended by biologists has been greatly influenced by one's conviction in the strengths of economics.

Landa and Ghiselin (1999) highlighted in their editorial to the first issue of the Journal of Bio economics. Rather than approaching the subject from the standpoint of one's profession, it is time to investigate what the two disciplines may provide to one another.

In this sense, bio economics demonstrates its significance by taking into account the possibility of reciprocal enrichment without raising the issue of an economic hierarchy.

4.6. CONCLUSION

In this chapter, a brief introduction to socio-biology and bioeconomics have been explained. This chapter also discussed about the evolution of socio-biology and its emotion. It provides highlights on the different theories of socio-biology such as inclusive fitness, the "problem of Altruism" and sexual selection and sexual conflict.

REFERENCES

1. Encyclopedia Britannica. (n.d). *Sociobiology*. [online] Available at: https://www.britannica.com/science/sociobiology (accessed on 21 September 2022).
2. Encyclopedia.com. (n.d). *Sociobiology* | Encyclopedia.com. [online] Available at: https://www.encyclopedia.com/science-and-technology/biology-and-genetics/biology-general/sociobiology (accessed on 21 September 2022).
3. Marciano, A., (2005). Book review: Peter Koslowski (ed.). 1999. Sociobiology and bioeconomics: The theory of evolution in biological and economic theory. *Journal of Bioeconomics, 6*(3), 317–327 [online]. Available at: https://www.researchgate.net/publication/5148681_Book_Review_Peter_Koslowski_ed_1999_Sociobiology_and_Bioeconomics_The_Theory_of_Evolution_in_Biological_and_Economic_Theory (accessed on 21 September 2022).
4. Nickerson, C., (2022). *Sociobiology Theory*. [online] Simplypsychology.org. Available at: https://www.simplypsychology.org/sociobiology.html#:~:text=Sociobiology%20is%20the%20systematic%20study,inheritance%20as%20widely%20as%20possible (accessed on 21 September 2022).
5. Science Direct, (n.d). *Sociobiology*. [online] Available at: https://www.sciencedirect.com/topics/social-sciences/sociobiology (accessed on 21 September 2022).
6. Studies in Economic Ethics and Philosophy, (1999). *Sociobiology and Bioeconomics*. [online] Available at: https://link.springer.com/book/10.1007/978-3-662-03825-3 (accessed on 21 September 2022).

CHAPTER 5

BIOECONOMY AND AGRICULTURAL PRODUCTION: CONCEPTS AND EVIDENCE

CONTENTS

5.1. Introduction 122

5.2. Significance of Agriculture for Bioeconomy 127

5.3. Challenges of the Agricultural Bioeconomy 130

5.4. Agricultural Production 132

5.5. Agricultural Production Economics 137

5.6. Primary Production Sustainability 138

5.7. Agricultural Sustainability 139

5.8. Agricultural Chains Sustainability Within the Concept of Circular Economy 141

5.9. The Challenges Toward the Adoption of a Circular Model in the Agri-Business 144

5.10. Evolutionary Scenarios for Agricultural Business Models 146

5.11. Unlocking the Potential of Agriculture with Evidence Based Production 147

5.12. Conclusion 149

References 151

Agricultural-production, often known as agriculture or farming, is the simplification of nature's food webs and the rerouting of energy for human cultivation and animal consumption. One might inquire. To put it simply, agriculture is reversing nature's smooth process of the food chain.

5.1. INTRODUCTION

The sun gives light to vegetation, which is the natural flow of the food web. Plants turn sunlight into sugars that offer sustenance for the plants (this process is called photosynthesis). Plants offer food for herbivores (plant-eating animals such as sloths), while herbivores give food to predators (meat-eating animals, i.e., jaguars) (Figure 5.1).

Figure 5.1. A farmer fertilizing maize plants.

Source: Image by PixaHive.

Plants and animals that have perished are broken down by decomposers or microorganisms. The nutrients from the plants and animals are released into the soil, and the cycle begins again. This connection is disrupted in agriculture.

Plants are preserved for human utilization rather than being eaten by herbivores. This indicates that not only plant eating animals but also carnivorous animals and even detritivores are excluded from the food chain.

When, on the other hand, a farmer plants maize to feed his cows, the animals consume the maize to plump up before being butchered for human use. Despite the fact that a herbivore (cow) is consuming the plant (corn), the web is broken when the animal is slaughtered for human consumption.

5.1.1. Organic Farming

In the early 1970s, environmental groups focused more on organic agriculture aimed at customers. Ever since, the expansion of organic agriculture outside the niche has had an influence on the entire food supply chain, as seen by the availability of associated markets to additional clients, the selling of organic foods in traditional stores, and the spread of organic supermarket chains.

On the supply side, developments are hurting small farms and family farming, which are expected to give way to big farms with greater mechanization, industrialized monocropping and vertical integration or contract farming.

> Several empirical studies on the evolution of organic agriculture undertaken in Germany, Italy, and Portugal as well as the Netherlands have advocated for such patterns. It is crucial to highlight that the uptake of organic methods appears to be independent of farm size indicating that farm size does not appear to represent an entrance barrier to organic food production.

Because organic farming does not use artificial fertilizers or chemical pesticides, yields are only 70–80% of what conventional cropping methods achieve. As a result, a push for substantial innovation to boost productivity is likely.

In turn, rising demand for organic fertilizers such as animal by-products, algae, and plant based composts as well as raw mineral phosphorus and potassium based fertilizers will promote the development of site specific fertilizer application systems (Figure 5.2).

Figure 5.2. Organic farming at the Batad rice terraces, Philippines.

Source: Image by Wikimedia Commons.

Soil and biological assessments will be used to maintain and improve soil fertility and crop protection. UGVs like automated mechanical mowing and crop harvesters may significantly minimize environmental effects, which is in line with the goals of organic agriculture. Guiding technologies and CTF, for instance, are already being utilized on organically arable and vegetable farms in the Netherlands as described by Vermuelen and Mosquera.

5.1.2. Urban Farming

In low income nations people spend a substantial portion of their money on food that is easily accessible to local markets. However, high housing and transportation costs make the essential food amount less affordable and most households may have limited access to the food safety nets that characterize rural agriculture.

These social structures safeguard disadvantaged households from harm to their livelihoods while also ensuring appropriate food intake and enhancing food security. Examples of safety net tools include the distribution of cash or food vouchers as well as the investment in public projects and employment insurance programs.

Urban agriculture is described as "small places within the city (e.g., unoccupied plots, gardens, verges, balconies, containers) for growing crops and rearing small animals or milk cows for private use or sale in neighborhood markets" and may provide a source of food and money for city inhabitants.

Food security may be attained in a variety of ways under this approach, as Poulsen et al. admit. Households, for example, can reduce food expenses, freeing up funds for other types of food, allowing for a more diverse and higher quality diet or meeting other requirements (Figure 5.3).

Figure 5.3. New crops-Chicago urban farm.

Source: Image by Wikimedia Commons.

Furthermore, urban agriculture can improve overall food security by boosting the variety, amount, and quality of perishable products in metropolitan environments. Dietary variety is acknowledged as a valuable indicator of family food security and micronutrient consumption in this respect.

In low income countries, family farming could even take many forms: production companies could depend on both crops and livestock or only crop production in the form of cultivated plots (home gardening, vacated good bit cultivation), ranging between different seasons and year round cultivation practices, even if conventional leafy greens are the most extensively grown crop (Figure 5.4).

Figure 5.4. A picture of home gardening.

Source: Image by Wikimedia Commons.

Many food producers in low income nations farm for their own use, with monetary benefits as a secondary goal. This concept is especially important in the urban context, which is typified by narrow regions that impede scale economies.

Numerous studies have discovered that the amount of food that is produced by urban agriculture is eaten by farmers instead of marketed. There is little evidence of innovative technology adoption in urban agricultural situations, as food production still mainly relies on old technology.

One exception may be the use of ICTs, which may play a big role in the future, supporting sustainable development through creative natural resource utilization; examples of technical breakthroughs include broadband infrastructures, better internet connectivity and phone applications.

5.2. SIGNIFICANCE OF AGRICULTURE FOR BIOECONOMY

The phrase "bioeconomy" has lately gained popularity. It has been the topic of several papers and solutions in numerous nations. The global economy is facing several issues, which has sparked interest in the notion of a bioeconomy.

These problems include sustainable natural resource management; sustainable production; public health improvement; climate change mitigation; integrated social and economic growth; and sustainable global development.

Many countries link the development of a bioeconomy with opportunities for innovation, economic growth, and job creation. Given the projected population expansion of 9 billion by 2050, it is apparent that the demand for natural resources would keep rising in the absence of a suitable development plan.

In view of the current issues, incorporating bioeconomy into the development strategies of EU member states appears to be a logical decision. The bioeconomy sector encompasses all activities related to novel production as well as the utilization and conversion of biological resources.

As a result, it includes agricultural, forestry, fisheries, food, and cellulose and paper manufacturing as well as components of the chemical, biotechnology, and energy industries. Bioeconomy is viewed differently in different countries and by different economic sectors. It is, however, cross-sectoral.

The notion is studied from the standpoint of innovativeness and the economic gains that may result from its development regardless of sector.

Bioeconomy refers to unique processing and value creation chain in which products from the sectors of original biomass production move across processing sectors, exchange, and distribution chains as well as reach consumers as food and biomaterials for further handling, industrial goods, and consumer products, forming an entire closed-circuit economic system.

These three aspects, namely biomass generation, processing, and production, and distribution and consumption, are linked by a system of developing and applying knowledge and technology (Figure 5.5).

Figure 5.5. SICET, biomass power plant in Ospitale di Cadore, Italy.

Source: Image by Wikimedia Commons.

Farming, which is the largest primary production sector, has become one of the bioeconomy's sectors. Agriculture impacts cultural landscapes by being involved with the deterioration of soils and water supplies and the deterioration of related products and environmental services.

This is responsible for the failure of biological diversity and 13.5% of global greenhouse emissions. As a result, bioeconomy agriculture must be sustainable in the future. It implies that agricultural output should be maintained in such a way that an adequate quantity of food and biomass is given to a growing population while conserving ecological functions and biological variety.

This sector, which produces food and items other than food, must strike a balance between human food supply and environmental protection. Agro-biotechnology (agricultural biotechnology) that aims to tackle the issues of the modern world connected to agricultural production is particularly significant in this context.

Agriculture and forestry play an essential part in the bioeconomy in the context of the EU's ambitious climate and energy goals until 2030. They also aid in the decarbonization of other industries (for instance, by decarbonizing transport through shifting to advanced biofuels and removing coal from the atmosphere). Agriculture generates the majority of the biomass raw materials for the bioeconomy.

As a result, the bioeconomy may not only help to sustain agriculture, but it may also be an important source of diverse income for farmers and a contributor to job opportunities, competitiveness, and growth in rural areas.

In terms of agriculture, the European Union's (EUs) bioeconomy plan states that the goal is to provide tools and skills for fruitful, resource efficient as well as adaptable systems for food, feed, and biobased natural resources, in conjunction with policies that help rural livelihoods without jeopardizing ecosystem services.

Furthermore, the suggested CAP revision for the period 2021–2027 requires member states to demonstrate whether their national strategic plans may contribute to more sustainable agriculture, environmental protection and mitigating climate change.

While assessing the bioeconomy is challenging, several initiatives have been devoted to estimating the importance of this sector at the national and European levels. Numerous scientific researches have been conducted on the bioeconomy.

However, the majority of them merely relate to the notion of bioeconomy and its political context. On the other hand, the importance of certain industries for the growth of the bioeconomy is cited less frequently.

For example, Fuentes-Saguar et al., for example, conducted such a study for the year 2010. Primary farming, food processing, biomass supplies, bioenergy, bio-industry, and non-bio-based activities were recognized by these writers.

Loizou et al., in turn, investigated the role of all bio-based industries in terms of employment and productivity. According to the Input-Output model Unfortunately, researchers solely looked at Poland's bioeconomy. Nevertheless, no research has been conducted on agriculture's role in the bioeconomy industry.

Agriculture scientists are constantly focusing on its variety in the EU, productivity, sustainability, and competitiveness. In comparison, rare studies discuss its possibilities and significance in executing the bioeconomy idea.

Therefore, the demand for widespread agricultural usage stems not only from the need to generate food and assure food security but also from the need to create biomass, which is a key source of renewable energy (RE) and biomaterials. It is projected that the burgeoning bioeconomy will necessitate an expansion in biomass supplies.

Unfortunately, not all of the biomass generated can be utilized. In Europe, the bioeconomy is inextricably tied to the Common Agricultural Policy (CAP), which relates to shared goals regarding food security and rural development.

The generation of non-food biomass from agriculture has been significantly promoted by the CAP particularly since the Agenda 2000 revision included the rural region's development plan as a policy priority.

Concurrently, the Revised Version of the EU Bioeconomy Strategic Plan and the Involvement of the Agricultural Sector observe that even the most essential performers in agricultural primary production (farmers) are not very well incorporated into the bioeconomy value chain, serving as biomass providers instead of bioproduct producers.

A redesigned bioeconomy plan should therefore prioritize farmer requirements and include steps to improve primary producer's roles in emerging bioeconomy value chains. It also stated that the sustainable growth of the bioeconomy in rural regions would surely be a significant positive element in addressing depopulation by producing employment and business possibilities based on current digital technology and creative business practices.

5.3. CHALLENGES OF THE AGRICULTURAL BIOECONOMY

Supplying more food with limited resources is a global challenge. This might be accomplished by the effective use of agricultural inputs on the one hand as well as the effective utilization of by products and waste recovery as a major component driving competition and sustainable value chains on the other.

Simultaneously, in order to revitalize rural economies, the notion of multifunctional farming must be incorporated into prospective rural business models. Increasing agricultural production would yield more promising results.

Yet, agricultural production is strongly dependent on two of the earth's most precious resources: land and water. It is universally recognized that the only way to get renewable biological material is through environmental protection.

This is certainly relevant for the optimum utilization of natural resources, notably water and soil bioeconomy being the main sources of conventional economic sectors (agriculture, forestry, aquaculture, and fishery).

Farming requires important and restricted materials to maximize biomass, like land, fertile, and functional soils, water, and healthy ecosystems, as well as weather factors in the form of resources such as mines and energy for fertilizer manufacture.).

All of these issues are intertwined with the environment, health, and food safety, as well as the customer's desire for healthy and local foods. Furthermore, they are caused by the expected increase in the global population, the shift toward animal protein-rich diets, the growing concern about antibiotic resistance, agricultural loss, and food scraps, particularly fruits, vegetables, and seafood.

In summary, the primary problems for establishing an agricultural bioeconomy are: ensuring food security; managing natural resources in a sustainable manner; and mitigating the impact of global warming.

Climate change mitigation using climate-smart technology is an important intervention area in the agricultural bioeconomy, with the goal of mitigating the negative effects of weather patterns on agricultural production. Several scholars believe global warming is having a significant influence on agricultural output.

Farming is today thought to be very energy demanding owing to the use of nitrogen fertilizers, chemical pesticides, irrigation, mechanization, and food for livestock production. The substitution of fossil based inputs with controlling and enabling ecosystem services promises to lessen agriculture induced consequences while narrowing yield discrepancies.

Agriculture's greenhouse gas (GHG) emissions have decreased over the past decade and are now about 25% lower than in 1990. Agriculture, on the other hand, accounts for about 10% of EU GHG emissions.

There is potential for a future decrease in this percentage by implementing particular efforts to enhance manure management, enteric fermentation, synthetic fertilizer usage, food loss monitoring and minimization, food waste recycling and meat processing.

The creation of local bio-economies might improve the resilience of vulnerable places, particularly isolated rural areas, in the future. Farm-scale biogas generators might minimize farmers reliance on electricity while also addressing the issue of manure management. Rural biorefineries might assist isolated rural regions in achieving energy and material self-sufficiency (Figure 5.6).

Figure 5.6. Measuring greenhouse gas emissions from smallholder systems.

Source: Image by Flickr.

5.4. AGRICULTURAL PRODUCTION

5.4.1. Resource and Efficiency in Traditional Agriculture

Nowadays, in India, like in many other developing nations with a strong agricultural past, the terms 'improved agriculture' and 'progressive agriculture' are becoming associated with the proliferation of high yielding varieties (HYVs) cultivated with ever increasing dosages of agrochemicals (Figure 5.7).

Figure 5.7. Application of agrochemicals including fertilizer to crop land and more nutrients entering water bodies.

Source: Image by Wikimedia Commons.

Time honored crop rotations, intercropping patterns and other vital characteristics of conventional agriculture have been brutally destroyed everywhere the new crop types have expanded.

- **Traditional Implements:** Farmers current plows as well as other tools were obsolete and require replacement. Using 'enhanced plows, the local farmer may forego the several plowings he provides to the soil.
- **Irrigation System**: The well-developed irrigation infrastructure was a key agent of traditional Indian farming. Watering by wellbores is both the most extensively used and the most productive technology for producing the greatest examples of attentive farming.

Furthermore, when it comes to wellbores, one cannot assist but be struck by the talent by which native farmers initially find a water supply, then by the building works of the wellbores, the types of wellbores and their appropriateness to the environment and means of the people and by different water-raising technologies, each of which has a specific rationale for its use, although the effectiveness of conventional irrigation systems was less fruitful (Figure 5.8).

Figure 5.8. Conventional sprinkler irrigation at leafy greens in the Salinas valley of California operated by farmer.

Source: Image by Flickr.

- **Crop Rotation System:** The methodical rotation technique of crop production was another significant feature of ancient agriculture. More than one crop may be observed on the same land at the same time, but it is easy to overlook the fact that this is an example of rotation in action. monocrop cultivation has been implemented recently, increasing land productivity and earning market surplus.
- **Soil-Mixing Practices:** In India, combining is common. The addition affects the consistency of the sand, making it more suitable for sugar cane and other garden crops that grow under irrigation.

The cultivator recognizes the significance of tank silt and in areas where these water reservoirs are widespread; they are meticulously cleaned out each year.

The silt that has accumulated in these tanks as a result of the washings of village sites and cultivated fields has some manorial value, and when applied at the rate of 40 bullock cart loads or more per acre, it contributes significantly to the composition of the soil.

- **Grain-Storage Practices:** Indian farmers food grain storing is no less enthralling. Indian farmers are masters of food storage and therefore will emerge unscathed from rough mud holes after twenty years.

Those who know the actual condition of fully ripe to which grain should be expected to exist in particular seasons; in other words, under different weather conditions to make sure it's preserving when thus stored; and the duration that, under transient conditions, this should lie on the open floor or ground to protect the very same item.

- **Scientists Farmer:** The Chhattisgarh area has highlighted the high level of abilities of farmers from distant tribal communities that have remained unaffected by government development projects.

Tribal groups continue to live independently; they obtained equivalent or even higher yields from indigenous rice varieties especially as opposed to the HYVs legally disseminated in other districts of the state. A further astonishment was the enormous variety of rice types planted by the farmers, who were well familiar with each of their holdings.

Some of those kinds were notable for their great yields, some for their superior cooking capabilities, still others for their scent and yet others for other treasured characteristics.

- **Grazing Land:** Originally, humans, wildlife, forests, meadows, and farmlands were interdependent and harmonious aspects of a single system. The villagers cared for the vegetation in the fields and also managed to ensure the communal grazing pasture.

Farmers cared for domestic cattle, often with the aid of a grazing assistant, and cultivated the crops, with or without hired labor or shared croppers. However, the water saving properties benefited the locals by contributing to the productivity of agricultural areas and giving shelter during summertime (Figure 5.9).

Figure 5.9. Grazing land at Bethlehem.

Source: Image by Wikimedia Commons.

- **Husbandry:** Cattle produced dairy products, that helped to supplement the nutritious quality of the villagers diets. Cattle waste supplied organic fertilizer to the crops, whereas fowl produced eggs and meat. The skins of deceased cattle were used to make footwear and other leather items, all of which were done in the community. Last but not least, bullocks mowed the land.
- **Crop Management**: Because of the diverse crop variety cultivated in conventional farming, excellent quality seeds must be made accessible to the people. Great field planning and assistance with manuring, seeding procedures, crop management, and post-harvest storing may result in higher crop quality and production.
- **New Agricultural Technology:** Modern agriculture technology in the form of tractors and fertilizers will once again favor the wealthier landowners, allowing them to enhance agricultural productivity and cash collections.

One's reliance on green manure and bullocks, on the other hand, is lessened, reducing the need for feed. All of these considerations may cause them to overlook the expansion and adequate upkeep of grazing areas.

- **Water Use:** The establishment of flat rate pricing for agricultural electricity is to blame for this bizarre current state of affairs. A farmer spends a predetermined fee per horsepower each month for power consumption under this approach. As a result, the marginal cost of lifting water is nothing. This results in energy waste, over-pumping, and ineffective crop selection.
- **Flat Rate Pumping:** further obscures growers genuine electricity costs. The tariff structure, as well as a bad combination of technology and management are to blame for water loss, unsustainable groundwater extraction and substantial energy losses linked with the distribution and end use of power in irrigation water pumping.
- **Energy Losses:** Major energy losses are linked with power distribution as well as inadequate choice, setup, repair, and operation of the electrical motor pump system. A vicious cycle operates two subsystems simultaneously the electricity supply system as well as the pumping stations system.

This vicious cycle is subdivided into three: the technical sub-cycle, the financial sub-cycle and the societal sub-cycle.

5.5. AGRICULTURAL PRODUCTION ECONOMICS

It is a sub discipline of agricultural economics that is engaged with the determination of cropping patterns and resource usage effectiveness in order to optimize the objective function of the farming village or the country within a system of restricted resources.

It is described as an adaptive strategy of research in which economic concepts are used to the utilization of resources such as land, manpower, money, and supervision in the farming business.

Agricultural production economics is concerned with the examination of production linkages and the principles of rational decision-making in order to maximize the use of farm resources on individual farms and to justify the use of farm inputs from the perspective of the overall economy.

The major focus is on bringing economic rationality to agricultural problems. The efficiency of farm inputs is the focus of agricultural production economics. As such, it is concerned with allocating resources, resource mixes, resource utilization effectiveness, strategic planning and resource management.

Agricultural production economics is the study of factor-product, element, and product-product interactions, farm size, returns to scale, loans, risk, unpredictability, and so on. As a result, agricultural production economics is concerned with any farmer issue that fits within the ambit of allocation of resources and marginal utility assessment.

5.6. PRIMARY PRODUCTION SUSTAINABILITY

Without a question, the phrase "Sustainable Development" or "Sustainability" is among the most popular subjects in recent decades. The fast environmental deterioration resulting from human activities prompted the shift toward a comprehensive approach to planetary conservation.

Nonetheless, in recent decades, the economic and social concerns confronting mankind have become as essential. As a consequence, there is a greater need to investigate solutions to preserve our world for future generations.

Although these difficulties appear to be modern, the concept of sustainable development dates back to 1713, when Hans Carl von Carlowitz articulated the simple, but no longer appropriate, notion that "one should collect just the same quantity of wood that matches the trees planted."

A few years later, Thomas Malthus emphasized the concern on overcrowding, stating that perhaps the population rate of expansion is significantly greater than the relevant resource production rate.

Decades of concerns and debate about concerns such as resource depletion, environmental degradation, poverty, and the relationship between ecology and economics resulted in the Brundtland Commission's 1987 definition of sustainable development.

Sustainable development, as per this concept, is "advancement that satisfies the requirements of the present without jeopardizing future generations ability to fulfill their own needs" (World Commission on Environment and Development, 1987). According to this concept, the component of sustainability is now taken into account in all facets of human action.

The intensification of primary production, although increasing output, has the unintended consequence of degrading food quality and thus the environment. It is worth noting that Rachel Carson noted the effect of pesticide use on crop health in the 1962 book Silent Spring, demonstrating that the negative effects of intensive agriculture had been observed years previously.

In recent times, nevertheless, there is a concerted effort to adapt to changing circumstances by dealing with climate change, poverty, hunger, and resource degradation and depletion. Farming can help with these efforts that are explicitly stated in the United Nations Sustainable Development Goals (SDGs) for 2030.

Several methods and techniques have been developed to increase crop production efficiency, whether economically or environmentally.

Such approaches often attempt to reduce the amount of input necessary, as with Variable Rate Application (VRA), or to finally eliminate the usage of crop protection chemicals, as the advent of Organic Farming.

Furthermore, as Information and Communications Technology (ICT) improve, agriculture is being mechanized and automatized, entering a new era of farming with the primary goal of addressing the negative consequences of agriculture while preserving quality and boosting output quantity.

However, the contribution of these technologies and processes to total agricultural sustainability should be evaluated and assessed, which is why a range of agricultural sustainability evaluation approaches have indeed been created.

These techniques are often built on widely used methods for analyzing sustainability in its three main pillars (Environment, Economy, and Society), with an abundance of indicators incorporated, each appropriate to the unique instance under consideration.

Because the approach utilized is frequently connected to the component of agricultural sustainability that has to be investigated, this work begins with a discussion of generic definitions of agricultural sustainability as they are indicated in the literature.

The notion of agricultural indicators is next introduced, followed by a detailed assessment of the indicators contained in the most recent agricultural sustainability bibliographies. The purpose of this work is to offer readers a comprehensive review of the various indexes that may be used to assess the effects of primary output.

5.7. AGRICULTURAL SUSTAINABILITY

The farming sector contains a large number of components and criteria that are involved in carrying out the necessary procedures. For that purpose, assessing its sustainability is a particularly difficult undertaking, not only because of its multidimensional character but also because the idea of agricultural sustainability has yet to get a widely agreed upon definition.

In most situations, its meaning is intimately related to the issue under consideration, and thus both the phrases "Agricultural Sustainability" and "Sustainable Agriculture" are presented in the literature. But these phrases are not synonymous because they play various functions in relation to the topic of the evaluation, resulting in a never ending controversy over the final meaning.

Agricultural sustainability, which is a sector-specific aspect of sustainability, refers to a set of sustainability principles (economic, environmental, and social) that should be used in farming activities. On the other hand, the phrase "sustainable agriculture" refers to all of the procedures, structures, and activities that contribute to and adhere to agricultural sustainability principles.

Hansen (1996) defined sustainable agriculture as an activity that adheres to a set of preset parameters across time (Hansen, 1996). Concepts of sustainable farming in the millennium that followed include those of Tilman et al. (2002), Chopin et al. (2016), and Sajjad and Nasreen et al. (2016).

Tilman et al. (2002) defined sustainable agriculture as "a practice which meets current and long-term societal needs for food and fiber as well as other similar concerns, whilst also maximizing economic benefits through conserving natural resources, the maintenance of other ecosystem services and functions, and long-term human development."

Accordingly, Chopin et al. (2016) defined sustainable farming as a paradigm that allows for the conservation of resources and ecosystems whilst maintaining production.

Finally Sajjad and Nasreen (2016) mentioned the three pillars of sustainability, stating that sustainable agriculture involves food supply systems that integrate social fairness, economic, and environmental wellness, and resource conservation (both human and natural).

Conway (1985) suggested a farm-level definition of agricultural sustainability, stating that agricultural sustainability is the system's ability to maintain production in the face of extreme suffering that can lead to significant disruption.

A few years later Lynam and Herdt (1989) expanded on Conway's concept of agricultural sustainability by including the use of reference standards in determining sustainability. Researchers defined sustainability as a system's ability to keep outputs that are about equal to or higher than historical normal values, with the amount of approximation dictated by prior volatility.

Many additional interpretations of agricultural sustainability and sustainable agriculture have been provided over the years and more will be given in the coming years.

Despite the fact that all these definitions contain various foci, the majority of researchers worldwide believe that the triple bottom sustainability principles must be applied in order to ensure the achievement of sustainability. As a result, agricultural sustainability may be defined as the result of the interaction of economic, ecological, and social issues.

Pretty (2008) suggested a sequence of foundational concepts to accomplish agricultural sustainability that included:

- The assimilation of biological and ecological procedures like nutrient cycling, nitrogen-fixing, soil renewal, allelopathy, competing, predation, and parasitism into food production processes;
- The reductions of using those non-renewable components that end up causing damage to the environment or even to the environment;
- The productive utilization of producers knowledge and expertise, so increasing self-reliance and replacing human resources for expensive external inputs, and
- The constructive use of people's collective skills to collaborate to address shared agricultural and natural resource issues, including insect, drainage, water, forestry, and credit control customers, avoiding harming natural resources or the ecosystem in general."

5.8. AGRICULTURAL CHAINS SUSTAINABILITY WITHIN THE CONCEPT OF CIRCULAR ECONOMY

The agri-food business is an important economic and political sector. Mostly within the EU, agri-food trade surpassed V250 billion, with food production and processing accounting for approximately 7.5% of total employment and nearly 4% of total added value in the EU and agriculture and food related industries and services employing roughly 44 million people.

Furthermore, the EU agriculture sector's production has been steadily expanding over the years, hitting V427 billion in 2017.

Furthermore, the agri-food industry is among the most governed and shielded industry sectors, with major implications for advancement (e.g., National Food Technology Platforms) and sustainability, including meeting human needs, supporting economic growth and job creation and reducing environmental impact.

It is heavily impacted by consumer habits and the evolving international socioeconomic and political environment. Rising environmental, social, and moral issues as well as enhanced understanding of the impact of food production and consumption on the natural surroundings has put pressure on agri-food businesses around the world among consumer organizations, environmental advocacy groups and lawmakers as well as several consumer groups.

Aside from the financial ramifications of agri-food production, corporations have progressively dealt with environmental and ethical challenges associated with their distribution networks across product life cycles from "farm to fork" in recent years.

The recent post-recession time has highlighted the importance of shifting the company goal from mere profit to growth and longevity.

The EU Strategy for Sustainable Development envisions divorcing economic development from environmental deterioration, as well as conformity with EU Directives, international agreements, and the UN SDGs (United Nations, 2020), The E.U. 2030 Food Agenda (e.g., HM Government, 2010; Diedrich et al., 2011; Levidow et al., 2012; Mc Cormick and Kautto, 2013; Maggio et al., 2015), the E.U.

Circular Economy Action Plan (European Commission, 2015, 2019b; McDowall et al., 2017; Leipold and Petit-Boix, 2018), and the 2030 Agenda for Sustainable Development. The United Nations (2015) and Lee et al. (2016) will necessitate an EU-wide reaction in terms of research, information sharing, and the integration of innovative, less hazardous agri-food manufacturing techniques.

5.8.1. Supply Chain Responsibility in the Agri-Food Supply Chains

Nowadays, social stakeholders expect supply chain accountability, in which all parts of the food chain participate, are fully engaged, exchange data and establish uniform standards and guidelines across the value chain.

Sustainable supply chain management expanded the definition of sustainability from the level of the organization to the point of the supply chain by providing companies with methods and technologies for enhancing their own and the sector's competitive nature, sustainability, and obligation toward stakeholders and customers expectations.

Agri-food enterprises must monitor multiple dimensions and criteria, including quality, sustainability, and cost, in response to calls for transparency, accountability, affordability, and multistakeholder participation. Set standards and reflect on the environmental sustainability success of their supply chain operations, as well as financial and social/ethical factors (Triple-Bottom-Line approach) (Figure 5.10).

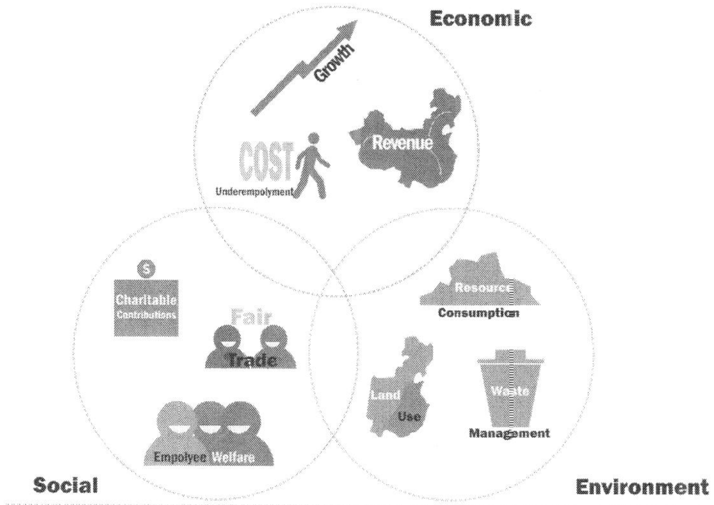

Figure 5.10. A graphic describing the some of the areas of study that the triple bottom line framework.

Source: Image by Wikimedia Commons.

Companies must also find places where environmental damage may be decreased in order to deliver an order winning plan for their food items while lowering production costs. Moreover, in order to unlock value, it is necessary to use an integrated strategy and make use of the underutilized potential of manufactured by-products and agri-food wastes for energy recovery and nutrient recycling.

Stakeholders are becoming more conscious that their success will depend on balancing the financial, societal, and environmental obligations across the supply chains using the "Triple Bottom Line" paradigm.

Only through the lens of the Triple Bottom Line could the agri-food industry's sustainable development be monitored and controlled. The Triple Bottom Line is a paradigm that helps firms balances their three obligations in order to achieve long-term success.

Such a system may serve as a foundation for companies and guarantee that they stay profitable while also preserving the natural environment and local-to-regional society (Moussiopoulos et al., 2017).

As a result, any contemporary restructuring in agri-food supply chain management must be cost-effective, ecologically, and socially responsible in aims to discuss particular concerns about Corporate Social Responsibility (CSR), stress to minimize waste and request to improve the quality of products and services.

5.9. THE CHALLENGES TOWARD THE ADOPTION OF A CIRCULAR MODEL IN THE AGRI-BUSINESS

A variety of problems exist for all areas of the economy in the quest to develop the circular economy model. The majority of the obstacles that companies confront are connected to the technology that must be used, as well as financial and quality issues.

In terms of technology, new processes must be established and/or existing processes must be altered, beginning with the reintroduction of trash, which is typical of lesser quality than virgin materials. Furthermore, in order to promote the circular economy, the economic viability element is crucial, particularly when addressing private for profit enterprises.

In general, industrial processes that employ waste or recycled resources are thought to be less effective than those that use raw resources. This is primarily due to the fact that extra processing stages are frequently necessary. In this respect, a method that uses recyclable materials should provide savings by reducing the use of natural resources while also taking into consideration relevant externalities.

Similarly, for the circular economy to succeed, the cost of "circular" items must be addressed. Because innovative procedures typically cost more than traditional manufacturing methods, material, energy or any other type of savings should be pursued so that goods become affordable when compared to the traditional manner of utilizing raw resources.

In contrast, such approaches replicability, scalability, and transferability would never be viable in the real business world. Last but not least, quality is an essential barrier to the circular logic of the agri-food business.

Evidently, at most, the very same degree of excellence would be essential in order to market circular economy items, despite the fact that markets will always be illiberal about these attempts. The latter emphasizes the importance of verification of the quality of secondary resources, which must be properly considered for the development of the circular economy concept.

The collaboration of all types of stakeholders is required to fulfill the goals of the circular economy in the agri-food industry. Producers, for example, must work very closely with animal farms, food processing factories, dairy businesses, butcheries, and biogas plants to name a few.

The mix of needed cooperation with the involvement of many stakeholders in various disciplines (academic and research institutions, policymakers, consultants, farmers, agri-food enterprises and so on) could be intimidating. In most situations, though, the following implementation stages are necessary to shift from conventional (linear) supply networks to circular supply chains.

The availability of data is suggestive of such implementation processes. possibilities for using secondary materials and raw material substitutes and long-term contractors for the exploitation of waste that can be used as raw material in other units (within or outside the same area) and closing the materials cycle through energy recovery.

In any instance, these measures and processes must attempt to maximize and utilize all available raw resources, products, and waste in order to conserve energy and minimize GHG emissions.

One of the most significant challenges that stakeholders who want to capitalize on the benefits of the circular economy and implement a circular model in their day-to-day operations confront is the fact that considerable investment expenses are usually involved (Araujo Galvao et al., 2018).

The latter frequently makes them hesitant to proceed with business reengineering and migrate to the new model. Despite the hefty investment cost, the potential is great, the economic possibilities are huge and interest in the exploitation of agri-food wastes is strong.

Although the commercial sector may begin the adoption of circular economy principles, it is the public sector (at the national, regional, or local levels) that may more effectively and efficiently promote the shift from the linear to the circular model.

As a result, the involvement of governmental authorities is crucial. These may help any relevant stakeholder by giving access to knowledge and information as well as providing networking possibilities.

Public authorities can enable actions and interact with national, regional, and local stakeholders, like companies, utility companies, academic institutions, research institutions and also civil society groups, by spreading information, organizing significant events and/or establishing information exchange portals.

5.10. EVOLUTIONARY SCENARIOS FOR AGRICULTURAL BUSINESS MODELS

Development and production continue to be important concerns in the world economy. According to some projections, agriculture product consumption is predicted to grow by 60% in 2050 compared to 2005–2007 annual data reflecting an average annual growth of 1.1%.

This will drive demand for more food production in terms of both quantity and quality, in an environment marked by various intervening circumstances, like the stalling of arable land growth, shortage of water supplies, a shrinking agricultural labor force and growing urbanization.

Advancements in agricultural technology, equipment, and practices are viewed as viable avenues for promoting long-term food production.

Smart farming, described as "a new management technology based on georeferenced information for the regulation of agricultural systems," is ushering agriculture into the digital and information era and is predicted to spark widespread societal changes affecting work conditions both in and out of farm limits.

Precision agriculture promotes optimum agricultural input management methods whilst allowing remote partnerships involving the entire food supply chain through the use of big data.

Under such conditions, technological advancement can support the implementation of the food supply chain to fulfill information needs on location and process-based, enabling the tracking of all actions conducted by the various actors in the supply chain.

Due to the fact that precision agriculture is more capital intensive than labor intensive, the agricultural area becomes an important component in investment amortization. As a result, corporate farming, or big scale agriculture on farms controlled by major corporations will be more likely to embrace highly specialized technology than smallholder groups.

Mergers in the form of plot consolidation will accelerate the accomplishment of economies of scale as well as the processing and utilization of big data, which can only be financed by huge enterprises, attracting financial investors (Corsini et al., 2015; EPRS STOA, 2017).

Small farmers' presence in this setting would become extremely risky; they would often be unable to repair or alter machinery, and they would undoubtedly face additional time and financial costs for necessary technological help.

The combined effect of the identified components higher productivity and demand for capital-intensive technology would therefore reorganize the role of small farmlands, forcing numerous set of retirees to sell their land (Corsini et al., 2015); several senior farmers have really no heirs in their family (the so-called "continuers"), whereas an increasing number of "newcomers" are trying to seek to gain entry into agriculture but without any previous experience.

Additional successions are reported in Belgium, France, Italy, Romania, Spain, and the United Kingdom, and are generally characterized by greater levels of creativity and various techniques, like organic agriculture and shorter distribution networks (Cavicchioli et al., 2015; Access to Land, 2018).

5.11. UNLOCKING THE POTENTIAL OF AGRICULTURE WITH EVIDENCE BASED PRODUCTION

Evidence based production (EBP) is expected to have a significant influence on agriculture supply chain efficiency. Proagrica defines EBP as "farming that gives more importance to technology and uses data to guide output."

EBP entails agronomists, advisers, input dealers, producers, and merchants using data to enhance productivity on farms and inside their own companies. It creates a safe, accessible, standardized, and seamless architecture for automatically transforming actual data into actionable information that brings actual advantages when deployed appropriately.

These advantages extend beyond crops in the field to possibly enormous profits from more efficiently utilizing current statistics to disclose new insights across the supply chain and to actually learn from mistakes.

Many farms have had several years worth of data kept in farm management information systems (FMIS) or in a complicated combination of outdated customer relationship management (CRM) tools, spreadsheets, and paper-based records. Farmers can now use FMIS to simplify and standardize data collection on their farms.

Yet, the massive potential of this information is badly underutilized in many firms. If present issues could be smoothly contrasted to comparable earlier experiences, such information would prove extremely beneficial to farmers, agronomists, and input providers.

For instance, if 2012 was a poor potato blight year and 2019 is similar, agronomists may readily look back at operations in 2012 to determine what methods were used effectively or unsuccessfully on farms.

Agronomists may readily integrate this data with the most recent meteorological, soil, sensor, and research results to give farmers tailored suggestions for specific field regions.

This may involve optimizing VRAs for each $10m^2$ field section based on current weather conditions and forecasts or employing automated systems to inform farmers of disease outbreaks (e.g., yellow rust) and advise them on the best course of action based on the amount of risk. This might involve just treating the most vulnerable parts rather than the entire field, limiting.

Merging anonymized and consolidated FMIS information with external data not just allows agronomists to deliver a markedly improved service to farmers, but also promotes superior decision-making elsewhere. Fertilizer companies, for instance, may swiftly identify locations that are prone to suffer from nutrient deficits using a mix of current applications, crop rotations, sensors, and satellite data.

This information is used to guide production and distribution planning, fertilizer application suggestions, and marketing messages. Grain traders can supply benchmarking data, which can assist farmers to fulfill product criteria more precisely when growing circumstances change.

Crop protection producers should proactively monitor real time consumption across many areas to spot possible product shortages locally and rectify such concerns as soon as possible.

Using EBP throughout a whole supply chain would make agricultural data far more powerful and versatile while allowing individual enterprises to keep complete control over whatever data may be exchanged with others.

Inputs are effectively turned into outputs across the sector, reducing waste. The significant savings realized by EBP will be critical in feeding an increasing world population with limited resources.

5.11.1. Redefining Industry Boundaries and Transforming Competition

Such simple but still very inventive solutions are available to all agricultural businesses, not just a chosen number. Harvard Business Review (2014) emphasized smart connected goods revolutionary ability to change competitiveness and redefine industry limits.

Presently, Proagrica gives all agricultural enterprises access to the power of connection, data from distant areas of the supply chain can be effortlessly gathered and analyzed to provide new insights that substantially boost production.

These developments not only improve main operations but can create totally new business models. Numerous businesses are already reaping major benefits from integrating their operations all without any need to invest in expensive technology or equipment. Proagrica links the lines to draw considerable value and actionable data from your existing policies and procedures.

5.12. CONCLUSION

To simplify, agriculture involves redirecting nature's natural flow of the food web. This means that not only are plant eating animals excluded from the food web, but also carnivorous animals and even decomposers.

Even though an herbivore is eating the plant, the web is interrupted when the cow is killed for human consumption. The natural flow of the food web is that the sun provides light to plants. Plants provide food for herbivores and the herbivores provide food for carnivores.

Instead of having herbivores eat the plants, the plants are protected for human consumption. However, if a farmer is planting corn to feed their cattle, the cattle eat the corn to fatten up and then are eventually slaughtered for human consumption.

Nutrients from plants and animals go back into the soil, and the whole process starts anew. Decomposers, or bacteria, break down plants or animals that have died. Plants convert sunlight into sugars, which provide food for the plants and the process is called photosynthesis.

REFERENCES

1. Bochtis, D., Achillas, C., Banias, G., & Lampridi, M., (2020). *Bioeconomy and Agri-Production: Concepts and Evidence* (1st edn.). [e-Book] Academic Pr. Available at: https://libgen.li/edition.php?id=138849452 (accessed on 21 September 2022).
2. Edwards, W., & Duffy, P., (2014). Farm management. *Encyclopedia of Agriculture and Food Systems*, [online] pp. 100–112. Available at: https://www.sciencedirect.com/science/article/pii/B978044452512300111X (accessed on 21 September 2022).
3. Patil, A., (2011). *Agricultural Production and Productvity*. [online] www.researchgate.net. Available at: https://www.researchgate.net/publication/220028128_AGRICULTURAL_PRODUCTION_AND_PRODUCTVITY (accessed on 21 September 2022).
4. Proagrica, (n.d). *Unlocking the Potential of Agriculture with Evidence-Based Production* - Proagrica. [online] Available at: https://proagrica.com/news/unlocking-the-potential-of-agriculture-with-evidence-based-production/ (accessed on 21 September 2022).
5. Rodino, S., (2018). *Bioeconomy Concept: Challenges and Perspectives for Agriculture*. [online] Hdl.handle.net. Available at: http://hdl.handle.net/10419/205087 (accessed on 21 September 2022).

CHAPTER 6

BIOECONOMICS OF FISHERIES MANAGEMENT

CONTENTS

6.1. Introduction 154
6.2. Why is Fisheries Management and Regulation Needed? 155
6.3. The Social Trap and Free-Rider Behavior in Fisheries 159
6.4. Fundamentals of Fisheries Bioeconomics 160
6.5. The Basic Bioeconomic Model 162
6.6. Age Structured Bioeconomic Model 162
6.7. The Fisheries Management Process 163
6.8. Historical Perspective on the Development of the Paradigm 168
6.9. Economic Analysis of Fishery Regulation 170
6.10. Bioeconomics of Ecosystem Interdependencies 173
6.11. Spatial Management of Fisheries 175
6.12. Dealing With Risk and Uncertainty in Fisheries Management 177
6.13. Conclusion 181
References 182

The integration of resource biology and ecology with the economic aspects that influence fishers behavior in space and time is necessary for systematic fisheries research that may create and evaluate hypotheses associating different management and regulation structures with fishery performance.

The chapter discussed about how management and regulation are different. When the same institution, management or regulatory framework is applied to fisheries with diverse resource and community settings, the results can vary.

This is because the fundamental qualities of fish resources produce different contexts of human interdependencies. These human interdependencies in resource usage are guided by the degree of resource mobility in location and time as well as the level of fishing autonomy of small-scale and industrial vessels.

6.1. INTRODUCTION

In marine fisheries, where most are often impacted by natural fluctuations, shifting dynamics of coastal ecosystem, and a lack of effective governance, modern fisheries bioeconomics can offer insight into establishing techniques to deal with the complexity of overcapacity and over exploitation.

The early theories in fisheries economics first developed in the mid-1950s by Canadian economists Scott Gordon (in 1954) and Anthony Scott, are closely related to the field of bioeconomics (1955).

Their theories drew on recent developments in biological fisheries modeling, particularly Schaefer's work from 1954 and 1957, which formalized a link between fishing activity and biological growth through mathematical modeling and empirical studies. Their work also had connections to ecology, the environment, and resource protection.

These concepts actually originated in Canada's multidisciplinary fisheries science community at the time. During a fruitful and creative period, particularly among Canadian fisheries researchers from diverse disciplines, fisheries science and modeling advanced quite rapidly.

In order to assess the biological and economic effects of various fishing techniques and fisheries management choices, economists were exposed to population modeling and fishing mortality and new interdisciplinary modeling tools became available for them.

6.2. WHY IS FISHERIES MANAGEMENT AND REGULATION NEEDED?

There are various reasons why fisheries utilization can be improved by controlling the activities of actual and potential participants. Given that this chapter is about bioeconomics, the absence of an acceptable system of property rights is particularly significant.

Although the subject cannot be covered in length here in this chapter, still it is generally agreed that property rights should contain the following features to encourage people to use resources for their highest-valued purposes (Randall, 1981; Schmid, 1987; Scott, 1989, 2008; Anderson and Holliday, 2007):

- **Exclusivity**: This refers to the extent to which the outputs created as a result of possessing and employing the resource for which the property right is defined are entirely within the owner's discretion to retain or renounce.

In a similar vein, the owner is in charge of paying any expenses connected with the usage of the resource. A crucial component of exclusivity is the ability to enforce these claims, and occasionally enforcement is specified as a separate quality.

- **Permanence**: The term "permanence" describes how long a right can be exercised. It matters because the length of time an owner can receive benefits or bear financial responsibility will determine the incentives for prudent use.
- **Security or Quality of Title**: The degree to which the right is secure from unauthorized seizure or invasion is referred to as the security or quality of title.
- **Transferability**: The ability to transfer a right to another person is referred to as transferability. The ability to utilize the resources to their best value depends on this as well.

These characteristics do not apply to fisheries resources, especially not to the extent that they do to other natural resources, as will be covered in more depth further in this chapter, even in situations when there is active government participation in fishery usage under open access.

In essence, this means that participants have little incentive to fully use a fish stock's reproductive potential. As will be established below, it can be quite effective to create regulation mechanisms that offer incentives similar to those provided by property rights.

Other characteristics and features of fisheries resources that can influence both short and long-term fishing behavior and exploitation patterns, with or without regulation, should be taken into account when thinking about fisheries management and regulation.

High transaction costs (information costs and enforcement costs), high exclusion costs, free-rider behavior, and a social trap condition must be recognized among these (Seijo et al., 1998; Caddy and Seijo, 2005).

High Exclusion Cost. It implies that it is difficult to restrict the usage of an existing fish stock to those who have the legal right to fish it. It is not necessarily possible to successfully exclude all other participants just because rules exist that limit fishing to a certain number of individuals.

Enforcing rights and rules in maritime fisheries is logistically challenging and extremely expensive because most fish resources are mobile and migratory. In order to confine participation to those who are authorized and fishing operations to those who are authorized, fisheries management and control necessitate substantial enforcement or police expenditures (Figure 6.1).

Figure 6.1. Women in fisheries: International maritime organization.

Source: Image by Flickr.

The areas that need to be policed are vast for oceanic (and many shelf) fisheries, and traditional patrol vessel operations are both expensive and inefficient. A non-enforceable right turns into an empty right in these situations.

High information costs. The significant uncertainties inherently present in natural systems, as well as a variety of additional biological, social, political, and economic aspects, raise the complexity of fisheries management, necessitating a precautionary approach to fisheries management (FAO, 1996).

Even in the context of stakeholders cooperation, it is generally highly expensive to get usable information adequate enough to know what the proper management strategy is and associated rules.

Additionally, interdisciplinary approaches connecting the scientific and social sciences are necessary due to the multidisciplinary character of the information needed to assess the status of the fisheries and the ecosystem that supports it (Figure 6.2).

Figure 6.2. Fisheries patrol vessel, Newlyn Harbor, FPV Saint Piran is operated by the Cornwall Inshore Fisheries and Conservation Authority.

Source: Image by Wikimedia Commons.

This book is an attempt to advance the subject of fisheries bioeconomics by fostering the necessary interdisciplinary analysis in fisheries through the fusion of biological and ecological concepts and methods with the economics of fisheries.

Effectively allocating property rights is severely hampered by the challenges of upholding the fundamental premises of exclusivity and cheap information and enforcement costs. The above mentioned characteristics of fisheries, which include their substantial exclusion and transaction costs, force us to look beyond the straightforward answer of "fair distribution of individual rights."

The issue that requires the most immediate change, both domestically and internationally, is the distribution of resources among stakeholders.

Fisheries play a significant role in society and the economy. Fishing related jobs are thought to employ 12.5 million people, and since the early 1990s, the value of fish traded worldwide has been estimated at US$ 40 billion annually. The total production oscillated and reached around a total mass of 100 million tons from capture fisheries and aquaculture during the same period (Figure 6.3).

Figure 6.3. Workers doing fishing job.

Source: Image by Flickr.

However, though many of the world's exploited fish stocks are currently exhausted, overfished, depleted, or in need of recovery, many are also harmed by environmental deterioration, especially in inland and coastal regions. Significant ecological harm, which may not always be repairable, and economic waste are frequently already obvious.

As a result of new technical advancements like Geographical Positioning Systems (GPS), radar, echo sounders, more powerful vessels, and better processing techniques (like surimi), fishers are now better able to exploit more live resources more intensely, which could make the issue worse.

The current state of the world's living aquatic resources is largely due to the inability of the current governance system for fisheries to ensure accountable and successful management of fisheries in the majority of countries.

The current unacceptable state of the world's fisheries and living aquatic resources must be accepted as a shared duty by fishermen, fisheries management authorities, and fisheries scientists, as well as those accountable for indirect effects such environmental degradation.

The States must see to it that coordinated action is taken to revers these tendencies. The next section outlines the steps that must be taken to adopt ethical fisheries management.

It's critical for fishery managers to understand that inaction in the face of over or irresponsibly exploited resources will have unfavorable effects down the road. Loss of potential advantages like food, income, employment, and others will occur as a result of reducing fish stocks to physiologically and ecologically detrimental levels, both immediately and over time.

Any stock at a very low level is likely to have negative effects on other dependent stocks and the losses may go beyond the stock that is directly impacted. In these situations, it cannot be immediately expected that reducing fishing pressure will result in a complete or quick recovery of the stock and its related environment. Losses occasionally may be substantial and long lasting or even permanent.

6.3. THE SOCIAL TRAP AND FREE-RIDER BEHAVIOR IN FISHERIES

Additionally, it should be acknowledged that the lack of exclusivity in the rights to open access to fish populations frequently puts fisheries in a condition known as a "social trap." According to Shelling (1978), a social trap arises when a fisher's micro motives in the short run conflict with the macro-outcomes he and other fishers want in the long run.

While maximizing economic output or maximum sustainable yield may be among the long-term targeted macro goals, the short-term micro motives comprise of capturing as many fish as possible in order to improve individual marginal gains.

Because future stock availability is uncertain, short-run marginal gains typically outweigh long-run outcomes. A sustainable yield from a fishery will typically only be possible when the number of fishermen is limited and they cooperate to apply some kind of effort regulation, taking into account seasonal variations in resource productivity and preferences of resource usage (Seijo, 1993).

A fisher, however, might be an unwitting free rider or a noncontributing user if the group is large. This sort of person typically manifests when the majority of the community members do not take voluntary collective action to stop the depletion of resources and when there is uncertainty over stock abundance (which is the usual case in marine fisheries).

6.4. FUNDAMENTALS OF FISHERIES BIOECONOMICS

One can think of a fishery as a stock or stock of fish and the businesses that have the ability to use them for profit. A fleet of comparable ships from a single port can exploit a single stock of fish in a relatively straightforward manner. Or it may be more intricate, with fleets from several ports employing various technologies to gather fish from numerous stocks that are connected ecologically.

In the course of this chapter, a wide range of fisheries are examined. But the goal is to offer a fundamental understanding using a very straightforward model. Even with their simplicity, the findings usually hold true for more complex but realistic systems.

First, the system's natural capital is the fish stock, as determined by its biomass. Its capacity for procreation and the production of new recruits, as well as individual growth rates, rates of natural mortality, and rates of fishing mortality, are all of importance.

If more biomass is added to the population through recruitment of new individuals and growth of existing ones than is lost to natural and fishing mortality, the stock will grow.

The fishing fleet, which harvests the stock to produce fish for the market and generate income for the fleet owners, is the second man-made component of a fishery. The production function (the correlation between stock size, vessel activity, and harvest), the cost of input, and the price of fish are the relevant factors.

> The fleet's size will typically change depending on the direction and magnitude of net returns. The fleet will generally grow if gross returns outweigh all manufacturing costs. In contrast, if the opposite is true, the fleet size would typically decline as owners sell their ships or go out of business.

Showing how a fishery would likely function in the face of these endogenous yet interdependent changes in the fleet and stock size is one of the goals of fisheries bioeconomics. It will be demonstrated that, if left to its own devices, a fishery would typically run in a way that produces "too much" fishing effort, which will have the effect of decreasing the size of the fish stock, or the "too low" effect.

The tale includes the standards for evaluating what is "too high" and "too low." The second goal is to develop and evaluate methods for controlling participants so as to achieve the appropriate degree of effort. It shouldn't come as a surprise that achieving the second goal requires constant effort. However, fisheries bioeconomics can be a crucial element in creating a useful fisheries management strategy.

It is vital to define the relationships that describe the aforementioned process in order to comprehend how a fishery really functions. What are the stock's population dynamics first? How will fishing affect the fish stock's evolution over time? Secondly, what role does the harvest production serve?

What connection exists between the inputs made and the quantity of fish that will be caught from a particular stock size? Finally, what is the relationship between the quantity of effort performed and net fishing returns, given prices and costs?

These connections will, as one might expect, vary from fishery to fishery based on the species of fish stock, the method used to harvest them, the final product, and the market where they are sold.

The relationships employed in the fisheries model must, to the extent possible given the availability of general knowledge and specific data, mirror those in the fishery in order to understand the behavior of a given fishery.

For teaching purposes, it starts with a few straightforward presumptions about these relationships. This will make it quite simple to make some crucial points. Hence, more, and more complicated relationships are introduced as the discussion progresses.

6.5. THE BASIC BIOECONOMIC MODEL

Now that a basic understanding of the biology underlying population dynamics has been established, one may add economics to the analysis. Discussions of the connections between effort, harvest, and stock size are made possible by the biological study.

However, it is vital to comprehend what degree of effort will actually be produced under specific conditions in order to comprehend how a commercial fishery operates. This is what a bioeconomic model is meant to do. Most people who engage in commercial fishing do so for financial gain.

It is feasible to create a model that can aid in forecasting anticipated levels of effort and output by include information on prices, costs, and how the profit level will fluctuate with output. The one presented here is quite straightforward and is based on the sustainable yield curve. Even though it is straightforward, it can be used to introduce the fundamentals of fisheries bioeconomics.

The bioeconomics of fisheries management is centered on two major issues. The first is: How much effort should be put forth? This inquiry implies an explanation of the origin of the word "should." How much work would likely be done by industry players if left to their own devices? Is the second query.

The significance of the questions comes from the fact that the two questions will have different correct answers. Fisheries bioeconomics is crucial in explaining why they differ and what may be done to shift a fishery toward the desired level of effort. In this paradigm, the questions can only be addressed in a static way, but this will do for the time being.

6.6. AGE STRUCTURED BIOECONOMIC MODEL

The biological model of Schaefer logistic growth that serves as the foundation for fisheries bioeconomic theory. Despite being a straightforward biological model, it is nevertheless used to examine some actual fisheries since the necessary data can be acquired with relative simplicity.

Because it permits simple mathematical manipulation and a smooth transition into simulation modeling, it is also helpful for analytical purposes. The fundamentals of stock size fluctuation both with and without fishing mortality are covered. As a result, it offers a valuable setting for illuminating the dynamics of open access, and optimal, and controlled utilization.

The Schaefer model encapsulates the growth and natural mortality of current individuals, as well as the net consequences of recruiting additional individuals. Furthermore, it is implicitly assumed that when tracking changes in stock size over time, both natural and fishing mortality can be taken into account separately.

Although it is more difficult, utilizing an age structured framework where recruitment, individual growth, and natural mortality are considered separately can provide a clearer view of population dynamics. Age structured models demonstrate how they might be incorporated into a bioeconomic model similar to those previously proposed.

The age structured model can be used to explain most of the economic and fishery management insights that may be learned from the Schaefer model. In fact, it is possible to create sustainable revenue and cost curves as well as population equilibrium and economic equilibrium curves, even if this needs numerical rather than analytical methodologies (PEC and EEC).

In that it enables the examination of age at first capture laws and for the uniqueness of various stock-recruitment connections, it does offer something extra. However, the main reason for introducing it isn't really to learn more about fishery economics. The goal is to ensure that the best bioeconomic model is capable of using economic principles.

6.7. THE FISHERIES MANAGEMENT PROCESS

Fisheries management consists of two distinct tasks. The first is determining, in light of the present biological and economic conditions, the desirable amount of harvest to be harvested. The second is to put restrictions into place so that the actual harvest matches the anticipated harvest, taking into consideration participant and management agency expenses as well as immediate and long-term implications on participant behavior.

As one goes forward, it will be vital to take into account the interrelationships between the two because they are not entirely mutually exclusive. Fisheries management and regulation are the terms used here for the first and second tasks, respectively.

6.7.1. Time Scales in the Fisheries Management Process

It is crucial to understand that fisheries management calls for objectives, studies, and actions on a range of temporal and political scales (from days to years and from local to intergovernmental). The activities at the various dimensions regularly overlap, and the same people, organizations, and institutions will frequently be involved in actions and choices that have an impact on many time- and political scales.

When it comes to transboundary fisheries, these various scales take place within a regional or larger context. When it comes to stocks that are restricted to a particular exclusive economic zone (EEZ) or local area, however, they take place at a national or local level.

There are three primary activities, occurring at and involving different scales, which should be explicitly considered by fisheries management authorities.

6.7.1.1. Fisheries Policy and Development Planning

Fisheries and the wise use of living aquatic resources are frequently crucial to national or local economy. They also interact with other social and economic activities that are physically close by or compete for the use of resources like habitat along rivers and coasts, water consumption, etc.

The macroeconomic and macro-policy backdrop dictates that national development planning plans must be taken into consideration when conducting fishing operations. Therefore, it is crucial that decisions about policies and planning are made with full awareness of the ramifications, expenses, advantages, and alternative resource use options.

These policy choices won't cover the specifics of everyday fisheries management tasks, including particular control techniques, but they should give broad guidelines for how the resources should be used and the priorities that should be set. The criteria by which access to resources is allowed would typically be included in the policy or policies.

The fishing policy could, for instance, specify whether small scale traditional fishermen or largescale industrial fisheries should be given preference in each fishery, or it could specify some other arrangement. The management authority and other pertinent government departments typically advise the government on the formulation and specification of policy. Regular policy reviews should be conducted (e.g., every 5 years).

6.7.1.2. Management Plan and Strategy

The general objectives and priorities to be pursued in the usage of a country's living aquatic resources will often be outlined in the fisheries policy.

The policy must be translated into a detailed management plan for each fishery, which should include the stocks under consideration, the agreed-upon biological, social, and economic objectives, the control measures and related regulations, specifics of monitoring, control, and surveillance, as well as other information outlining how the fishery will be managed.

The management authority should design the management plan and strategy with full involvement from the recognized interest groups and it should be reviewed and evaluated every three to five years, including a performance "audit."

6.7.1.3. Management Implementation

The management plan outlines the specifics of how and by whom the fisheries will be managed. It should have a management procedure that explains in fully how management choices are to be made in response to changes in the fishery, particularly in reaction to yearly variations in resource status.

The management procedure would then specify how the total allowable catch (TAC) is to be calculated annually, for instance, on the basis of stock assessment using commercial catch and effort statistics and the findings of a fisheries independent survey. For instance, the management plan might specify management by a TAC.

The action and decision making required to guarantee that the management plan is implemented and operates effectively comprise management implementation. It therefore includes duties like licensing of fishermen, monitoring, control, and surveillance as well as liaison with interest groups on the status of the fishery and resources in relation to the management plan.

It also includes gathering the data required to make resource and fishery control decisions, such as determining the annual TAC in accordance with the management procedure.

6.7.2. The Paradigm of Modern Fisheries Management

The twin questions of what stock size to aim and how to get from the current stock to the target stock were covered in our discussion of dynamic optimal utilization. These concerns are crucial to the paradigm that forms the basis of contemporary fisheries management policy, despite the fact that those questions originated from a tightly constrained model that was intended to maximize the present value of net income.

This paradigm was established through an evolutionary process, and a large portion of it is based on the Food and Agriculture Organization (FAO) of the United Nations' Precautionary Approach to Fisheries Management.

When reduced to its most fundamental components, fisheries management's objective is to reach a target stock size on a sustainable basis. A policy decision about the size of that stock must take into account social and political management goals while also being bound by the stock, fleet, and environmental conditions at the time. It is not, in, and of itself, a scientific conclusion.

The goal stock size should be determined by the amount of sustainable "benefits" that may be obtained from using the fish stock. Along with the sheer amount of advantages, it may also be important to take into account the stock's resilience to environmental shocks.

And all else equal, the higher the stock size, the more resilient the stock will be. There may be a trade-off between increasing resiliency and accepting lower sustainable benefits.

However, how should these advantages be described? Conceptually, linkages between several significant variables, such as employment and the balance of payments, and stock size should be able to be plotted.

This would provide a wider view on options than simply identifying the locations of maximum yield and net economic returns. Additionally, it might be conceivable to assign relative weights to each of these factors so that the curves can be added to reveal the stock size at which the weighted sum of advantages is greatest.

However, quite frankly, the bulk of fishery management plans in place today stipulate that the best course of action is to increase long-term sustainable catch. In other words, XMSY, the stock that will give the highest sustainable yield, should be the target stock size.

The Magnuson Stevens Act in the United States basically mandates this. Repeating that XMSY is probably not the stock size that will enable the maximization of net returns in sustainable or present value terms should not be necessary.

Furthermore, the maximum economic yield (MEY) or the dynamic maximum economic yield (DMEY) cannot be defined only in terms of the stock size that is provided. Maximizing the profits from harvesting the sustainable harvest from that stock size is also required.

Australia recently released a policy declaration that identifies XMEY as the required target stock size, which is a significant move that suggests the results of bioeconomic analysis may be flowing into practical policy. Given that the objective is to maximize sustainable net returns, it is obvious that they are referring to what has been referred to as static MEY.

Additionally, it states that XMEY is bigger than XMSY, which is in line with recent studies that demonstrate that this is the right relationship even when taking DMEY into account. The need to maximize net returns at the desired stock size is also part of Australian policy.

Given the stochasticity of the marine environment, the practical limitations on the accuracy and availability of data, and the current political-institutional framework for decision making, the general procedure is to use a target/limit approach to set an ongoing policy designed to achieve or maintain the target stock size, once the target is specified.

A predetermined goal stock size, a predetermined limit stock size, and predetermined harvest control criteria are the paradigm's fundamental elements. When the stock is close to or below the limit stock size, further urgency and stringency are provided.

An algorithm known as a harvest control rule determines the desired harvest limit based on stock size at any given time. Target/limit framework is also used in the harvest control rule's application. A limit harvest level is produced by the harvest control rule and is intended to be the maximum amount that may be taken while still allowing the target stock to be maintained or reached in a reasonable amount of time.

The target harvest level is set below the limit harvest level when recommending the actual allowable harvest as a safety measure to maintain the stock on the desired growth trajectory in the event that the estimates of the current stock size, predicted stock change, or the limit harvest level are off.

Using objective and limit reference points, this management paradigm takes both stock and flow aspects of management into account. A foundation for assessing the stock's current condition is provided by the goal and limit stock sizes.

The stock may be deemed overfished if it is now below the limit stock size. If the stock is being overfished in the sense that it is not moving in the desired direction toward the target stock size, then the current harvest rate is over the limit harvest rate.

The standard procedure would be to use a target/limit strategy to set an ongoing policy intended to attain or maintain the target stock size after the target stock has been determined. This is because of the stochastic nature of the marine environment, the limitations imposed by the real world on the accuracy and availability of data, and the pre-existing institutional political structure for decision-making. The paradigm's fundamental elements are an established target stock size, an established limit stock size, and established harvest control rules to keep the stock at or moving toward the target size, with provisions for increased urgency and rigor when the stock is close to or below the limit stock size.

6.8. HISTORICAL PERSPECTIVE ON THE DEVELOPMENT OF THE PARADIGM

By describing the environment in which this paradigm was created, the constituent elements can be more clearly understood. It was primarily created to combat the nearly unanimous belief that the current fisheries management and regulation initiatives were ineffective. Two particular causes of the failure were mentioned among others. Data restrictions were the first, and managerial politicization was the second.

Limited data is a widespread and multifaceted issue. For starters, there is a dearth of sufficient biological data pertaining to stock status and the anticipated effects of fishing. In other words, it is not always practical to create biological models of real fisheries using age class models or the Schaefer surplus production model.

Information is lacking in certain instances, and the majority of times it is incomplete. There are numerous sorts of uncertainty, even when there is information. An overview will be helpful here, even if a more thorough consideration of risk and uncertainty in fisheries management and regulation is necessary. Uncertainty comes in many different forms.

For instance, there is a lot of variability even though recruitment may be connected to the biomass of the spawning stock. The accuracy with which these things can be predicted has its limits. We refer to this as process uncertainty.

Empirical biology studies on stock population dynamics involve three related but conceptually separate types of uncertainty. Observation uncertainty refers to measurement and sampling inaccuracy in the data collection process.

The selection of the appropriate functional form for a given connection is a factor in model uncertainty. Does the equation accurately reflect the way nature works? Which equation the Beverton-Holt or the Ricker equation best describes recruitment? The procedure of estimating the parameters of the equation that is chosen also entails some uncertainty. Scientific uncertainty is a term that is frequently used to refer to these two different types of uncertainty.

The management process itself is not without its own uncertainties. Setting a maximum harvest is one thing; ensuring that the crop is kept within that limit is quite another. This is known as implementation uncertainty, and it is based on how well participant actions can be seen and managed.

The perception of inherent flaws in the institutional framework for fisheries decision-making served as the second driving force behind the development of this paradigm. Making the right choices for safe permissible harvest levels when there was a trade-off between stock maintenance and employment in the fishing sector frequently presented difficulties (Figure 6.4).

Figure 6.4. Fisheries management meeting, Solomon Islands.

Source: Image by Flickr.

Because the advantages of conservation are long-term and speculative while the drawbacks are immediate and certain, politicians frequently make shortsighted decisions. In other words, they are aware that any move to reduce harvest will have an immediate and certain impact on employment for individuals that are easily identifiable and who make up their constituency.

The advantages, however, are perceived as coming in the form of vague stock improvement, are of ambiguous magnitude, and will affect industry participants ambiguously. Additionally, they won't happen for a while, and it's difficult to pinpoint who will benefit. As a result, allowed harvests frequently exceeded what could have been wise.

6.9. ECONOMIC ANALYSIS OF FISHERY REGULATION

The process of managing fisheries, which entails selecting a goal stock size and a harvesting schedule to attain or sustain it, can be challenging and complex. However, the issue of fisheries regulation, which entails figuring out how to manage harvest such that desired and actual catches in any given year coincide is just as problematic.

Events that take place at one time can have effects down the road. Regulations also fit within this category. The effectiveness of the current regulations and the overall capacity to limit harvest in the future can be impacted by actions taken in a single time on the stock and fleet sizes. It's actually so complicated that it's difficult to know where to start when explaining it.

Consider the following condensed history of how regulation to control annual harvest levels has changed over time. It heavily references the Commission on Ocean Policy Report. How participants reacted to the regulations, what effects this had on their effectiveness, and how regulation programs were changed to make changes are key components of the plot.

Of course, fisheries were not at all managed in the beginning. Anyone who wanted to fish could. According to how it was used above, there was open access. Early on, though, governments would demand that participants get a permit for their boats or equipment.

The licenses were freely available for a little cost, and their main function was to retain records and possibly, in some situations, to contribute to the government's coffers. The licenses were not transferable, but that wasn't a problem as they were freely available.

Managers would start putting controls on the behavior of the existing participants if things became bad enough. No attempt was made to question the established practice that everyone should be allowed to fish if they so desired. Any attempt to restrict the number of participants would have been absurd.

Instead of utilizing output restrictions to limit harvest, input controls were used to describe the types, quantities, and methods of gear that were permitted as well as to restrict the accessible fishing grounds or seasons. This sort of restriction raises fishing costs by limiting operators options and encourages them to alter their fishing methods in order to increase catch in light of the limitations.

This has the dual effects of lowering the regulation's biological effectiveness and raising the price of fishing. Additionally, managers implemented output controls, such as establishing total allowed catches (TACs) or trip caps for specific fishermen (Figure 6.5).

Figure 6.5. Fishermen sorting fish for market.

Source: Image by Pixabay.

In order to catch more fish despite the rules and to do so faster than other fisherman, before any overall limit is reached, fishermen are incentivized by these management strategies to invent new types of gear or tactics. Since any uncaught fish is likely to be captured by someone else, neither input nor output controls offer incentives for individual fishermen to postpone or skip fish collection.

As a result of the incentive to maintain one's individual catches as high as possible, which is a necessary component of both input and output restrictions, management, and fishermen are forced to play a frustrating game in which the fishermen always have the ultimate say.

Fishermen are motivated to create new, legal fishing techniques that permit them to raise their effort while still adhering to the rules in reaction to each new restriction intended to limit the amount of fishing effort they are permitted to produce. The resultant rise in harvest drives managers to enact more stringent regulations, while fishermen come up with increasingly creative ways to get around them.

For instance, if boat length is restricted by managers, fishermen may widen the boat if doing so increases fishing power. Fishermen have an incentive to change inputs to catch fish faster than their rivals do rather than aiming to build boats and design equipment that can harvest efficiently with total output controls. If input restrictions are used, fisherman will try to circumvent them.

Because it takes time for fisherman to change their gear or behavior, such rules may be ecologically beneficial in the short term. However, the brief rise in stock size only serves to finance more advancement, such as the development of fishing-powered boat designs. The "race for fish" has been used to describe this phenomenon.

Along with environmental issues, the scramble for fish can also have safety implications. Fishermen could feel pressured to work in hazardous weather circumstances rather than give up harvests to rivals by waiting for better weather if they have a sharply reduced amount of time to harvest.

Managers began to regulate total catch or effort as the next step in the evolution of contemporary fishery management programs by limiting the number of participants through limited access programs. They are widespread now, but when they were originally introduced, they caused a lot of controversy because they went against the widely believed notion that everyone should have the right to fish.

Limited license programs were the prevalent name for these schemes. The licenses that were granted as part of limited access schemes have undergone a number of significant alterations.

In the first place, they were by definition not generally available but rather provided in small quantities to particular participants, typically to a subgroup of current members.

Second, they were often a permission to operate a certain size and style of boat rather than just a broad License to fish. Thirdly, they were typically transferrable. A permit had to be purchased from an existing owner in order to take part in limited access fishing.

Finally, there were guidelines for changing an allowed vessel when the permit was sold or if the owner desired to buy a new boat. The true goal was to restrict the allowed fleet's capacity to take catches, not merely the quantity of licenses.

6.10. BIOECONOMICS OF ECOSYSTEM INTERDEPENDENCIES

The provided single species and single fleet, distributed uniformly over space in a deterministic setting, analytic, and numerical bioeconomic development assumptions are eased. Relevant ecological interdependencies across species along the trophic web must be taken into account when managing fisheries with environmental considerations.

Over time, it can be crucial to comprehend their dynamics in order to better comprehend fishermen behavior. The extent to which multispecies and their bioecological interdependencies should be incorporated into bioeconomic modeling and analysis, however, will depend on the pertinent fisheries and ecosystem management questions raised to address stock recovery and sustainability strategies within an ecosystem framework

The availability of bioecological and economic data for serious parameter estimation of increasingly complex mathematical models needed to address the identified. We must take into account the variability of fleets targeting or incidentally collecting the chosen species through time in the context of the pertinent ecosystem.

Expanding the single-species bioeconomic approach to take into account the biological and financial interdependencies inherent in the ecosystem was acknowledged as being important by Sinclair and Valdimarsson (2003) and Van den Bergh et al. (2007). We will be working with the ecological and technological interdependencies of multispecies and multi-fleet fisheries in order to address these ecosystem elements.

The physical and bioecological characteristics of an ecosystem are spatially characterized and understood, and this is crucial for managing fisheries using an ecosystem approach.

When attempting to design spatial management strategies of single stock populations and/or metapopulations, understanding the geographical borders of population distribution over space, their needed habitat, and potential interdependent stocks of the same species connected together by dynamic oceanographic patterns during their life cycle, becomes a major input.

In this ecological context, functions that disperse recruits in areas of variable density take the role of functions that assume a homogeneous recruitment distribution over space. The management of fisheries that target stationary or low mobile species will benefit the most from this.

To comprehend and describe fishers spatial behavior over time, it is necessary to expand costs and revenues functions over space when introducing the spatial dimension to the bioeconomic study.

Understanding how fishermen choose their next fishing location based on past visits and the most accurate dynamic fishing community information of resource availability in alternative fishing sites is essential to this analysis. When calculating steaming and fishing expenses over space and time, port locations and the related distances to fishing locations become essential inputs.

Therefore, estimating quasi-profits of variable costs at alternate fishing sites throughout earlier time periods is necessary to depict spatial fishing behavior across time. The circumstances of single-stock fisheries and metapopulations, as well as the bioeconomics of spatial dynamics in marine fisheries.

6.10.1. Current Challenges of the Ecosystems Approach to Fisheries (EAF)

When creating EAF in coastal states, the following major concerns will need to be resolved (Seijo, 2007): In changing management practices to implement an EAF is likely to result in disagreements with stakeholders; this needs to be taken into account and accommodated when creating an EAF for a particular fishery.

In developing coastal states where it is already challenging to implement appropriate:

- Data collection for single species, obtaining scientifically valid data in support of fisheries management using an ecosystem approach could present major challenges;
- Data collection requirements are higher with the EAF than with single target species analysis of fisheries;
- In improving coastal states where it is already hard to implement sufficient data collection for single species;
- Information costs may need to be covered by the various users of the ecosystem in order to meet the fundamental requirements for implementing an operational EAF;
- Expenses of building and maintaining data collection and analysis systems for full marine ecosystems and their users (i.e., small-scale, and industrial fishers, ecotourists, and non consumptive users) are likely to be important;
- The management of fisheries must take into account the limited knowledge and growing uncertainties regarding biotic, abiotic, and human components; and
- The emphasis cannot be solely on biological monitoring but must also take into account the human dynamics involving institutional, economic, and social dimensions.

As acknowledged by Cochrane et al. (2004), the adoption of EAF is likely to be gradual and many nations, organizations, and people are still working to comprehend and interpret exactly what is meant by the word EAF. Because there are so many more unknowns and dangers, the transition to EAF would frequently be managed incrementally and adaptively.

The time needed to learn and gain knowledge and the requirement to properly evaluate the distributional effects of EAF initiatives are two factors that demand consideration. EAF goals and guiding principles need to be updated and broadened to more accurately reflect institutional, social, and economic ramifications.

It would be helpful to know the fundamental steps required to set up a frugal bioeconomic ecosystem approach to fisheries within this complex framework.

6.11. SPATIAL MANAGEMENT OF FISHERIES

The last ten years have seen research and experimentation into the spatial management of single stocks and metapopulations as fisheries management shifts from single species to ecosystem concern. The results of adopting marine protected areas (MPAs) for fishery spatial management along with adjacent and distant alternative port options.

The implications of an MPA's location in a setting involving a metapopulation with a source sink arrangement are then demonstrated.

6.11.1. An MPA to Manage a Single-Stock Population Under Open Access

When using an MPA to manage a single stock population spatially, it is important to take into account how its location may affect the fleet's target port. When the MPA is situated either close to a port or far from the port of origin, we see the spatial fishing behavior.

The model correctly anticipated that there would be no fishing in the MPA's alternate locations. Due to the high expense of steaming, effort is typically distributed closer to the fleet's port of origin in both situations.

We have the species biomass trajectories with regard to time and effort in three scenarios under open access:

- Without an MPA;
- Having an MPA close to a port; and
- Without a port and with an MPA

In the absence of MPAs, species biomass declines more quickly than in situations where MPAs are developed. As is to be expected, over time, biomass tends to be at higher levels when the MPA is close to a port as it approaches bioeconomic equilibrium (Figure 6.6).

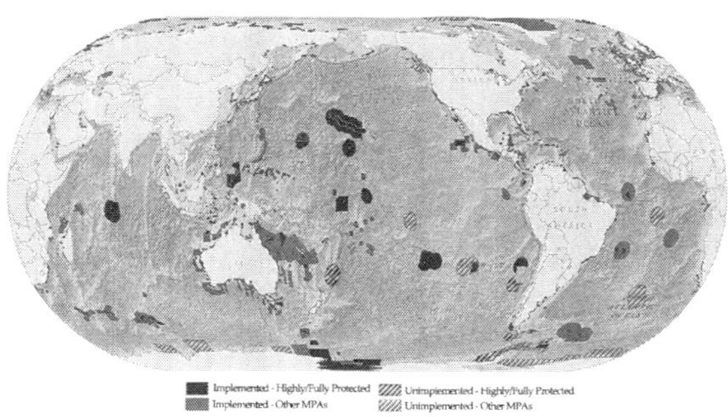

Figure 6.6. MPAs globally in 2020.

Source: Image by Wikimedia Commons.

The cost of steaming the particular fishery will determine the relative differences in the bioeconomic equilibrium biomass for the three scenarios. The cost-benefit analysis of the usage of MPAs as a spatial management technique for marine fisheries and an overview of recent modeling work to understand their potential performance under various resource and effort dynamics over space and time are provided.

6.11.2. Spatial Management of Fisheries: MPAs

An MPA is a geographical region with clearly defined borders that has been set aside to help with marine resource protection. The formation of marine reserves is frequently mentioned as a novel technique for stock and ecosystem recovery.

The creation of marine reserves may have the following advantages:

- Preservation of spawning biomass
- Supplying a source of recruitment for the neighborhood
- Supplemental restoration of fished areas through emigration
- Preservation of natural population age structure
- Preservation of untouched habitat
- Protection from poor management in fisheries areas, and

Through two direct mechanisms, MPAs help neighboring fisheries:

- Net emigration of adults and juveniles from protected to unprotected areas; and
- Export of eggs and larvae that will eventually settle in unprotected habitats where fish can be caught.

The first will be influenced by the degree of resource mobility throughout space (e.g., Walters, 2000), and the second by mechanisms involving the dispersal of eggs and/or larvae. According to Gell and Roberts (2003), populations tend to grow in size, people live longer, and their reproductive contribution increases in marine reserves.

6.12. DEALING WITH RISK AND UNCERTAINTY IN FISHERIES MANAGEMENT

Fisheries management has discovered in recent years that elements about which information is typically lacking have an impact on population dynamics. It is frequently unknown or difficult to determine the significance of bioecological factors that affect population dynamics and bioeconomic analyzes.

Interdependencies between species and fleets on both an ecological and technological level may have an impact on fishing a particular target species.

The study of the dynamics of oceanographic processes dispersing larvae that will eventually settle in source or sink areas, where habitat and food availability are crucial for defining the dynamics of metapopulations, is another aspect of understanding and estimating resource and fisher behavior over space and time.

It is important to take into account how much the local ecology affects population dynamics because this relationship is frequently highly complex. It is not well understood how changes in the environment affect fish populations locally or globally.

In addition to the observable periodicities of these oscillations and the recently discovered correlations with fish harvesting, the underlying cause-effect mechanisms have not yet been pinpointed with a high degree of certainty.

In order to effectively prevent or mitigate overexploitation and overcapacity, fisheries management must acknowledge that not only are fish population dynamics complex and affected by factors that are frequently unexplained, but that it is also difficult to predict fishers behavior over time and space.

The detailed understanding of the factors that affect fishing behavior, which can differ based on fishermen's cultural background and context, the technology used for fishing, as well as perceptions and strategic behavior affecting compliance with the regulatory scheme in place, is necessary for managing fisheries by prescribing fishing effort.

It is essential to have knowledge of the population dynamics of the species, the interaction of fish populations with their environment and other species of the ecosystem, and the dynamics of fishers behavior in order to manage fisheries successfully.

This is in addition to getting adequate institutional systems in place that evidently allocate property rights to resource users (individual or community rights) and selecting regulatory input and/or output controls.

As was already said, this is a complicated undertaking with stochasticities and complexity that calls for responsible management that accepts the uncertainties and uses decision making techniques that explicitly and methodically take them into account.

In addition to Fogarty et al. (1992) and Berkson et al. (2002), Hilborn, and Peterman (1996) have noted a variety of sources of uncertainty related to stock assessment and management methods, such as uncertainty in resource abundance, model structure, model parameters, resource user behavior, future environmental conditions, and future economic, political, and social conditions.

In order to manage fisheries in light of these many uncertainties, it was advised at the Lysekil meeting (FAO, 1995) to use Bayesian and non-Bayesian decision theory as well as the incorporation of limit and target reference points. The list of sources of uncertainty stated above has lately become more complex due to mounting evidence of the effects of climate change on fisheries.

The distribution of marine and freshwater species is changing as a result of climate change, according to data presented by Cochrane et al. (2009). In addition to experiencing changes in the size and productivity of their habitats, species are being forced toward the poles.

Additionally, it is anticipated that ecosystem production will tend to rise at higher latitudes and decline in tropical and subtropical ocean regions. The physiological processes of species may be impacted by higher temperatures in freshwater and marine habitats, which could have both positive and negative consequences on fisheries.

This will depend on the species' sensitivity to variations in salinity and temperature as well as their ability to move about and choose settings that are more conducive to them. The aforementioned study notes that seasonality of important biological systems is being impacted by climate change. Additionally, it is changing food webs, with unpredicted effects on fisheries production.

One possible effect of climate change on marine fisheries is a change in species abundance, which can affect reproduction, recruitment, and individual growth patterns. Another potential effect is a change in ecosystem productivity, which supports fisheries. A possible third effect is the change in species availability and their patterns of spatial distribution.

6.12.1. Climate Change Increases Uncertainty in Marine Fisheries

Fish production is becoming more uncertain as a result of climate change, creating additional difficulties for risk assessment, which is often predicated on the likelihood that future occurrences will occur. The scant historical evidence of potential impacts of climate change on fisheries may not be sufficient to set expectations for the future.

Possible outcomes and adaptive responses of fishers to gradual climate shifts, as well as synergistic interactions among climate change and other stressful situations affecting coastal fishing communities, are some of the increased uncertainties that fisheries supervisors and other decision makers are facing today (in addition to those reported by Hilborn and Peterman, 1996) Eutrophication, overfishing, runoff pollution from subpar agricultural and animal production techniques, and the capacity and resilience of fishing communities to combat these numerous stresses on coastal environments are a few examples (Figure 6.7).

Figure 6.7. Fish aggregating device (FAD) being deployed in Gwanatafu, a fishing community in North Malaita, Solomon Islands.

Source: Image by Flickr.

There are straightforward methods for including these expanding uncertainties and related risks in the bioeconomic analysis of fisheries. An application of decision theory to a stock fluctuation fishery, with, and without mathematical probability, to address specific management concerns, illustrates the uncertainty prevalent in crucial biological and economic characteristics of most fisheries.

Through the application of Monte Carlo analysis to a scenario with numerous fleets impacting various population structure components, the risks of surpassing limit reference points of biological and economic performance characteristics are demonstrated.

For the estimate of the dynamic bioeconomic consequences of alternative stock recovery procedures to meet LRP criteria, a multi – fleet age-structured model, similar to the one established, is utilized. The fishery bioeconomic model, management choices, and decision criteria with and without mathematical probability are linked with the relevant risk analysis, which uses the Monte Carlo method to evaluate the likelihood of exceeding limit reference points.

6.13. CONCLUSION

In this chapter, the role of bioeconomics in fisheries management have been explained. It also discusses about the importance of fisheries management and various regulations needed to manage it. In this chapter, different fundamentals of fisheries bioeconomics have also been discussed. It provides highlights on the basic bioeconomic model. Towards the end of the chapter, it discussed about the economic analysis of fishery management, spatial management of fisheries.

REFERENCES

1. Anderson, L., & Seijo, J., (2010). *Bioeconomics of Fisheries Management.* [online] https://www.wiley.com/. Available at: https://www.wiley.com/en-us/Bioeconomics+of+Fisheries+Management-p-9780813817323 (accessed on 21 September 2022).
2. Fao.org. (n.d). *Introduction to Fisheries Management.* [online] Available at: https://www.fao.org/3/w4230e/w4230e05.htm (accessed on 21 September 2022).
3. Gatto, M., (1988). Bioeconomic modeling and fisheries management. *Ecological Modeling, 42*(2), 161, 162. [online] Available at: https://linkinghub.elsevier.com/retrieve/pii/0304380088901147 (accessed on 21 September 2022).
4. Gupta, A., (2016). Bioeconomic fishery management. *Advances in Environmental Engineering and Green Technologies*, [online] pp. 261–281. Available at: https://www.igi-global.com/gateway/chapter/140575 (accessed on 21 September 2022).
5. Shepherd, J., & Clark, C., (1986). Bioeconomic modeling and fisheries management. *Biometrics,* [online] *42*(3), 683. Available at: https://www.jstor.org/stable/2531229?origin=crossref (accessed on 21 September 2022).

CHAPTER 7

BIOECONOMICS OF AQUACULTURE

CONTENTS

7.1. Introduction 184

7.2. Bioeconomic Modeling and Salmon Aquaculture 186

7.3. Preliminary Analysis of the Culture Potential of the Freshwater Angelfish: ***Pterophyllum Scalare*** ***192***

7.4. Farm Design 197

7.5. Marketing and Economic Considerations in the Production of ***P. Scalare*** ***203***

7.6. The Bioeconomics of Recirculating Aquaculture Systems 204

7.7. Bioeconomic Model 207

7.8. Technology and the Bioenergetic Model 210

7.9. Limitations and Future Research Directions 214

7.10. Conclusion 215

References 216

Natural fish resource depletion would have a significant impact on future global food supply. Fish is presently the principal protein source for around one billion people or one-fifth of the world's population.

The one and only option to alleviate the rising shortage of global fish supply in the future would be to effectively manage fish stocks, embrace ecologically friendly technology, and promote equality in the distribution of fish supply through policy-driven initiatives as well as through farming.

7.1. INTRODUCTION

At the moment, the significant growth in fish supply is mostly owing to an expansion in aquaculture. Fisheries is the fastest expanding agricultural sector. And over 220 finfish and shellfish species are grown. It currently accounts for around 27% of all seafood consumed by people globally and 18.5% of total global seafood output.

Since 1984, global aquaculture production has nearly quadrupled, and the UN Food and Agricultural Organization projects that by 2030, aquaculture will control fish production, with farming producing more than half of all fish consumption. Species richness, such as shrimp and salmon, are two of the major marine cultivated species (FAO, 2003).

Although aquaculture intensification has enormous promise for boosting global fish supply and resolving issues around food security, poverty, livelihood, and income, it has not been without consequences and disputes. Many marine and diadromous finfish are raised in movable net cages close to shore, for instance.

The aquaculture industry's rapid expansion and technological advancement have frequently outpaced society's ability to regulate the growth of this diversified and dynamic sector. This has had a negative influence on the ecosystem, limiting the expansion and development of aquaculture and affecting other resource users.

As a result, pollutants and other wastes are released, ecosystems, and wild populations are degraded and destroyed, genetic, and biological diversity is lost, disease outbreaks occur, and resource use disputes occur.

While the salmon and shrimp industries provide economic advantages, they also provide the most visible instances of the potential and actual detrimental environmental repercussions of aquaculture when management is poor (FAO, 2003).

Because of the complexities of aquaculture production environments and the numerous constraints presented by the industry's fast expansion, extensive modeling initiatives are required to offer growers technical capabilities and policy choices for decision-making.

As per Cuenco (1989), there are numerous reasons why aquaculture systems should be modeled, such as:

- Modeling is an excellent tool for developing, testing, and improving hypotheses and theories.
- Models could make sharp predictions about the effects of various management techniques on the system.
- Models could provide informed predictions about the impacts of multiple management techniques on the system.
- Modeling offers a functional tool for swiftly conducting several 'what if' tests and makes it easier to examine the ramifications of different assumptions or management techniques for large and complex intensive aquaculture systems, that are rarely practicable in their naturalistic setting.
- Models act as a means for identifying what is unknown by arranging whatever is learned within the context of the models.
- Models aid in the assessment of complicated connections in aquacultural systems.
- Modeling increases the application of much more quantitative and exact methodologies in aquaculture research.
- These models may combine knowledge from theory, laboratory, and field investigations into a cohesive whole in order to discover knowledge gaps, sparseness, and/or inconsistency. Models can integrate knowledge from theoretical, laboratory, and field studies into a consistent whole so as to identify areas where knowledge is lacking, sparse, and/or inconsistency.

Bioeconomic models are an excellent analytical technique for studying the interaction of the various components (biological, physical, technical, economic, and institutional) of aquaculture systems. Bioeconomic models may address concerns about financial viability, optimal configuration design, optimal operating methods, and recommendations for further research (Leung, 1994).

Nevertheless, bioeconomic models cannot be directly generalized between species since each species development is governed by unique circumstances and characteristics.

7.2. BIOECONOMIC MODELING AND SALMON AQUACULTURE

7.2.1. Bioeconomic Models: An Overview

Bioeconomic models, in particular, relate to the application of mathematical approaches to simulate the efficiency of 'living' production systems subjected to economic, biological, and technical limitations (Allen et al., 1984).

Bioeconomic models are concerned with the systematic integration of biological performance and physical systems and their relationship to economic factors such as market pricing, allocation of resources, and institutional restrictions (Cacho, 2000).

In comparison to traditional output function analysis, bioeconomic modeling provides an alternate technique for representing the production process. Because biotechnical linkages may be more thoroughly specified, it enables evaluations of a broader variety of environmental variables than would generally be achievable with solely economic models (Leung, 1994).

Bioeconomic models may help producers and decision-makers find optimal production system designs, operation management methodologies, and alternative development and policy initiatives. Effectively built bioeconomic models can give guidelines for planning private aquacultural plans in reaction to optimal sector development and regulatory problems.

Bioeconomic models for aquacultural systems are still very restricted in comparison to agricultural systems. The use of bioeconomic models for marine resources began with dynamic optimization analysis for the growing population (Clark and Munro, 1975).

Such models were subsequently modified to include optimal control and current capital theory (Clark, 1985). Bioeconomic models of aquaculture production have concentrated in the last decade on producing a more accurate representation of biological systems and establishing relationships with more complicated economic models (Cacho, 2000).

Aquacultural output presents a serious challenge to economic modeling. Because the aquaculture farmer cannot clearly see the development of the "crop," management choices must be based on indirect and subjective metrics of productivity.

Even though resemblances with the growth of other constrained animals have been an important initial step for some aquacultural production models, critical discrepancies especially in pond ecology, tracking, and feed utilization necessitate appropriate analysis sophistication that the other animal production processes do not need (Hatch and Kinnucan, 1993).

Because aquatic species are sensitive to heat as well as other climatic influences (e.g., soluble oxygen, ammonia, salinity, and pH) the complex combination of ecological parameters could have a considerable influence on aquaculture development and profitability. The most critical management concern is undoubtedly the tracking issues linked with the growth of a population that can't be seen or controlled.

At any particular time, the producer has no control over the number of animals or their condition. To reflect the difficulty to track the growth of the animal population during the growth period, aquaculture models can frequently require a stochastic feature.

Feeding usage is yet another source of confusion for aquaculture production facility managers since the quantity of feed actually ingested by the fish can be noticed qualitatively when fish rise to the top to feed, but cannot be determined with so much accuracy.

7.2.2. Application of Bioeconomic Models to Aquaculture: A Review

Leung (1994) discovered how mathematical optimization models, particularly risk programming, were widely employed in fisheries during the 1984–1993 era in the investigation of the usage of bioeconomic models in aquaculture.

Leung believed that such a phenomenon might be attributed to a number of factors, along with an increment in the whole farm implementations, the acceptive of risk programming mostly in the early 1980s, as well as the accessibility of more potent computer software, like generalized additive model (GAMs), which encouraged the use of non-linear as well as integer model types.

The increased usage of dynamic programming models from 1984 to 1993 was mostly caused by an increase in production planning systems. Because of the change in modeling species in less predictable culture conditions, there were fewer optimum control applications throughout the same time span.

More budgetary simulations were created between 1984 and 1993, while very few process simulators were created. This development might be attributed to a shift toward more comprehensive agricultural evaluations.

Given the scarcity of works focusing on bioeconomic models since 1993, it was thought worthwhile to revisit prior publications focusing on salmon aquaculture bioeconomic studies. Some of the important elements of four articles. The first one is by Johnson (1974), who used linear programming analysis to plan launch dates and stock selection at a hatchery.

Gates et al. (1980) used dynamic linear programming to optimize fish culture decisions in a water reuse system and a financial feasibility study for salmon grow-out in the United States' New England area. Bjorndal (1988) used optimal control theories and the Faustmann model to optimize salmon cultivation harvests in Norway.

Bjorndal (1990) was the first to release a book on the economics of salmon aquaculture that provided extensive data and analysis on salmon farming. Sylvia and Anderson (1993) used a dynamically multilevel programming model to optimize public and private net pen salmon farming operations in the US Pacific Northwest. This final study is particularly relevant in this context since it created data for both private and public salmon aquaculture policy plans, addressing concern for the environment.

7.2.3. Issues in Bioeconomic Modeling of Salmon Production in Chile

Since it began salmon aquaculture in 1986, Chile has been the world's leading producer of exported salmonids. Chilean salmon exports totaled \$973 million in 2001, accounting for 3.5% of total Chilean exports and nearly 35% of fishing-related exports. During that year, the nation's total export revenues were \$1.8 billion (EMS, 2003). Southern Chile is an excellent site for aquaculture. Its mild temperature, large undiscovered fresh and salt-water regions, lack of ice cover on its enormous lakes, and protected saltwater sources provide year-round secure growing habitats for farmed fish (Figure 7.1).

Figure 7.1. Aquaculture fish farming in the fjords south of Castro, Chile.

Source: Image by Wikimedia Commons.

Nonetheless, current aquaculture techniques in Southern Chile have sparked a heated debate about the environmental impacts of both freshwater and saltwater systems (Buschmann et al., 2002).

Benthic pollution, water column pollution, disease control, genetic impacts, exotic introductions, toxicants, and antibiotics, effects on marine animals and birds, noise pollution, and esthetics are all causes of environmental issues.

Furthermore, forestry, mining, as well as urban and industrial wastewater discharges, compete for resources and/or damage salmon aquaculture. (Weber, 1997; Goldburg and Triplett, 1997).

In land-based tray systems, salmonid aquaculture begins with a three-to five-month egg hatching period. When the young salmon hatch, they are put in freshwater net enclosures until they reach smolt stage 2. The young fish are subsequently transported to grow-out enclosures in the sea.

Both the freshwater and saltwater phases of salmon aquaculture result in the release of nutrients (particularly nitrogen and phosphorous), organic compounds associated with feeds (carbon, growth hormones, anti-parasitic drugs, disinfectants, and antibiotics), and bioaccumulated metals or hydrophobic organic compounds associated with the fishmeal-based food fed to hatchlings but not retained in fish tissues.

Most of these environmental consequences have been widely recorded across the world. However, the underlying ecosystem systems are poorly understood and need more careful study (Buschmann et al., 2002).

As a result, the long-term viability of such techniques and the long-term effects of expanding aquaculture on freshwater lakes and coastal zones remain unknown. Environmental issues in the salmon aquaculture sector across the world, as well as in Chile, have had an influence on firm-level production plans and regional industry growth.

In certain situations, environmental issues have imposed direct production costs on producers by influencing salmon growth or death rates, or indirect costs as a result of the implementation of public laws aimed at reducing environmental externalities by regulating private production practices.

Many nations have adopted different tactics to deal with the environmental impacts associated with salmon farming. Environmental regulations in Japan and Norway have encouraged the spread of salmon aquaculture, but Scotland, Ireland, the United States, and Canada have imposed stronger limits on business behavior in order to prevent or minimize environmental externalities.

Salmon aquaculture has caused tension and controversy in the United States, leading to moratoriums and in some cases, outright bans of the sector. (Goldburg et al., 2003).

Environmental challenges have been particularly difficult in net pen salmon aquaculture. These issues have had an influence on firm-level production tactics, production costs, and the evolution of the industry.

The Chilean salmon aquaculture sector has a critical problem in determining how to co-develop both private and governmental policies that might decrease externalities while allowing responsible aquacultural growth to occur in a cost-effective manner.

Economic models that are strategically built can provide information for building private aquaculture plans in response to optimal industry development and regulatory concerns. Information may be utilized to make firm-related choices and to enlighten decision-makers about different development scenarios and the possible consequences of various regulatory policies.

The most relevant piece of research addressing these concerns in our evaluation of bioeconomic models relating to aquaculture is by Sylvia and Anderson (1993), which is detailed below. To assess the feasibility of different technologies and policy alternatives for balancing ecosystem management, the models generated must incorporate biological, technical, and environmental processes.

These models should allow for a cost-benefit analysis of different production tactics and locales, as well as legislative and regulatory frameworks. The findings could help to shape both management approaches that maximize economic benefits to producers and legislative proposals that safeguard the aquaculture industry's long-term development.

The findings should also be included in instructional programs dealing with aquaculture's long-term growth. Because real mathematical algorithms capable of resolving all significant concerns have yet to be fully established, a multi-level and multi-objective model addressed by a multi-stage approach may be the most appropriate.

When environmental externalities are an issue, the use of multi-level and multi-objective analysis can address both private and public concerns in aquaculture. To investigate the possible costs and advantages of different production processes or different production techniques could be assessed in light of various governmental regulatory frameworks.

Furthermore, instead of focusing primarily on biological or environmental implications, the multi-level method may be utilized to address social and economic concerns as well as the effects of various regulatory regimes. In the Pacific Northwest of the United States, Sylvia, and Anderson (1993) created a multi-level and multi-objective model for net pen salmon farming.

The model assumed that producers would maximize profits, while public policymakers faced four policy objectives: revenue, benthic quality, profits, and tax revenues. The study's policy tools include the number of authorized sites and the effluent fee. Different site locations, optimal methods could be devised.

Environmental concerns in the aquacultural systems have resulted in increasing company production costs as well as indirect expenses due to environmental liabilities. The problem for Chile's salmon aquaculture business is to figure out how to build both private and governmental solutions that decrease externalities while supporting safe aquacultural growth.

Economic systems that are appropriately built could give data that can be useful for making effective firm-related choices as well as advising policymakers regarding different growth scenarios and the possible implications of alternative regulatory measures.

If environmental externalities are a concern, bioeconomic modeling can help the salmon aquaculture business manage both private and public policy challenges. Bioeconomic models may address concerns about economic feasibility, optimal system design, and optimal operational procedures.

Optimal strategies can be devised by investigating the possible costs and advantages of different production processes or alternative site locations. This sort of information may be used to direct the policy process, focus discussion, and assess policy choices in order to maximize both private profits and societal welfare.

By including biological, environmental, economic, and institutional components, multi-level, and multi-objective bioeconomic models can help to solve private and public policy challenges influencing salmon production. Long-term expansion in the aquaculture business needs both environmentally sound techniques and resource management that are sustainable.

Such methods may be fostered through the establishment of best management practice standards for salmon aquaculture in Chile that encompass best practices in site selection, technology, and management. Commercial producers productivity and fish health, as well as government policy and regulatory measures. The outcomes of bioeconomic models may help to generate such effective management practice suggestions.

7.3. PRELIMINARY ANALYSIS OF THE CULTURE POTENTIAL OF THE FRESHWATER ANGELFISH: *PTEROPHYLLUM SCALARE*

The aquaculture sector has always been characterized by a high failure percentage of new initiatives and poor profits, and the Australian aquaculture industry is no different.

Even though these are to be anticipated in any growing sector, failures in several instances could be linked to a lack of understanding of the basic connection between a species' biology, the physical culture system, site selection, the economics of the culture system, as well as the market opportunity of the aquaculture species (Allen et al., 1984; O'Hanlon, 1988; Pillay, 1990; Shepherd and Bromage, 1988; and Logan and Johnston, 1992). That lack of knowledge frequently results in poor farm layout, planning, and administration (Figure 7.2).

Figure 7.2. Freshwater angelfish – Pterophyllum scalare.

Source: Image by Wikimedia Commons.

Aquaculture has been practiced for over 2,500 years in various regions of the world. Yet, it is only lately that this has started to progress from a primitive, survival scale of output and innovation towards a more complex, fully advanced procedure (Allen et al., 1984; Pillay, 1990; and Shepherd and Bromage, 1988).

However, aquaculture has still yet to undergo the same level of growth as agriculture, which has led to the domestication of several plant and animal species (Raven and Johnson, 1990). Aquaculture, on the other hand, may claim just a few cultivated food fish species, like common carp, *Cyprinus carpio*; channel catfish, *Ictalurus punctatus*; rainbow trout, *Oncorhynchus mykiss*; Atlantic salmon, Salmo salar; and Tilapia spp. (Allen et al., 1984; Pillay, 1990; and Stickney, 1994).

In comparison, the ornamental fish market contains several varieties of fish that have been tamed over several years via ongoing breeding. The goldfish, Carassius auratus, is the most well-known of these species, having been raised in China for 2,500 years (Penzes and Tolg, 1986).

Guppies, *Poeclia reticulata*; angelfish, *Pterophyllum scalare*; gouramis, Trichogaster sp.; platys, Xihophorus variations; and Siamese fighting fish, *Betta splendens* are instances of additional domesticated species present in this sector.

The ornamental fish business is broadly accepted around the globe (Bedford-Clarke, 1993; Bassleer, 1994), with global wholesale sales of ornamental fish of more than US $4 billion per year (Winfree, 1989) and retail sales totaling more than US $7.2 billion in 1986. (Andrews, 1992).

The market is well developed in Australia, having ornamental fish or aquarium fish hobbyists accounting for around 11% of Australian homes (Humphrey, 1989), and at least one million enthusiasts (O' Sullivan, 1991). Every year, the Australian ornamental fish company sells aquarium fish worth more than $80 million in market cost. (pers comm, R. McKay, 1991).

The vast majority of ornamental fish marketed in Australia are sourced from ranches and wild fisheries in other countries. The total number of ornamental fish imported into Australia in 1993–1994 was 7,872,909 tails, with a total value of more than $2,720,000 landing in Australia (Australian Bureau of Statistics Foreign Trade Data).

Because of the rise in import costs, it has become more cost-effective and appealing for Australian hobbyists and farmers to breed numerous species commercially, particularly the more skilled, higher-value lines of tropical ornamental fish.

The sector is predicted to develop further over the rest of the 1990s and to have promising future opportunities, with output for 1994–1995 projected to be valued at about $10 million (O'Sullivan, 1991). In Australia, around 20 ornamental fish species are now grown on a commercial basis (McKay and Reynolds, 1983).

One such case is Pterophyllumscalare (Lichtenstein) (Pisces; Cichlidae), a popular medium-priced cichlid. In Australia, commercial production of this species is presently modest, and the ecological, marketing, and economic elements of commercial production are little known.

P. scalare is a famous species with a well-established reputation in the aquarium business, with a variety of color variants and also long and short varieties generated by selective breeding. The Australian industry for angelfish is projected to be in the range of 320,000 tails every year, including sales of 56% tiny, 40% medium, and 4% giant-sized (per comm, R. Datodi, 1992).

The bulk of *P. scalare* marketed in Australia is currently sourced from Asian farmers. This shows that there is still a good possibility for *P. scalare* production in Australia as an import substitution. These species' supplies are no longer dependent on wild fisheries (Brown and Gratzek, 1982), and their needs must always be fulfilled by farm-produced fish (Axelrod and Burgess, 1979).

Nonetheless, little information is known on the economics of growing this species, or any other attractive fish species, in Australia or elsewhere. With ongoing demand in the ornamental fish business and inconsistency in supply, *P. scalare* has the possibility of being an aquaculture species.

7.3.1. Bioeconomics in Aquaculture

Historically, assessing the cultural capacity of a particular species has been basic and one-dimensional, depending on a species commercial potential (Allen et al., 1984). Nevertheless, in addition to analyzing a cultured species commercial potential, careful consideration should be given to its biology and the physical culture technique to be employed.

A more integrated approach has lately been embraced, with these characteristics being considered as a dynamic, interconnected system. This approach recognizes the dynamic interdependence that exists between different fields.

Bioeconomics refers to the integrative links that exist between the biological and economic aspects of a physical production process, Allen, and colleagues (1984); O'Hanlon (1988); Shepherd and Bromage (1988).

Instead of the typical one-dimensional vision of a farm as merely a series of accounts, this connection considers the aquacultural operation as a fully dynamic system that takes into consideration the interrelationships of three domains. Bioeconomics evaluates a species' culture in three functional domains as outlined by Allen et al. (1984). These functional categories are as follows:

- The biological properties of the cultivated species,
- The physical culture system layout and administration; and
- The economic outcomes of the physical culture system and the commercialization of cultivated species.

From a bioeconomic standpoint, the biological properties of the cultivated species are likely the most basic and restricting aspect. For effectiveness, a thorough grasp of reproduction, growth, and development, nutrition, and physiological functioning is required (Allen et al., 1984).

Basic scientific study on the selected species' biology is required to do this. These properties could be represented as the biological sub-model of the culture system. The technological component of a culture system is equally crucial, and it must be built to fulfill the biological properties of the cultivated species as well as the economic objectives of the culture idea (Allen et al., 1984; O'Hanlon, 1988).

The physical system, represented by the physical sub-model, must fulfill the physical needs for maximum biological function at the most cost-effective rate (O'Hanlon, 1988). To achieve this purpose, the physical model must supply the following: adequate water purity, enough area for progress and expansion, and a proper feeding schedule.

The physical system governs and manages the fish culture environment, supplying the abiotic elements required for reproduction and development (e.g., space for fish, water of good quality, food, oxygen, and waste removal). The fish in the system respond by living to reach commercial size, at which point they are sold and the network is refilled.

The concepts of economics must apply to the fish farming operation, as they do to any other commercial activity, both on a microeconomic (i.e., farm) and a macroeconomic (i.e., industry, home market, international market) level (O'Hanlon, 1988), with changes in the external environment just as relevant as changes in its internal environment (i.e., availability, input costs, pricing, and need of product).

The following economic elements must be addressed while assessing the output of a specific cultural system:

- Market evaluation or anticipated market for the product;
- Production costs;
- Projected sales and revenue;
- Cost and revenue fluctuation; and
- Business profitability.

Operational expenses (i.e., the expenses of inputs such as heating, pumping, food, chemicals, and labor), capital costs (i.e., the money necessary to develop plants and machinery, structures, and space), and marketing expenses are all examples of expenses.

For any particular production situation, the functional relationships between both the biological and physical elements result in a distinctive set of economic situations (i.e., cost of production, net cash flow, earnings), with adjustments in any facet of the biological and/or physical parts that have a small extent of effects on the overall result (Allen et al., 1984).

The economic sub-model represents such economic situations. As a result, in order to set up an appropriate structure, the aquaculture farm architect must be able to analyze information from fields such as biology, construction, and finance.

7.4. FARM DESIGN

From the biological, physical, and economic sub-models, a farm design for farmed species may be constructed. This design may be used to specify both the relationships between parts of the system and the functionality of the system, while also allowing for comparability with other systems (Allen et al., 1984; Huguenin and Colt, 1986; and Logan and Johnston, 1992).

Creating a farm design is a difficult process that requires a broad range of biological, technological, marketing, and economic data. The interaction between the system's numerous components adds to the system's complexity (Allen et al., 1984; Huguenin and Colt, 1986; and Logan and Johnston, 1992).

The farm layout shall quantify the connections between the physical system and the species' capacity to adapt in order to ascertain the following elements:

- The enterprise's economic feasibility;
- Opportunities for additional study;
- The technology to be employed;
- Production management;
- Marketing management; and
- Location selection.

7.4.1. Product Definition

Anon (1979) provided a product definition to aid in the formulation of a cultural system's output goals. The product definition is a concise explanation of the item which will be created by the system, explaining the product's objective or application in line with consumer demands (Forteath, 1993).

The above comments would preferably be used to select the most effective species and culture system to match the product characteristics and customer needs because this is rarely possible with several farm models that are premised on scarce biological factors and culture requirements, or on methods and biological metrics for other types of fish.

Until now, product definition has indeed been employed sparingly in the design and construction of aquaculture facilities, resulting in poor performance of aquaculture species and, in many cases, costly adjustments to production plants or production methods (Forteath, 1993).

Aquaculture plants provide a wide variety of goods, including oyster spat, Kurama prawns, and table fish. Microalgae, macroalgae, zooplankton, and sea urchins are some of the less common varieties of products produced by aquacultural firms.

A typical fish farm's goods might well be classified as follows:

- Commercial or gaming fish (ranching);
- Food fish;
- Decorative or aquarium fish;
- Fodder and baitfish; and
- Egg production.

It should also be noted that when comparing local and international market marketplaces, the product definition for a specific species may differ; in fact, in accordance with fundamental marketing principles, the product definition may frequently differ depending on which foreign market is evaluated (Logan and Johnston, 1992).

Farms may also generate many types of products or a line of products of related items. This results in the farm having over one product specification. For instance, a hatchery and grow-out facility could offer eggs, fingerlings, and plate-sized fish. This expanded product line distributes the risk, providing flexibility and the capacity to adjust to shifting markets (Logan and Johnston, 1992).

As a result, while managing the farm becomes more difficult, there are intrinsic benefits to having greater flexibility in changing production tactics to changes in market demand or the demands of different market groups.

Once the product description for the cultivated species has been defined, and a clear grasp of the marketing aims and qualities has been established, a more rigorous technique may be utilized to further examine the economic benefits of growing the species.

It could be used as a starting place for farm layout, for instance, to assess culture system size, suitable culture technologies to be employed, and process and packing requirements. According to Forteath (1993), product definition is seldom utilized in aquaculture to assess these criteria.

7.4.2. Biological Sub-Model

The biological sub-model must be as realistic as feasible since it serves as the foundation for the physical and economic sub-models and indicates the system's constraints and requirements. This sub-data model might originate from a variety of sources. Allen et al. (1984) classifies these data sources into four categories.

Aquaculture-oriented experimental data is the earliest, most trustworthy and most important source. These investigations, which can range from laboratory to pilot-scale production system trials, are frequently aimed at providing specialized knowledge on cultural subjects. The second source of data comes from trials that are not related to aquaculture.

Such information, which might be generated from physiological or ecological studies of fish, is valuable in improving knowledge of environmental impacts on qualities like fish reproduction and growth.

However, when basing models on non-aquaculture experimental results, Parker (1981) and Huguenin and Colt (1986) take note of the following:

- The goals of study and design may differ;
- The time factor: studies are frequently short-term, whereas farming is long-term;
- Despite the fact that biological and behavioral processes are normally scale-dependent, the effects of scale are sometimes overstated or overlooked;
- Information from other species, even strongly linked species, might not provide an accurate representation of another species' effectiveness; and
- Intrinsic disparities between laboratory and outdoor circumstances.

The third collection of information comes from current metabolic and growth models built via earlier research. Nevertheless, their utility might well be diminished owing to changes in the cultural and experimental conditions and also variances in the intent or objective of the current model and the cultural model. The fourth data source is the utilization of energy or mass balance relationships.

These connections can be used to predict upper and lower bounds of experimental values and may be useful as alternatives to experimental data (if available) or as a check for experimental results. As a result the researcher may have access to a diverse set of biological data in order to create the biological sub-model.

Normally, this information is derived from a mix of all four sources. Such sources frequently enhance one another by filling gaps in information from each unique source (Allen et al., 1984).

Even so, it is critical that the biological sub model accurately reflects the real nature of the interactions of the cultured species with their environment in order to construct an appropriate physical sub model and determine the system's economics.

As a result, pilot scale studies are frequently done prior to commercial scale manufacturing to eliminate inaccuracies in the biological sub model. When the biological sub model is complete, work on the physical sub model may begin.

7.4.3. Physical Sub Model

All of the culture facilities and management practices needed to hatch, grow, harvest, and prepare the cultured species are included in the physical sub model.

Such facilities are typically developed with only a subset of the biological requirements of the cultured species in consideration and thus are frequently based on the biological requirements of other species owing to an absence of biological data (Forteath, 1993).

Preferably, the facilities will be built to meet the product definition's product and marketing objectives and also the biological needs of the cultivated species.

The farm architect has access to a wide variety of culture systems, from enormous pond systems to more intensive culturing like raceways, cages, and recirculating systems (Wheaton, 1977; Muir, 1982; Laird and Needham, 1988; Shepherd and Bromage, 1988; Pillay, 1990; and Stickney, 1994). The sense of influence over the cultural environment must rise in tandem with the intensity of production.

Tilapia production, for instance, has long been practiced on a subsistence basis, including little more than the gathering of fish from ponds (Allen et al., 1984). As a result, there is essentially little control over tilapia production, and output from such a system is modest and rather erratic (Pillay, 1990).

A shift toward more intensive tilapia production has resulted in increased capital expenditure (i.e., pumps, aerators, tanks, etc.) and increased inputs (i.e., more labor, artificial food, etc.), resulting in an increase in operational expenses. Nevertheless, with this increased degree of control and intensity, a far larger output is feasible. (e.g., 64 kg/m^3) (Balarin and Haller, 1982; Stickney, 1994).

Location selection is also a crucial aspect that may have a significant influence on the farm's technology selection, operation, and profitability (Shepherd and Bromage, 1988; and Stickney, 1994). Previously, site selection was primarily dependent on the availability of adequate amounts of appropriate water for expected usage and future growth.

Nevertheless, additional issues like electrical supply, road access, and so on must be addressed. Access to trade is another critical component that is sometimes disregarded. If a sufficient local market for the product is not accessible, access to transportation must be addressed.

The trade-offs between access to a suitable location and excellent quality water must be balanced against the expenses of accessing markets and distributing the product to customers.

As a result, several factors influence the selection of systems and technology, such as the ones that follow:

- Biological requirements of cultivated species
- The availability of land
- The availability and quality of water
- Climate
- Infrastructure (such as utilities and roads)
- Labor and technical abilities
- Feeds
- Level of capital investment.
- Product distribution (harvesting, warehousing, processing, and transportation); and
- System marketing and economic objectives

When the physical sub-model is finished, an economic sub-mod must be created.

7.4.4. Economic Sub-Model

The economic goal of aquacultural firms is often profitability for the proprietors. It must be acknowledged, even so, that the venture might need to meet a variety of social, political, legal, technical, and environmental goals (Huguenin and Colt, 1986), like stockpiling native fish stocks, supplying seed stock for subsistence farmers, fish farming research institutions, and public aquaria.

A variety of elements must be considered while designing this sub-model:

- The level of capital investment;
- The level of gross income;
- The level of operational costs; and
- The level of debt and equity.

Managerial choices on such elements are made as part of uncertainty in input supply and demand for the ultimate output. The heterogeneity in the biological reactions of the cultured species and their interaction with the physical system adds to the complexities. As a result, economic considerations must be examined alongside physical and biological sub-models.

Advertising in aquaculture is frequently overlooked (Shaw, 1986; and O'Hanlon, 1988). Most farm designs have been predicated on producing items that the designers believe the market wants, instead of necessities. However, proper manufacturing and marketing strategies may be devised to meet market demand through market research and subsequent product definition.

This strategy is driven by the market rather than the product. According to Shaw (1990), due to the increased degree of competition in the business, this market-driven strategy is becoming increasingly vital for aquaculture producers. Farmers must also build and retain competitive advantages over other farms in order to sustain market share and profitability.

7.5. MARKETING AND ECONOMIC CONSIDERATIONS IN THE PRODUCTION OF *P. SCALARE*

Australia has good potential for ornamental fish production since it already has a well-established infrastructure for the aquaculture and ornamental fish sectors. The Australian ornamental fish cultivation sector is now experiencing significant development, with the need for ornamental fish likely to rise (Treadwell et al., 1992).

The cultivation of ornamental fish in Australia is predominantly for import replacement (Lee, 1991), with the competitiveness of domestic producers, both locally and globally, increasing due to the increasing cost of foreign ornamental fish and the global interest in top-quality fish. (Bassleer, 1994).

7.5.1. Market Assessment

With the evident growth in ornamental fish culture prospects comes the need to identify and analyze suitable species for commercial production. As noted previously, one such contender is *P. scalare*, which no longer relies on wild fisheries for supplies.

A widespread scarcity of *P. scalare* has emerged as a result of a disease epidemic termed "Angelfish sickness" in 1987. (Lambourne, 1991; and Gratzek et al., 1992). The causal agent is considered to be viral, despite extensive study since its discovery has provided little knowledge about the disease's specific mechanics (Lambourne, 1991; Gratzek et al., 1992).

The illness has significantly impacted *P. scalare* output worldwide, generating demand for disease-free stock (Gratzek et al., 1992). According to industry sources, the market size and value for *P. scalare* in Australia is roughly 350,000 fish each year, worth $300,000 (per comm R. Datodi, 1992). Nevertheless, due to a lack of market information, it is hard to corroborate industry estimations.

There is scant data on the nature of *P. scalare* home production or ornamental fish production in general. As a result, an exploratory industry survey is required to assist in developing patterns and a comprehensive image of the business.

7.5.2. Economics

The continued demand for *P. scalare* in the ornamental fish business, along with the inconsistency of supplies (due to angelfish sickness), suggests that *P. scalare* has commercial potential. This possibility looks to be especially promising for the Australian ornamental fish sector. However, statistics on the economics of producing *P. scalare*, as well as other ornamental fish species in Australia and elsewhere, are missing.

The cultural capacity of a species may be examined by creating several financial scenarios due to the interactions between the generated biological and physical sub-models.

Financial information like cash flow, profit, and loss, and accounting information could be used to depict these financial situations. This data is beneficial in displaying a cultural candidate's economic potential and may be used to compare it to other species and other forms of economic investment.

7.6. THE BIOECONOMICS OF RECIRCULATING AQUACULTURE SYSTEMS

Parallel to agriculture's "Green Revolution" in the 1970s, aquaculture had a phase of extraordinary growth in the 1980s known as the "Blue Revolution." Increased returns were achieved as a consequence of advancements in feed ingredients, nourishment, water quality, illness prevention, and screening for economically attractive characteristics.

With the introduction of these technologies, there was a greater emphasis on intensive culture operations that utilized closed, artificial confinement levels with high input levels. The most significant benefit of closed systems versus open pond production is that the system's environment can be precisely regulated. (Spotte 1979).

The mechanical and biological components of water filtration have been the topic of studies on closed, or re-circulating, technology (Brune and Tomasso 1991; Lawson 1991). Nevertheless, the aquaculture sector has recognized that complicated and novel system designs cannot guarantee the economic viability of recirculating systems (Figure 7.3).

Figure 7.3. Recirculating aquaculture system.

Source: Image by Wikimedia Common.

As a consequence, researchers are starting to concentrate their attention on the economics of recirculating technology. Of fact, conducting economic tests on commercial-size recirculating systems may be challenging and costly. One option to overcome this difficulty while still generating the required info is to employ bioeconomic models that adequately represent the system's underlying bioenergetic activity (Allen et al., 1984).

To investigate the temporal dynamics of species development in pond aquaculture systems, bioenergetic models have been frequently employed (Paloheimo and Dickie 1965, 1966a, 1966b; Machiels and Henken 1986; Cacho 1990). Some studies have even looked at the effect of metabolic feedback on growth (Cuenco, Stickney, and Grant, 1985).

Bioeconomic models that integrate bioenergetics and producer strategic planning are less prevalent. Cacho, Kinnucan, and Hatch (1991) employed a bioeconomic model of pond catfish production to establish the most cost-effective feed regimens.

Researchers have also utilized bioeconomic models of various sophistication to investigate open system rearing of shrimp (Karp, Sadeh, and Griffin, 1986), carp (Talpaz and Tsur, 1982), lobster (Botsford, Rauch, and Shleser, 1974), and tilapia (Botsford, Rauch, and Shleser, 1974). (Liu and Chang, 1992).

To the best of our knowledge, no research has investigated a recirculating production system from a comprehensive bioeconomic perspective, taking into account not only actual metabolite-constrained growth over time, but also the economic restrictions imposed by profit-seeking providers.

7.6.1. Conceptual Model

Considering the main characteristics of a recirculating aquaculture system, a conceptual model may be created using bioenergetic interactions and ecosystem dynamics concepts. Despite being simplified and without particular references to pH and water pollutants, this model incorporates the essential environmental components that impact the operation and administration of a recirculating system.

Taking just the bioenergetic element of the model into consideration, flows in the system are predominantly driven by fish weight as mediated by metabolism and hunger. Water temperature and feed amount and quality, which are theoretically within producer control, can be utilized to change the different fluxes and hence the temporal route of fish growth.

Feeding and development produce waste products and need oxygen, but most bioenergetic models believe that all these elements are digested or provided by the surrounding environment. Nevertheless, due to their possible influence on individual fish development, mortality, and the overall expansion of total biomass in the production system, recirculating system models must adequately account for such feedbacks.

In the conceptual model, metabolic waste materials are divided as solid or total ammoniacal nitrogen (TAN). Unionized ammonia nitrogen (UAN), a poisonous component of TAN, is one element of a review system which can restrict fish development, causing changes in appetite or, if big enough, induce fish mortality.

Biological filtration, according to the conceptual model, limits the accumulation of UAN, but it also contributes bacterial respiration to the biological oxygen demand (BOD) produced by fish respiration and solids breakdown.

Open flow-through systems reduce UAN and BOD accumulation by water exchange, whereas recirculating and certain pond systems require oxygen to satisfy BOD requirements through mechanical or liquid oxygen aeration.

Mechanical filters remove suspended materials from the system. The performance of biological and mechanical filters is crucial to fish development and the sustainability of a recirculation system during the grow-out cycle.

There would be no development or death reviews until both filters are totally effective. In the absence of proper filtration, emergency water exchange may be used to limit the impacts of UAN and BOD, depending on the laws and regulations regulating a given species cultivation.

7.7. BIOECONOMIC MODEL

The move from a conceptual biophysical model to an empirical bioeconomic model, which can be utilized to quantitatively explore optimal system function, necessitates the selection of a specific culture species. Given a decent imitation of the natural surroundings, almost any fish may be cultivated in a closed environment.

Commercial usage of recirculating systems, on the other hand, necessitates consideration of economic concerns.

In brief, a viable species for food-fish culture must have a well-established market, a sufficiently high value, be legal for culture, be tolerant of poor water quality and high stocking densities, thrive on pelleted food, have an effective feed conversion ratio (FCR), and be accessible locally (hatcheries or wild stocks). Given these characteristics, tilapia is a fish with significant potential for usage in closed systems.

7.7.1. Tilapia as a Culture Species

Tilapia is the collective name for a group of warm-water fishes of the Cichlidae family. Tilapia, a native of Africa, has been farmed in open systems in the Middle East for generations, a method that has spread throughout the world.

While tilapia cultivation has traditionally been forbidden in many regions of the United States because of worries about exotic species introduction, tilapias developed in tropical and subtropical climates and are not cold-resistant (Bowen, 1982. Many states have authorized tilapia farming in recent years, with the Louisiana legislature legalizing tilapia culture under stringent restrictions in 1991.

One of these constraints was that tilapia can only be grown in closed systems. Due to their resistance to overcrowding and poor water quality, researchers have selected tilapia as a preferred species to be used in recirculating systems (Drennan and Malone 1990).

Tilapia are also suitable for closed systems because of their capacity to adjust to salinity, temperature, and dissolved oxygen (DO) variations. In addition to its toughness, tilapia has high economic potential due to its capacity to replace a variety of high-valued fish. Tilapia has mild white meat that restaurants and retail customers may substitute for rare sea trout, redfish, and snapper.

Tilapia quickly achieved commercial popularity in the United States because of its flexibility. Domestic tilapia output was expected to be 6,818 metric tons (live weight) in 1994, a 20% increase over 1993.

The domestic produce was worth around $15.7 million at the farm gate. Furthermore, tilapia imports into the United States totaled 14,585 metric tons in 1994, representing a 29% increase over 1993 and a $25.6 million-dollar value (Figure 7.4).

Figure 7.4. Farmed Nile Tilapia in a fish market.

Source: Image by Flickr.

7.7.2. Economic Framework

The application of the conceptual model to a recirculating tilapia production system necessitated the development of particular bioenergetic and metabolic feedback sub-models, as well as their integration within a larger economic framework. The section explains the economic framework and the general management issues that recirculating system operators confront.

Over the course of a growth cycle, management has the most impact on the variable costs involved with short-term decision making. In addition to the obvious monetary costs of stocking, feeding, and electrical power use, indirect costs might develop when a system's technology fails to entirely eliminate metabolic wastes. Such variable expenses manifest themselves as decreased fish development and higher mortality.

7.7.3. UAN Feedback Sub-Model

Nitrogenous chemicals accumulate in recirculating systems caused by the buildup of expelled metabolic waste, fecal waste, and uneaten feed. Speece (1973) demonstrated that TAN production can be calculated directly from the feed ratio using an ammonia conversion constant.

According to Fivelstad et al. (1990), the converting factor would be affected by the ration quality, especially the quantity of protein in the diet. A simple method for calculating the converting factor for warm-water systems is that each kilograms of fish excretes the same number of grams of TAN each day as the percentage of dietary protein in the ration (Meade 1973; Liao and Mayo 1974; Drennan and Malone 1990).

However, TAN is only of relevance due to its link to UAN, the most poisonous form of aqueous ammonia. UAN levels are typically computed as a proportion of the TAN in the system and are largely dependent on system pH and temperature (Drennan and Malone 1990).

At a constant pH of 8.0 and a temperature of 30 degrees Celsius, roughly 7.5% of the TAN will be in the unionized form (Emerson et al., 1975).

7.7.4. BOD Feedback Sub-Model

The use of oxygen in a recirculation system, or BOD, is caused by three processes: fish respiration, ammonia compound oxidation by autotrophic bacteria, and organic material degradation by heterotrophic bacteria (Wheaton, Hochheimer, and Kaiser 1991).

The BOD created by fish respiration is typically calculated by adding the oxygen necessary for active and normal metabolism. The nitrifying autotrophic microorganisms that colonies accessible substrate in a recirculating system could be related to increased oxygen consumption.

Mostly, nitrification happens within the biological filter because of its high specific surface area.

Nitrifying bacteria require roughly 4.65 grams of oxygen for every gram of TAN oxidized (Wheaton 1977), or approximately 0.062 grams of oxygen per milligrams of unionized ammonia.

The combination of BOD, aeration, and the resultant leftover dissolved oxygen (DO) in the system ultimately determines the amount by which BOD influences tilapia development and mortality. Tilapia are one of the most resistant warm-water fish, with survival documented at DO levels as low as 0.1 mg per liter (Caulton 1982).

More cautious estimates indicate that tilapia development begins to slow at DO concentrations of less than 5.0 mg/liter, with death beginning at 1.5 mg/liter (Caulton 1982). Unfortunately, nothing is known about the functional link between DO and growth rates in tilapia or comparable species.

7.8. TECHNOLOGY AND THE BIOENERGETIC MODEL

Considering that the emphasis in recirculating system management is on filter technology, it was expected that for optimal system performance regardless of feed quality, flawless biological filter (BE = 1.0) and mechanical filter (SRE = 1.0) functioning would be necessary.

The model reveals that production of 700 g of tilapia generates a terminal system fish density of around 50 grams/liter (0.4 pounds/gallon) over a wide range of feed quality when run without inefficiency. Nevertheless, the harvest duration is reduced from 270 days using a 20% crude protein diet to 230 days using a 40% crude protein diet.

This 40-day difference can be attributable to the higher growth offered by the high protein diet, despite the fact that tilapia in natural systems eat relatively low in the food web and are not typically considered to be fast growing.

In closed systems, high protein diets are often supplied. Proteinrich feeds may also be advantageous in such ideal filtration circumstances since no metabolic feedback will be available to reduce the potential benefits of employing high protein diets.

Though theoretically possible, few, if there are any, agricultural production systems have been observed to function at peak efficiency. Therefore, given less-than-perfect filter operation, it is helpful to examine the implications of altering feed quality on days-to-harvest and maximum system density.

The number of days to harvest increases as filter system effectiveness diminishes, whereas maximum system density drops, with the impacts being more evident for good quality feeds.

In conclusion, the simulated biological trajectories are affected by feed amount, grade, and the degree of technological inefficiency. Furthermore, the form and divergence of system biomass density and FCR trajectories indicate that feed quality/technology interactions are strongly reliant on fish size, with metabolic feedback dampening UAN and DO accumulation.

While decreasing growth is the primary cause of density loss in these models, future increases in target harvest weight and/or lower filter performance might eventually result in large mortality reactions.

7.8.1. Isoquant Analysis

The total feed eaten was plotted against the relevant dietary protein in an optimized simulation of the limited biological development model to yield isoquants. The lines on the isoquants show the feed amount and quality mix needed to generate a specific size of tilapia at three degrees of biological filter efficiency.

Whereas the isoquants do not depict the actual distribution of feed during the growth season, a few findings are worth mentioning.

First, the isoquants' negatively sloping, convex section shows that there is some substitution between feed amount and quality.

Second, the substitutability range changes greatly with harvest weight as well as biological filter effectiveness.

For example, biological filter effectiveness of BE = 1.0 yields an isoquant for a 700 gram tilapia with replacement options for feed size ranging from 1800–2300 grams and feed quality ranging from 10–37% crude protein. Beyond a protein level of 37%, the isoquant begins to slope upward, indicating that the protein's marginal product is negative.

Cacho, Kinnucan, and Hatch (1991) obtained comparable findings in pond simulations with unconstrained 600 gram catfish generation. This overall pattern remains true for the generation of smaller-sized tilapia, with replacement options happening throughout a smaller proportion of feed amount.

The isoquant finding suggests that there is a certain amount of replacement between feed amount and feed grade in the 20–40% crude protein range, with the precise magnitude influenced by variations in filter performance as well as fish size.

In addition to lowering overall system density, smaller tilapia's greater protein absorption efficiency may help to prevent the buildup of metabolites from higher protein diets at less-than-optimal levels of biological filter operation.

Nevertheless, as fish grow in size, the range of advantages of a greater protein diet is heavily influenced by filter effectiveness. Only with low to medium quantities of dietary protein is replacement of feed quantity and quality practicable in the most inefficient example presented (BE = 0.90).

7.8.2. Economic Impacts

Considering the model's structure, the best BE-SRE pairing for all dietary protein intake happens because there is no inefficiency or when BE and SRE both equal one. Under these ideal management parameters, a 700-gram tilapia may be developed in 265 days with a 20% dietary protein. The net returns for this combination are 8.9 cents per liter, or 0.034 cents per liter every day.

Moving away from this ideal management arrangement diminishes returns, but the rate of reduction for SRE drops is rather moderate. For example, when BE=1.0 and SRE=0.5, the time required to acquire a 700 gram fish increases by two days, resulting in a 3% fall in daily returns to 0.033 cents/liter/day.

This drop is strongly attributable to inefficient solid removal and the requirement for additional aeration. Additional drops in SRE produce significant increases in production time and declines in net returns, with an SRE of 0.25 resulting in a 15% fall in net income.

Whereas declines in SRE result in reduced returns for any given amount of BE, these changes are minor. Nevertheless, decreases in BE for any given amount of SRE cause significant changes in returns and production times.

The seeming sensitivity of the model to variations in BE can be attributed to the nature of the BE-SRE feedback linkages in actual recirculating systems.

Inadequate biological filtering results in the progressive accumulation of poisonous UAN.

Although UAN can cause death, there is a very large UAN concentration range where the most immediate consequence is to slow development to allow the biological filter to digest the waste products. As a result, the slowing of growth rates tends to reduce the increase in UAN levels. As a consequence, increased UAN levels may happen without triggering a catastrophic system crash, except at the expense of ultimate return loss.

When management competence is insufficient to maintain zero levels of UAN, optimal production time approaches necessitate less than satiation feeding and hence longer production times. Due to the obvious model structure, decreases in SRE levels may not have the same effect since they may be at least partially countered by boosting aeration power.

This latter technique effectively resembles how producers run more aerators during low DO seasons. The observed slight reductions in returns for higher values of SRE at any given BE, on the other hand, imply that the optimal amount of power input to the system is significantly less than the quantity required to eliminate the effects of DO increase.

Aquaculture producers must make a variety of decisions that will influence the long-term profitability of their business. As producers increase their resource utilization, managerial competence becomes increasingly important. Small management errors are perhaps the most dangerous for recirculating systems.

Because of the nature of recirculating system operation, manufacturers must fill their systems at high densities in order to offset the higher fixed and variable costs involved with closed system operation. Increased levels of biomass, on the other hand, suggest smaller margins for error in controlling the biophysical environment.

As an outcome, managerial competence becomes critical to the financial viability of recirculating system operations. The model's findings suggest that the lack of management skills can erase the typical benefits of employing a high protein diet.

Although greater protein levels promote quicker development and are thus often employed in the industry, the increased direct feed costs and indirect expenses owing to metabolic feedback result in poorer daily returns if protein-rich feed use has been maintained throughout harvesting.

This model demonstrated that effectiveness in solid removal had a negative influence on returns, although the majority of the negative effects were associated with reductions in biological filter performance. As the effectiveness of biological filters decreases, the time to harvest increases and returns decline.

Furthermore, when stocking density grew, direct gains in returns were guaranteed only if no metabolic feedback occurred. Higher stocking density may also result in economic failure if the filter technology is operated inefficiently.

Thus, there is a tradeoff between stocking density and management competence, with the tradeoff being significantly influenced by dietary protein levels. In essence, economically acceptable tradeoffs between dietary protein and stocking density occur over rather small management ability ranges.

The biological reality of re-circulating systems may limit lucrative system operation in the absence of highly skilled and effective management. Such simulated findings might help to understand why recirculating systems have failed to achieve broad commercial success.

7.9. LIMITATIONS AND FUTURE RESEARCH DIRECTIONS

The constraints present in the foregoing assessments, like all simulation studies, may be traced all the way back to the hypotheses that were used to create the model, the selection of input and output variables, and the necessity to create a simplified representation of the complexity of the real world system.

As a conclusion, the method cannot be viewed to represent real system efficiency for any individual recirculating aquaculture operation. Rather, the model might be regarded as a broad, simplified recirculating system in which the focus of research is on the interactions between important factors that influence economic results.

Finally, these sorts of simulation studies can be useful in determining the critical variables that influence system productivity. As a result, suggestions on how to enhance the systems for commercial production purposes are made. The model, in its current form, must not be used to decide on particular production activities undertaken by individual aqua culturalists.

This research might be developed in a variety of ways to better investigate the influence of management on the economics of recirculating system operation. Further comprehensive characterization of the metabolic sub-models needs further biological study.

However, modeling of the biological filtering process might be enhanced by integrating growth relations for the local bacterial colonies. This inclusion is recommended by the frequently expressed industry remark that recirculating system operators produce at least two crops at the same time: harvestable fish biomass and no harvested, yet crucially significant, bacterial biomass. This original study modeling methodology might also be used to investigate the economic impact of biological production shocks (acute over-feeding, disease-related mortality) and pricing risk (both input and output) under different levels of managerial skill.

Modifying the model to study the usage of non-divisible inputs, varying proportional returns to scale, and differential development among individual fish might be another worthwhile area of research.

7.10. CONCLUSION

At present, the rapid increase in the fish supply is due in large part to an upsurge in aquaculture. However, bioeconomic models cannot be directly extrapolated between species as each species growth is determined by specific factors and parameters.

Bioeconomic models are an excellent methodological approach to studying the interaction of the various components (biological, physical, technological, economic, and institutional) of aquaculture systems. Depleting the world's natural fish stocks will significantly impact the world's food supply in the coming years.

This has led to adverse environmental impacts that often constrain the growth and development of aquaculture and affect other resource users. About one billion people rely on fish as their primary source of protein. That's one-fifth of the global population.

While aquaculture intensification holds great promise for increasing global fish supply and addressing concerns about food security, poverty, livelihood, and income, it has not been without consequences and conflicts. Bioeconomic models can answer economic feasibility questions, optimal system design, optimal methods of operations, and research direction (Leung, 1994).

REFERENCES

1. AgEcon Search, (1996). *The Bioeconomics of Recirculating Aquaculture Systems*. [online] Available at: https://ageconsearch.umn.edu/record/31681/ (accessed on 21 September 2022).
2. Anderson, J., Asche, F., & Garlock, T., (2019). Economics of aquaculture policy and regulation. *Annual Review of Resource Economics, 11*(1), 101–123 [online]. Available at: https://www.annualreviews.org/doi/10.1146/annurev-resource-100518-093750 (accessed on 21 September 2022).
3. Jolly, C., & Clonts, H., (2020). *Economics of Aquaculture.* [online] Available at: https://www.taylorfrancis.com/books/mono/10.1201/9781003075165/economics-aquaculture-curtis-jolly-howard-clonts (accessed on 21 September 2022).
4. Pomeroy, R., Ureta, B., Solis, D., & Johnston, R., (2008). Bioeconomic modeling and salmon aquaculture: an overview of the literature. *International Journal of Environment and Pollution, 33*(4), 485 [online]. Available at: https://www.inderscienceonline.com/doi/abs/10.1504/IJEP.2008.020574 (accessed on 21 September 2022).
5. Willis, S., (1995). *Bioeconomics in Aquaculture: Preliminary Analysis of the Culture Potential of the Freshwater Angelfish Pterophylluai Scalare.* [online] Eprints.utas.edu.au. Available at: https://eprints.utas.edu.au/14512/2/willis_chp1-3.pdf (accessed on 21 September 2022).

CHAPTER 8

BIOECONOMIC OF INVASIVE SPECIES

CONTENTS

8.1. Introduction ... 218

8.2. Invasion Process and Feedbacks Between Biological and Economic Systems ... 230

8.3. Bioeconomic Impact of Existing Policy on Invasive Species ... 232

8.4. Integrating Economics and Biology for Invasive Species Management ... 234

8.5. What are the Disciplinary Impediments of Ecological Economic Modeling? ... 236

8.6. Trait-Based Risk Assessment for Invasive Species ... 238

8.7. Management of Invasive Species in the Great Lakes ... 240

8.8. Conclusion ... 247

References ... 248

An introduced organism that becomes overpopulated and harms its new environment is an invasive species. These invasive species adversely affect habitats and bioregions, causing ecological, environmental, and/or economic damage, although most introduced species are neutral or beneficial with respect to other species.

8.1. INTRODUCTION

For native species that become harmful to their native environment after human alterations to its food web, the term can also be used for example along the northern California coast due to over harvesting of its natural predator, the purple sea urchin (*Strongylocentrotus purpuratus*) which has decimated kelp forests, the California sea otter (*Enhydra lutris*). Invasive species have become a serious economic, social, and environmental threat since the 20th century (Figure 8.1).

Figure 8.1. The purple sea urchin (Strongylocentrotus purpuratus).

Source: Image by Wikimedia Commons.

A natural phenomenon is an invasion of long established ecosystems by organisms, but human facilitated introductions have greatly increased the geographic range of invasion, rate, and scale. For millennia, as both accidental and deliberate dispersal agents humans have served, beginning with their earliest migrations, accelerating in the age of discovery and accelerating again with international trade.

Notable examples of invasive plant species include the kudzu vine, yellow star thistle, Andean pampas grass, English ivy and Japanese knotweed. The New Zealand mud snail, feral pig, European rabbit, gray squirrel, domestic cat, carp, and ferret are the examples of invasive animals.

As a serious side effect of international trade invasive species are now recognized worldwide. They damages increase over time and often spread irreversibly. To prevent the arrival of species or eradicate them early in an invasion, control their local abundance once they have become established, or slow their spread, to reduce such damages, private, and public investments are increasing in an effort.

Most often, however, as a new cost of doing business the damages of invasive species are accepted, and to minimize the impact humans change their behavior. In this chapter, to guide policy development in support of more cost-effective management, it is argued that integrating ecological and economic analyzes is essential.

A key goal is to provide answers to such questions as how many dollars should be invested in prevention versus control, and to describe quantitatively the feedbacks between economic and ecological systems and what benefits are derived from such investments.

The impacts of some high-profile invasive species described by this chapter, looks to epidemiology for a model of how research and management could be better integrated to inform policy and explains the extent to which ecological and economic systems are integrated.

Experts and the public have recognized two important things about many anthropogenic environmental changes in the last two decades: these changes are increasingly global in scope is the first thing, and the second thing is, they are hard to reverse. It is referred to as "invasive species" throughout this book, these characteristics apply with special force to harmful nonindigenous species.

To increasing our understanding of this problem, both the global scope and the difficulty of reversing invasions impart considerable urgency. Invading organisms reproduce and spread, even if it is ceased introducing more individuals.

Without management the problem of harmful invasive species gets worse. Especially to those of us in universities, research to better understands an invasion comes naturally to scientists and social scientists.

It is also, however, believed that to natural resource managers and policy makers, given society's explicit desire to reduce the current it is urgent to focus our research on questions important and future damages caused by invasive species.

We want our research and its implementation to increase social welfare. Because invasive species are, by definition, driven by human activities, usually commercial enterprises using the perspectives and tools of economists is appropriate. In industry practices and consumer behavior solutions will derive from changes.

That humans move around the globe humans are as much the target of our study as the species. It must be conducted collaboratively by natural and social scientists, and in the context of possible management and policy responses to invasive species, if research is to inform natural resource management and policy.

After considering some specific examples of invasive species it is elaborated on these general points, their environmental and economic costs, and societal responses to them.

The damages caused by that subset and the costs of controlling them can be substantial, while the subset of introduced species that become invasive is small. At the economic damages non-native species cause, methods economists often use to measure those damages, and tools used to assess invasive species policies an in-depth look taken by this chapter.

To put the problem in perspective, for activities related to invasive species Federal agencies reported spending more than half a billion dollars per year in 1999 and 2000 ($513.9 million in 1999 and $631.5 million in 2000 (U.S. GAO 2000)). On prevention approximately half of these expenses were spent.

On managing non-native species several states also spend considerable resources; for example, the Great Lakes states spend about $20 million each year to control sea lamprey (*Petromyzon marinus*) (Kinnunen 2015) and Florida spent $127.6 million on invasive species activities in 2000 (U.S. GAO 2000).

As actual damages costs to government may not be the same, which generally fall disproportionately on a few economic sectors and households.

For example, in damages to the equine industries in Colorado and Nebraska alone, the impact of the 2002 outbreak of West Nile virus exceeded $4 million (USDAAPHIS 2003) and in public health damages in Louisiana more than $20 million (Zohrabian et al., 2004) (Figure 8.2).

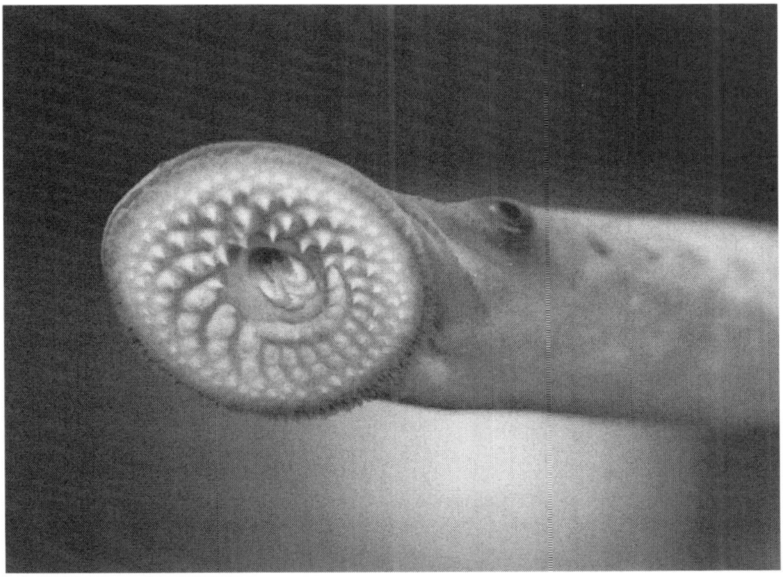

Figure 8.2. Sea lamprey (Petromyzon marinus).

Source: Image by Flickr.

Zebra mussels (*Dreissena polymorpha*) cause $300–$500 million annually in damages to power plants, water systems, and industrial water intakes in the Great Lakes region (Great Lakes Commission 2012) and are expected to cause $64 million annually in damages should they or quagga mussels *(Dreissena bugensis)* spread to the Columbia River basin (Warziniack et al., 2011).

Global biodiversity is threatened by invasive species and an estimated $120 billion per year cost the United States economy. While providing insights into ecological and evolutionary processes, understanding invasions will aid in their prevention.

Their relative lack of evolutionary history with species native to the invaded region is a defining characteristic of invasive species. Both on the invasive species and on native species in invaded communities, this article reviews evolutionary characteristics of successful invasive species and evolutionary impacts.

During invasions novel selection pressures can drive rapid evolution, and observed phenotypic changes often represent the combined effects of evolution and phenotypic plasticity.

For informing policy decisions studies on economic impacts from invasive species vary in their rigor and usefulness. How to distinguish impact studies that were done well from those that were done poorly, and appropriate use of values calculated in impact studies economic impacts and methods used to calculate them.

Key contributions of economics to invasive species science and provides a quick overview of behavioral and economic responses to invasive species risk discussed by this chapter. According to four main themes the chapter is organized.

Human-mediated vectors of introductions and development of trade and regulatory policies that prevent the movement of invasive species into uninvaded areas economic research on the introduction of species has focused on people's understanding of invasion risk and potential impacts and how they respond to that risk.

In their new geographical areas invasive species aggressively invade new continents so that these species become dominant. Invasive species include plants, mammals, birds, fish, amphibians, reptiles, arthropods, mollusks, and plant and animal diseases, benign components of their original habitats.

The black rat, house sparrow, and kudzu included in North America by some invaders familiar (Rattus, *Passer domesticus, Pueraria montana var. lobata*).

Invasive species cause many problems for humans in that they degrade natural communities once established, and damage agricultural species with pests and diseases. By cutting off boating along rivers invasive aquatic species can cut off local commerce, and by clogging the operation of hydroelectric dams cause local electricity emergencies, particularly in countries such as New Zealand that rely on hydroelectric power.

Those species that arrived on continents after the sixteenth century after human global travel, commerce, and migration increased are known as invasive species.

Global travel by humans has greatly accelerated the rate of intercontinental movement of species while species have moved between continents for millennia. The associated explosion of invasive species has caused the decline of native species on their continents of origin and new invasions of species have paralleled the movements of humans worldwide (Figure 8.3).

Figure 8.3. House sparrow (Passer domesticus).

Source: Image by Wikimedia Commons.

A 'New Pangea' has been created by this relatively free movement of biota between continents. The displacement and extinction of native species caused by invasive species ultimately will cause the overall worldwide number of species to decrease continents will have more species as new invasive species arrive.

Although there are many documented cases of species being transported to other continents for horticultural or agricultural purposes, exotic species introductions are generally unintentional.

Invasive species are not just another species for the species-richness list because invasive species cause environmental degradation, from a philosophical perspective. The perspective that invasive species are problematic in natural areas is fundamentally dissimilar to the idea that foreigners can cause harm to a society, while some authors criticize the furor over invasive species as being akin to xenophobia.

Due to the damage invasive species cause to natural plant communities, ecologists over invasive species are concerned.

8.1.1. Caulerpa: Successful Eradication

Sometimes tire of the organisms under their care and release them, aquarium keepers, like owners of all sorts of plants and animals. In two Southern California coastal embayment, populations of the invasive seaweed *Caulerpa taxifolia* were discovered in 2000. This species has been sold widely in aquarium shops because it is fast growing, hardy, and beautiful including a very invasive strain (Walters et al., 2006) (Figure 8.4).

Figure 8.4. The green algae Caulerpa taxifolia – aquarium cultivar.

Source: Image by Flickr.

A well-documented history of harmful invasions caused by some of these same characteristics. Including the Mediterranean Sea, commercial, and recreational fishing, recreational activities like scuba diving, and tourism have all suffered in various invaded marine ecosystems (Meinesz 1999).

A consortium of private and government agencies launched a concerted eradication effort using chlorine applications under anchored tarps when the species was discovered in California. The effort cost at least $3.7 million over 5 years (Woodfield and Merkel 2005), and it was successful.

However, the need for many similarly expensive management situations would probably occur in the future as other naive aquarium owners dispose of unwanted plants without policy responses to prevent additional Caulerpa introductions (Walters et al., 2006).

To declare the Mediterranean aquarium strain of *C. taxifolia* a federal noxious weed the U.S. Department of Agriculture (USDA) used its authority under the Plant Protection Act of 2000. In interstate commerce such a designation gives the USDA authority to prohibit importation, exportation, or movement of the species.

The state of California went a step further and made it illegal to possess *C. taxifolia* and nine other Caulerpa species in 2001. Nevertheless, to purchase in all states various species and strains of Caulerpa remain easy (Walters et al. 2006).

Then, is a success story, the story of Caulerpa eradication near San Diego. Referred to as "early detection, rapid response, and eradication," supported by additional efforts (of minimal success thus far) to prohibit future introductions, it is an example of successful implementation of a strategy.

8.1.2. Sea Lamprey: Successful Control

The construction of the Well and Canal by-passed Niagara Falls and allowed sea lamprey (Petromyzon marina), across the continent and about a century earlier, along with ships and barges, access to the upper Great Lakes. Despite the fact that most sea lamprey previously lived their adult lives in the Atlantic Ocean, large, and self-sustaining populations soon thrived in the upper lakes.

While the increased navigation fostered commercial activities that were beneficial to humans, the invasion by sea lamprey was not. Adult sea lamprey is parasitic on other fish species, using their rasping and sucker-like mouth to feast on the blood of commercially valuable species such as lake trout (*Salvelinus namaycush*) and whitefish (*Coregonus spp.*). The result was declining fisheries and a public outcry.

Fortunately, larval sea lamprey is confined to the tributaries of the Great Lakes, before assuming their adult bloodsucking habits where they reside for about 7 years. A chemical discovered in 1955, the larvae are easy to locate and are highly susceptible to 3-trifluoromethyl-4-nitrophenol (TFM). TFM kills sea lamprey larvae with acceptably low effects on other species, when applied at appropriate concentrations in tributaries (Figure 8.5).

Figure 8.5. Salvelinus namaycush fish.

Source: Image by Wikimedia Commons.

Since 1956 on monitoring and poisoning sea lamprey the United States and Canada have together spent about $15 million annually. With these continuous expenditures sea lamprey populations plummeted, and harm to the fisheries is kept tolerably low.

The ongoing expense of which is justified by even larger benefits in the protection of Great Lakes fisheries, the management efforts directed at sea lamprey constitute a remarkably successful "control" effort.

8.1.3. Gypsy Moth: Successfully Slowing the Spread

In 1869, gypsy moth (*Lymantria dispar*), escaped an unsuccessful attempt at silk production in Massachusetts, which had been imported from its native range in Europe. Thus began an invasion of North America that is ongoing today.

Gypsy moth infestations can completely defoliate vast forests of oak and other trees and can achieve such abundance that their excrement and bodies are sometimes a serious nuisance in urban areas (Figure 8.6).

Figure 8.6. Gypsy moth (Lymantria dispar).

Source: Image by Wikimedia Commons.

With an aggressive integrated pest management program outbreaks of gypsy moths are often controlled. When the periodic population outbreaks are treated with pesticides expenditures to keep their populations acceptably low are very high, in areas where the gypsy moth is now a permanent resident. The best that can be hoped for in these areas is successful control, not eradication, as for sea lamprey.

Therefore, future control costs will be high (perhaps forever) if pesticide treatments are chosen, for every acre that becomes infested as the invasion progresses. Otherwise, to the periodic damage to urban and natural forests, humans must simply adapt (sensu economics, not evolution).

The USDA and states from Wisconsin south to North Carolina spend about $12 million annually to slow the southwestward march of gypsy moths across the country because of the damage and/or control costs once gypsy moths become established.

From about 13 miles per year to about 6 miles per year a combination of trapping, aerial spraying of insecticides, and mating disrupting pheromones has slowed by 50% the advance of the invasion front (Sharov et al., 2002).

In the area in advance of the invasion front—an area of roughly 9,000 square miles (1,500 miles × 6 miles), although this effort is expensive, it is cost-effective because damages are avoided, at least for a year. The avoided damages are much higher than the costs of the slow-the-spread program (Sharov 2004).

Overall, the scientific and management responses to the gypsy moth are a successful example of a slow-the-spread strategy, preventing long-distance, especially human-mediated, dispersal ahead of the advancing invasion front remains a challenge for this program.

8.1.4. Most Other Invasive Species: Uncontrolled Damages and Unchecked Spread

Stories that end in at least some level of success— slowing the spread of the gypsy moth, control of sea lamprey, eradication of Caulerpa—are rare and unfortunately are vastly outnumbered by harmful invasions that proceed apace to a grim and often irreversible outcome. Some of the most widespread visible, and dramatic examples come from forests.

In the United States, previously two of the dominant trees in eastern natural and urban forests, respectively (Burnham 1988; Gilbert 2002), a combination of nonindigenous insects, fungi, and other parasites and pathogens have essentially extirpated American chestnut (*Castanea dentata*) and American elm (*Ulmus americana*).

From ongoing invasions many other beloved and valuable species seem likely to face a similar demise: flowering dogwood (*Cornus florida*), has declined in abundance by more than 90% in some forest types over the last two decades, destroyed by the anthracnose pathogen (Holzmueller et al., 2006).

Eastern hemlock (*Tsuga canadensis*) is declining as the hemlock wooly adelgid spreads across the East and Midwest; American beech (*Fagus grandifolia*) is succumbing to beech bark blister; by butternut canker, butternut (*Juglans cinerea*) invariably dies after infection, in the Northeast and Midwest which is common and spreading (Ostry and Woeste 2004); mortality of ashes (*Fraxinus spp.*) across the Midwest hovers near 100% as the emerald ash borer advances (BenDor et al., 2006); and several species of oak (*Quercus spp.*) are vulnerable to sudden oak death, the spread of which has only recently begun but has already jumped from the West Coast to the East Coast in the nursery trade (Gilbert 2002) (Figure 8.7).

Figure 8.7. American chestnut (Castanea dentate).

Source: Image by Wikimedia Commons.

With many arriving in the United States as hitchhikers in shipments of plants, wood products, or wood packing material.

It is not just accidentally introduced pests and pathogens that damage forestry production and damage natural and urban forests. Such as the kudzu vine (*Pueraria lobata*), deliberately introduced plants, are also outcompeting native vegetation for light, nutrients, and space. And they can seem like a good thing at first like the gypsy moth.

At the 1876 Centennial Exposition in Philadelphia, the American public first saw the fast growing, attractively purple flowered kudzu vine from Japan (Forseth and Innis 2004). For decades thereafter, it served well as an ornamental plant that also provided summer shade under overgrown porches, particularly in the southeastern United States.

Later, as justifiable concerns grew about the severe soil erosion and nutrient depletion that accompanied intensive cotton agriculture, especially during the first half of the twentieth century, the U.S. government distributed 85 million seedlings, paying southern farmers to plant them (Forseth and Innes 2004).

Especially as other economic forces caused the decline of row cropping and livestock operations that had included management of kudzu, as for so many introduced species, only later did the downsides to kudzu become apparent. Millions of kudzu plants began to escape control altogether (Forseth and Innes 2004).

The costs of kudzu had become painfully obvious, by mid-century. Kudzu now occurs from Texas to Florida and north to New York, covering over 3 million hectares, which increases by about 50,000 hectares per year (Forseth and Innes 2004).

A 6-year effort was required to eradicate kudzu from the Chickamauga and Chattanooga National Military Park and forest productivity losses are between $100 million and $500 million per year, power companies spend about $1.5 million annually to control kudzu.

As the species continues to expand its geographic range from the southeastern United States, the best that can be hoped for is locally successful eradication efforts, whose long-term success depends on continued monitoring and control.

Unfortunately, including hundreds of species, the list of deliberately introduced plants like kudzu that have become very harmful to livestock, forestry, agriculture, and natural ecosystems is long. It also continues to grow.

8.2. INVASION PROCESS AND FEEDBACKS BETWEEN BIOLOGICAL AND ECONOMIC SYSTEMS

Following the vignettes above, with thousands of additional examples it is continued to illustrate the issue of invasive species, replete with idiosyncratic biological details. Such catalogs of examples, however, the processes that are common to all invasions can be obscured. To prescribing appropriate management responses understanding the processes, in turn, is essential.

In a vector species are carried, which transports the species either incidentally (e.g., insect pests in lumber shipments, ballast water of ships, viruses carried by humans themselves or overtly (e.g., the pet and horticultural trades) (Figure 8.8).

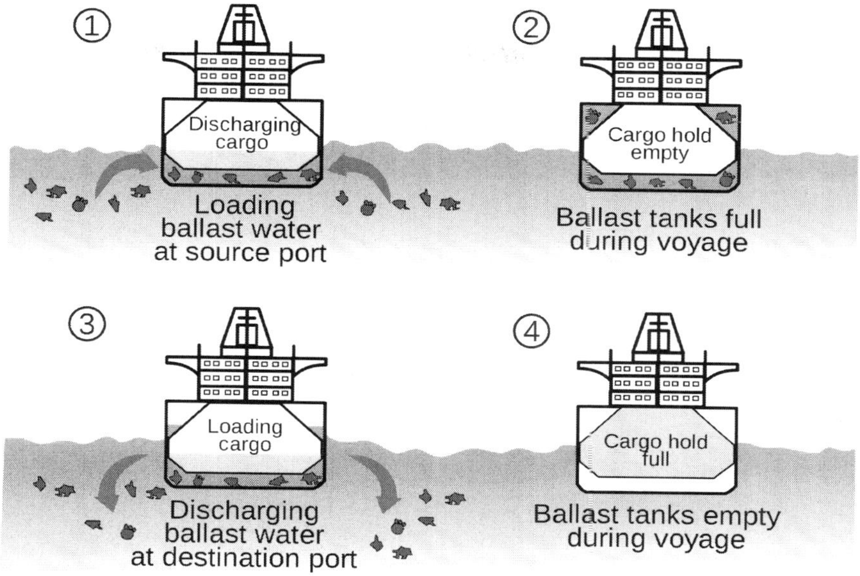

Figure 8.8. Diagram showing the water pollution of the seas from untreated ballast water discharges.

Source: Image by Wikimedia Commons.

Some proportion of the organisms may be alive when they are released or escape at a location outside their native range, depending on the traits of the species, conditions, and the duration in the vector.

Many to most species subsequently go extinct in a new location, depending on the taxonomic group of organisms, but a proportion—up to 50% for animals (Jeschke and Strayer 2005) and on the order of 5% for plants (Keller et al., 2007) — establish a self-sustaining population.

Perhaps not even detected by humans, a proportion, again about 5–50%, spread widely and become abundant at many new locations, while some of these established species remain localized. Such species—up to 25% of introduced animals (as calculated from the numbers above) and roughly 0.3% of introduced plants—cause undesirable environmental and/or economic changes and are categorized as invasive. Invasive species, which are a subset of nonindigenous species by definition, are bad.

When these underlying processes and probabilistic transitions during invasion are recognized policy and management implications become clear. As any invasion progresses the possible human management responses narrow.

> Before a species arrives in a new range or at the point of entry a vignette, prevention is possible only early in the process. Eradication depends on the rapid convergence of appropriate resources, technology, and political will.

Eradication is costly and sometimes impossible once a species is well established. Only two options remain: control of populations in selected locations, and adaptation by humans when the opportunity for eradication has passed.

Adaptation has been vastly more typical than any other response, in most countries, including those in North America, except when pests or pathogens have threatened either humans directly or highly valuable agricultural crops. It is suffered the consequences of invasions apart from these exceptions.

In the United States and elsewhere increasingly feature prevention efforts, in the last decade, however, investments in eradication, control, and finally prevention have increased for natural ecosystems, and policy discussions.

8.3. BIOECONOMIC IMPACT OF EXISTING POLICY ON INVASIVE SPECIES

International agreements are important arenas in which the feedbacks between the biological and economic systems are adjudicated. Great damage can be caused by movements of species within countries (Perry et al., 2002), but a large focus of ongoing policy development is international.

Dispersal to neighboring countries and to countries strongly connected by trade becomes much more likely once a species is introduced to one country. The interests of many other countries affected by decisions about importation or exportation by one country.

International agreements are the usual venue for more explicitly recognizing steps that should be taken to prevent exportation, as well as importation, of harmful species while national policy often focuses on importation.

Although more than 50 international and regional legal instruments address invasive species, few of these are binding (Shine et al., 2005).

Of these, at preventing environmental harm the binding agreement most directly aimed, is the Convention on Biological Diversity (CBD), ratified by more than 170 parties (not including the United States).

However, the obligation for compliance lies with each signatory country under the CBD, and the repercussions for noncompliance are virtually nonexistent.

In contrast, because the costs of noncompliance are high, international trade agreements have exerted the strongest influence over invasive species policy. Although the following comments apply also to binational and regional agreements, such as the North American Free Trade Agreement, globally, the most relevant agreements are those based in the World Trade Organization (WTO).

There is an inherent tension between promoting trade and preventing the introduction of invasive species, because the overarching goal of WTO is to increase international trade (which increases the probability of biological invasions).

Under WTO, if a nation wishes to reduce the introduction of invasive species, the International Plant Protection Convention specifies standards (through the Agreement on the Application of Sanitary and Phytosanitary Measures [SPS Agreement]) that national laws must meet (Hedley 2004).

To invasive species of all kinds these standards applied, including plant pests, plants, animal parasites and animals. The impact on trade must be minimized by any regulations to reduce the introduction of unwanted species.

On the importing nation there is the initial burden in demonstrating the need for protection, which must demonstrate via a scientific risk assessment that an import is likely to cause a harmful introduction. It remains largely unclear what constitutes a scientific risk assessment that can meet the SPS standards, while the role of scientific risk analysis appears preeminent in the SPS Agreement.

In favor of the exporting country most of the cases that have been adjudicated have been decided (Pauwelyn 1999). Rather than take precautionary measures to prevent the introduction of invasive species, countries are under pressure to quickly open their borders to imports.

To prevent invasive species while not unduly hindering the high-speed, high-volume international flow of goods by the difficult balancing act, not yet achieved, is to provide adequate safeguards (Jenkins 2002).

Most national policies, to the threat of invasive species have responded very little, for at least two related reasons, including those in the United States.

First, as S120 billion annually for the United States while the costs of invasions have been estimated (Pimentel et al., 2005), to parse with respect to policy options for specific vectors such aggregate estimates are certainly incomplete and are difficult, and few specific rigorous economic analyzes exist (Lovell et al., 2006; Olson, 2006).

Second, if their costs (in lower trade or shifts in the economy) outweigh their benefits (in decreased damages from invasive species) policy responses aimed at reducing invasions and increasing human welfare could instead lower human welfare and cause unanticipated economic distortions (Lovell et al., 2006; Olson, 2006).

As damage costs of invasive species, the vectors by which they move around (otherwise policies might be misdirected), and the costs of alternative policies, rational policies depend on better quantification of the externalities of trade manifesting. Fortunately, at this nexus of biology, economics, and policy research progress is rapid.

For example, a recent analysis demonstrated that the Australian Weed Risk Assessment, under which any plant proposed for importation into Australia is allowed only if it survives a risk assessment, brings net economic benefits to Australia (Keller et al., 2007). In this book, circumstances of costs, benefits, and spatial scales alternative policy and management strategies are warranted, it is explored in more detail.

8.4. INTEGRATING ECONOMICS AND BIOLOGY FOR INVASIVE SPECIES MANAGEMENT

When it triggers costs that outweigh any attendant benefits an established species is considered invasive. In the past, an approach that assumes the economic system and the ecosystem affect each other in a one-sided way, many researchers have used, to separate risk assessment from risk management caused by them.

A change in the ecosystem is viewed as only changing the economic system or a change in the economic system is viewed as changing the pressure on the ecosystem. The idea of the two-way interactions and feedbacks between human and natural systems does not address by this approach (e.g.,

Clark 1976; Crocker and Tschirhart 1992; Heal 1998; Barbier 2001; Brown and Layton 2001; Wätzold et al., 2006).

In the economic system ecosystem changes alter human behavior and productivity. Either by adapting the environment or by adapting to the environment, people recognize the change in their productivity, and they adapt to this change. leading to further changes in the ecosystem, when people adapt, they alter the pressure they put on the ecosystem, (Swallow 1996; Sohngen and Mendelsohn 1998; Perrings 2002).

By integrating ecological and economic modeling into a single cohesive framework this idea of a bioeconomic system can be addressed. To get more precise estimates of invasive species damages on human and natural systems is the motivation behind integration. Integration accounts for interdependencies, or feedback loops.

Traditionally, using dynamic models' economists have captured the notion of feedback loops. At most, one or two feedback loops and operate at a relatively aggregate level with a few exceptions, most standard bioeconomic models consider into the underlying problem at hand such models can provide the needed insight.

However, to help avoid the unintended consequences of poorly advised policy more ecological or economic detail is needed, in other cases. With more realism this challenge of balancing model tractability is not new in science, but when addressing the economics of invasive species management, it matters significantly.

8.4.1. Why Integration?

Why bother to go through all the trouble to integrate economics and biology for invasive species management? The straightforward answer is that it can provide more environmental protection at less cost and it will give us "better science" for policy.

We account for the impact of economics on biological systems and vice versa by integrating, and the feedback loops between the two systems are captured. If we can do so,

- Predictions of human behavior and species population densities will be less biased as we generate better risk assessments, and
- Through more efficient expenditure of scarce public and private resources for prevention and generate greater net benefits we support better risk management.

For instance, the case of fishing pressure or harvest effort is considered. As a constant treating fishing pressure or harvesting effort— among the systems unaffected by feedback how humans adapt to a change in the fishery does not account for. As fish populations fall due to an array of biological considerations, the harvest of fish also falls with constant fishing effort. Integrating economic systems and ecosystems via fishing efforts captures this initial change. What it does not capture is how a change in one system can lead to a change in behavior in the other.

When the fish species declines, will the fishing effort be constant? Many economic factors cause humans to transfer their efforts from one fish species to another or from fishing to other activities when the fish population declines. If no account were taken of these feedbacks, this shift in behavior could lead to a different ecosystem steady state.

8.5. WHAT ARE THE DISCIPLINARY IMPEDIMENTS OF ECOLOGICAL ECONOMIC MODELING?

By challenges and pitfalls ecological-economic modeling is necessary and feasible in principle and is accompanied. Ecological-economic modeling combines the knowledge and concepts of two disciplines using a particular methodological approach—modeling.

Combining these three disciplines requires:

- An in-depth knowledge of both disciplines by the researchers involved,
- Adequate identification and framing of the problem to be investigated, and
- A common understanding of modeling and scales between economists and ecologists.

Typical for ecological-economic modeling, in fulfilling these requirements, impediments, and pitfalls, which are likely to come up? Learn what other disciplines can do for you, and what you can do for them, as stated succinctly by a colleague.

8.5.1. Deep Knowledge of the Two Disciplines

The average economist's awareness of what ecologists do is not well developed and *vice versa*. By ecologists with business or finance economics is sometimes confused.

Failing to appreciate that land management is a major issue in such subdisciplines as landscape ecology and conservation biology, some economists think ecologists are solely interested in collecting and studying plants and animals for their own sake.

To researchers who have limited or no experience with the other discipline such confusions or prejudices are probably restricted. For numerous reasons, including the benefits of specialization, but even scientists who closely work with colleagues from the other discipline often do not have a profound understanding of this discipline.

When a scientist lets his or her own narrow focus assume that simplified views represent a complete picture of the other discipline's concepts, ideas, and methods, limited knowledge of the other discipline becomes an issue. To make full use of the richness of knowledge that exists in the other discipline, then, she or he misses the opportunity.

The full knowledge available in economics has been automatically incorporated may miss essential aspects of a problem, ecologists who assume that, by integrating costs of conservation measures into their models.

Examples include risk aversion of economic agents, transaction costs, asymmetric information between policy makers and land users, and property rights. Similarly, economists are often unaware of the knowledge ecologists have about the temporal, functional structure of ecosystems, and spatial, and restrict themselves to simplified—spatially homogeneous, static or scalar—descriptions of ecological systems and processes.

To examine real-world phenomena in different ways in addition, ecologists, and economists are taught. They identify different factors they consider mattering, formulate different research questions, and set up different research projects when looking at the same biodiversity management problem.

Also, some researchers will learn from each other in both disciplines, and they will acknowledge the depth of the other discipline. During the entire project, some will agree on the overall aim of the research, and they will cooperate.

How to handle questions of spatial and temporal scales that frequently differ across the two disciplines: a key issue remains that strains communication and integration of ecological and economic knowledge. From eight ecological and economic journals, Drechsler et al. (2007) surveyed 60 models related to biodiversity conservation that were randomly selected.

Economic models are conceptual, and are static and do not address uncertainty, are formulated and solved analytically; most ecological models are solved numerically or through simulation found by them.

Specific to a particular species and a geographic region, constructed on rules simulated step by step to model the dynamics of the system and consider various uncertainties, ecological models are more specialized. In some fields of economics, we are aware that computational models are applied; they are rare in the economic analysis of biodiversity management and less common relative to ecology.

8.6. TRAIT-BASED RISK ASSESSMENT FOR INVASIVE SPECIES

From one of two points, the political justification of environmental policy generally begins. To protect the environment a strict biocentric view supposes that society has an obligation on the one hand. To include the qualification that this protection can come at some cost but not too severe a cost to society a pragmatic bio centrist would amend this assertion.

The obligation to protect the environment only obtains insofar, an anthropocentric view supposed that on the other hand, as it serves the long view aimed at protecting the safety and security of citizens, the general welfare of humanity and options on society's future productivity.

The premise of this book is that only the pragmatic biocentric and anthropocentric views are acceptable to society, a book about invasive species and human economic systems.

Thus, as the bottom line arbiter of acceptable policy in either case we are concerned about human welfare and costs to society: there are some benefits we wish to maximize, and similarly, we wish to minimize costs. Therefore, a process of weighing costs and benefits is environmental policy.

An approach to incorporating both the cost-benefit trade-off that seems a natural basis for decision making and the fact that at any point in time the state of scientific understanding is known as risk analysis, in other words the ability to predict future events can only be asserted probabilistically.

At its core, assigning probabilities to these states (such probabilities may depend on events we control), and assigning costs and benefits to these states and risk analysis consists of defining possible future states of the environment.

We obtain a distribution of future costs by multiplying the costs associated with conditionally independent future states and their probabilities of occurrence. To determine what actions should be taken to maximize welfare over the predicted distribution of future costs, rules (i.e., a decision theory) can later be introduced.

At the introduction stage over other available options there are a number of reasons to prefer risk assessment.

First, to prevent the introduction of invasive species risk assessment informs efforts, once established or adapting socially and economically to pests making it a more efficient way to reduce impacts than either eradicating species. Indeed, for avoiding the economic and environmental costs of invaders preventing introduction is usually the only option, eradication is often impossible.

Second, while preventing the arrival of high-risk species, and can thus present a lower detriment to commerce than other methods for preventing the arrival of invaders accurate risk assessments will permit introduction of low-risk species (e.g., eliminate entire trades).

Third, to make a scientifically defensible decision about whether a species should be allowed for trade nations, can use risk assessments. Regulating trade standards of WTO (World Trade Organization 2005) are met by following this approach and without posing any risk producers can sell their products to international markets.

Introduction of nonindigenous species can be done using application of risk analysis methods. Species have been transferred intentionally and accidentally by the engagement of human societies in trade and travel.

Number of species being introduced increases significantly as globe is occupied by humans and societies become more connected (Lodge et al., 2006). To become established a proportion of these species go on and undesirable impacts are the result of a subset of these.

Including the potential to alleviate poverty (Sachs 2005), because trade brings great benefits to society, as an unavoidable externality of activities that increase social welfare these invaders have generally been regarded. However, two groups of species have been aggressively managed. These are known pests of agriculture, such as human parasites and diseases and foot-and-mouth disease.

By large potential costs of invasion without the potential for benefits efforts to prevent the proliferation of such species are warranted. In contrast, species that are intentionally introduced are chosen for their potential to produce economic benefits for the importer, such as garden plants. In these cases, if the species becomes invasive society must weigh the benefits against the costs that will accrue.

8.7. MANAGEMENT OF INVASIVE SPECIES IN THE GREAT LAKES

To the nature of food webs in the Great Lakes, biological invasions have clearly wrought irrevocable changes and how humans interact with those resources. The total economic loss due to invasive species is estimated to exceed $120 billion annually in the United States alone (Pimentel et al., 2005). For the monetary costs of invasive species in the Great Lakes no clear estimates exist, but the annual total must certainly be in the billions of dollars. Furthermore, on Great Lakes ecosystems and society the impact of invasions clearly has ecological and nonmarket costs and the latter are difficult to quantify.

For example, by zebra mussels of native unionid mussels (some of which are of conservation concern) the extirpation in many inland lakes has left "graveyards" of unionid shells in place of once-thriving native mussel beds; using the frameworks such ecological costs could be quantified although this has not yet been done (Figure 8.9).

Figure 8.9. The Zebra mussel is an aquatic invasive species found at Diamond Lake in Umpqua National Forest in Oregon.

Source: Image by Flicker.

The system serves as a useful model to illustrate how NIS management can benefit from an adaptive tiered approach with the attendant ecological and economic impacts of NIS in the Great Lakes.

The inherent value of prevention of invasions is recognized by all experts virtually (see Ruiz and Carlton 2003; Lodge et al., 2006), and in the current chapter, identity of species, the focus on vectors, timing, and invading the Great Lakes has a number of strengths.

First, to be focused on mechanisms most important in transmitting NIS to the lakes, explicit prioritization of vectors allows funds and efforts. By the many species we are aware of and by others not yet identified but that may use a particular vector, the value of this approach is that we can prevent invasions both. Second, by prioritizing and eliminating the strongest vectors to the lakes the number of species invasions prevented is most likely to be maximized.

Thus, management of ballast of ships arriving from Europe should reduce the risk of future invasions recent patterns of invasion to the Great Lakes indicate that. The efficacy of current management programs as well as direct future programs is informed by the analyzes of invasion timelines and the identity of particular NIS can help.

For example, for transoceanic ships although midocean BWE policies reduces the risk of introducing species intolerant of high salinity, for sediment dwelling species or those capable of producing resistant resting stages these strategies appear to have been less effective.

For example, since ballast water control policies were implemented in 1993 nine NIS were likely introduced in ballast sediment. In the Great Lakes this information underpinned recent programs aimed at the management of no ballast on board (NOBOB) residuals (U.S. Coast Guard 2005; Canada Shipping Act 2006). As of 2008, before entering the Great Lakes all vessels from non-North American source ports must flush ballast water and/or ballast residuals while operating on the Great Lakes if they intend to perform any ballast discharges.

New introductions of European or Asian species should be effectively eliminate by this policy via the ballast vector. However, it might be some time before the efficacy of this policy can be assessed because of time lags.

To identify whether specific organisms pose an invasion risk based upon assessments of life history attributes, environmental suitability, propagule pressure, or a combination of these approaches, modeling efforts may be useful (e.g., Kolar, and Lodge 2002; Muirhead and MacIsaac 2005; Herborg et al., 2007).

To only those species for which excellent background information exists these approaches will likely be limited and as potentially problematic that are perceived (e.g., Chinese mitten crabs *Eriocheir sinensis*, Asian carps).

However, to discriminate between NIS that may or may not establish and spread managers can utilize the output of these models, and those likely to have large versus small impacts. To guard against introduction of those NIS most likely to survive and become problematic in the Great Lakes, management efforts could be tailored accordingly.

To prevent all invasions a focus on prevention cannot be expected. Early detection is desirable, although often difficult, in such cases. Scientific risk assessments are required to determine the appropriate management response once new NIS incursions are detected (Figure 8.10).

Figure 8.10. A picture of Eriocheir sinensis.

Source: Image by Wikimedia Commons.

As having little potential for establishment, spread, or impact following establishment, some NIS may be perceived. By examining the life history attributes of the species (e.g., Kolar, and Lodge 2002) and interspecific interactions and economic impacts in regions where the species is established, these assessments can often be made.

It might be appropriate to take no further action other than managing the vector that was responsible for the introduction if the risk of establishment and/or the risk of adverse impacts is deemed to be low. Additional actions may be warranted if the species is deemed a moderate to high risk.

A control-the-spread strategy is applied when these actions which consist of eradication fails. Aquatic NIS that has successfully eradicated the number of invertebrates is quite small (e.g., black striped mussel Mytilopsis in Australia and green alga Caulerpa in California; Bax et al., 2002; Williams and Schroeder, 2004.

Detection of the incursion at a sufficiently early stage is a central problem that the population size and range of an NIS are very small and relatively easy to manage.

Nevertheless, to establish monitoring programs to facilitate early detection of nascent invasions when it is economically advantageous, cases may occur, particularly where the threat to native biodiversity is great or the potential for biofouling is large.

Although the recent establishment of quagga mussels in the western United States highlights the difficulty in eliminating vector activity, the 100th Meridian Project was designed with this in mind. Including the electrical field barrier in the Chicago sanitary and ship canal (CSSC), creation of barriers to dispersal, is an example of a control the spread strategy that may be effective not only for target species (e.g., silver, and bighead carp) but other NIS, as well.

Managers and society must adapt to life with the established NIS when prevention and eradication are ineffective. As for the case with dreissenid mussels in the Great Lakes, managers are essentially helpless with respect to distribution of the NIS at this point.

Here, by controlling its local or regional abundance management efforts may consist of limiting damage associated with the NIS, through chlorination of water intake pipelines to reduce mussel biofouling, and application of biocides to specific streams to reduce recruitment of sea lamprey as is done on the Great Lakes.

For reducing abundance and economic or ecological impact, in a limited number of cases, new markets may be created to exploit the NIS, as has been done by instituting a bighead carp fishery on the Upper Mississippi River. In summary, as a critically important form of human-mediated global change the introduction of NIS has emerged.

With society bearing the economic impacts of those invasions the Great Lakes have been highly receptive to NIS and are now greatly disturbed by them. Most evidence points to a small number of vectors, especially ballast contents, as the predominant source of new NIS to the Great Lakes.

To manage NIS in the Great Lakes appropriate strategies is developed, and is clearly a work in progress, but from previous invasions both within and outside of the basin much can be learned, to shape management programs of the future (Figure 8.11).

Figure 8.11. Upper Mississippi River National Wildlife and fish refuge.

Source: Image by Wikimedia Commons.

8.7.1. Risk Assessment for Invasive Species in the Great Lakes

After being intentionally introduced for commerce in the aquarium/pet, live food, horticulture, bait, and other trades many invasive species become established. There is great potential for managing the vectors of introduction to prevent the arrival of likely invaders because these species are specifically identified for commerce.

Species are likely to become invasive if released this in turn requires predictions. Because the theory and methods for ecological predictions are still being developed, preventing the intentional introduction of invaders is a natural area of overlap between policy/management and academic ecology.

Here, as part of my Ph.D. dissertation (Keller 2006) I describe two projects that were conducted and one ongoing project that I am involved in as a result of my Ph.D. research. From the trades in live aquatic organisms the first two projects of my dissertation investigated invasion risks.

I spent two summers sampling organisms from the aquarium, water garden, live food, bait, and biological supplies trades, in one, funded by a federal grant that my adviser, David Lodge, and a previous graduate student, Cindy Kolar, brought into the lab.

To purchase and identify samples of all live aquatic species being sold the project was limited geographically to the southern basin of Lake Michigan and involved two summers of visiting stores. The trades sell many species that are currently invasive or that appear likely to become invasive in the future is our principal findings.

In addition, aquatic plants sold almost always had "hitchhiker" species and species identifications used by the trades are often wrong or ambiguous (Keller and Lodge 2007). To those of other research teams these results are similar (e.g., Reichard, and White 2001; Maki and Galatowitsch 2004; Padilla and Williams 2004; Rixon et al., 2005; Weigle et al., 2005; Cohen et al., 2008).

Conducted in the Lodge lab and by others my second project built upon previous work (Kolar and Lodge, 2002 (e.g., Richardson et al., 1990; Veltman et al., 1996; Reichard and Hamilton 1997; Pheloung et al., 1999; Champion and Clayton 2000).

The Laurentian Great Lakes and the 48 contiguous United States are the two nested geographical scales where I gathered trait data for all nonnative mollusk species. To accurately discriminate between species that have and have not become invasive we found that annual fecundity is sufficient (Keller et al., 2007a), and for both geographic scales that this relationship holds.

As part of a decision tool for determining which species pose an acceptable risk in this relationship could be used, for example, the aquarium and live food trades.

Only the first has had any direct policy impact to date, although each of these projects has policy implications. When Lodge was invited to meet with the mayor of Chicago this came about. He and I spent several hours putting together two pages of recommendations, in preparation for the meeting, to reduce the risks from aquatic invasive species for actions the City of Chicago could take.

A list of species that we considered high risk is included, from the findings of the first project it was drawn partly (Keller and Lodge, 2007.

By the Chicago Department of Environment, we were contacted and invited to participate in a working group with the aim of identifying aquatic species in trade that pose a high risk to Chicago waterways, including Lake Michigan, six months after Lodge's meeting.

We assembled a list of 13 animals and 13 plants working with other stakeholders, from my previous work and other work from the Lodge lab again drawing in part (Kolar and Lodge, 2002. These species have subsequently been prohibited from live sale within Chicago by action of the Chicago City Council (Invasive Species Control Ordinance of 2007).

Although it has contributed to the ongoing federal debate about invasive species policy, in contrast, to the best of our knowledge, the mollusk risk assessment is not being considered directly for policy.

It is likely that a side project I took on toward the end of my dissertation will have a greater direct impact while some of the species identified in my dissertation work are now regulated in Chicago. In 2006 the invasive aquatic plant Hydrilla was discovered in Indiana.

To assemble a working group with the aim of determining the invasion risks posed by aquatic plant species in trade prompted the Indiana Aquatic Invasive Species Coordinator. I was invited to be involved and I had previously presented my work at Indiana Department of Natural Resources meetings.

To devise a system for assessing risk I have met regularly with managers, policymakers, nongovernmental organization representatives, and plant retailers over the last 18 months. We hope that in the next 12 months we will be able to submit a list of the species in trade, annotated for their invasion risk in Indiana, to the Indiana Department of Natural Resources for consideration, this project continues to make progress.

8.8. CONCLUSION

In the conclusion of the chapter, it explains about the bioeconomics of invasive species. This chapter also discussed about the invasive process and feedbacks between biological and economic systems. It provides highlights on the bioeconomic impact of existing policy on invasive species. Towards the end of the chapter, it discussed about the trait – based risk assessment for invasive species. In this chapter, management of invasive species in the great lakes have also been discussed.

REFERENCES

1. Cacho, O., (2006). Bioeconomics of invasive species in aquatic ecosystems. *Aquaculture Economics & Management, 10*(2), 107–124 [online]. Available at: https://www.tandfonline.com/doi/abs/10.1080/13657300600695616 (accessed on 21 September 2022).
2. Keller, R., Lodge, D., Lewis, M., & Shogren, J., (2009). *Bioeconomics of Invasive Species: Integrating Ecology, Economics, Policy, and Management*. [e-Book] Pdfdrive. Available at: https://www.pdfdrive.com/bioeconomics-of-invasive-species-integrating-ecology-economics-policy-and-management-e159312329.html (accessed on 21 September 2022).
3. Mooney, H., (2010). In: Reuben, P. K., David, M. L., Mark, A. L., & Jason, F. S., (eds), Bioeconomics of invasive species: Integrating ecology, economics, policy, and management. Oxford and New York: Oxford University Press. $99.00 (hardcover); $49.95 (paper). xv + 298 p.; ill.; index. ISBN: 978-0-19-536798-0 (hc); 978-0-19-536797-3 (pb). 2009. *The Quarterly Review of Biology, 85*(1), 96, 97 [online]. Available at: https://www.journals.uchicago.edu/doi/10.1086/650237 (accessed on 21 September 2022).
4. Rossi, S., (2022). Ecotourism and invasive species. *A Journey in Antarctica*, [online] pp. 147–152. Available at: https://link.springer.com/chapter/10.1007/978-3-030-89492-4_19 (accessed on 21 September 2022).
5. Warziniack, T., Haight, R., Yemshanov, D., Apriesnig, J., Holmes, T., Countryman, A., Rothlisberger, J., & Haberland, C., (2021). Economics of invasive species. *Invasive Species in Forests and Rangelands of the United States*, [online] pp. 305–320. Available at: https://link.springer.com/chapter/10.1007/978-3-030-45367-1_14 (accessed on 21 September 2022).

INDEX

A

Advanced materials 72
algae 123
altruistic behavior 92, 103
amphibians 222
anatomy 105
Andean pampas grass 219
animal consumption 122
antibiotics 189
anti-parasitic drugs 189
aquaculture 184, 185, 186, 187, 188, 189, 190, 191, 192, 193, 195, 197, 198, 199, 202, 203, 204, 205, 206, 214, 215, 216
arthropods 222
artificial fertilizers 123

B

Benthic pollution 189
Bio-based products 72
Bioeconomic models 185, 186, 192, 205, 215
Bioeconomy 1, 2, 3, 4, 5, 6, 7, 8, 9, 10, 11, 13, 14, 15, 16, 17, 18, 19, 20, 21, 23, 30, 61, 62, 63, 68, 75, 84, 90
Bioeconomy Contribution Index (BCI) 45
bioeconomy policy 62
bioeconomy regulations 62
bioenergy 2, 3, 7, 14, 17, 19, 25, 26, 27, 29
biofuels 2, 8, 9
biological diversity 184
biological waste streams 72
biomass 2, 3, 4, 6, 8, 10, 11, 14, 16, 17, 21, 22, 24, 25, 26, 27, 29
Biomass resources 3
biomaterials 127, 130
bioproducts 2, 10
biotechnology 2, 3, 4, 5, 7, 8, 9, 14, 15, 16, 18, 19, 21

C

carbon footprints 72, 89
Castanea dentata 228
cellulose 127
channel catfish 193
chemical pesticides 123, 131
chemicals 62, 68, 69, 72, 89
climate change 32, 46, 50, 58

Common Agricultural Policy (CAP) 130
composite fiber sand medicines 72
Coregonus spp. 225
Cornus florida 228
crop production 125, 134, 139
cross-laminated timber (CLT) 70
Cyprinus carpio 193

D

Darwin's theory 95, 103, 112
Data collection 175
disease control 189
Dreissena polymorpha 221
dynamic maximum economic yield (DMEY) 167

E

ecology 154, 178
economic system 116
ecosystem 68, 73, 75, 81
energy security 32
enforcement costs 156, 158
Enhydra lutris 218
environmental degradation 3
Environmental Life Cycle Assessment 40
environmental technology 3
environmental threat 218
esthetics 189
evolutionary psychology 94, 101
Evolution theory 115
exclusive economic zone (EEZ) 164

F

Fagus grandifolia 228
feed 2, 3, 11, 16, 26, 28, 72, 77
fisheries management 154, 156, 157, 159, 161, 162, 164, 166, 168, 175, 178, 181, 182
fish supply 184, 215
Food and Agriculture Organization (FAO) 166
food chain 122, 123, 142
food security 32, 42, 46, 184, 215
food web 122, 149
forestry 2, 3, 25
fossil resources 72
free-rider behavior 156

G

game theory 96, 100, 104
genetic impacts 189
Geographical Positioning Systems (GPS) 159
greenhouse gas (GHG) 2
greenhouse gas (GHG) emissions 36

H

healthcare 32, 45
herbivores 122, 123, 149
high exclusion costs 156
High transaction costs 156
human cultivation 122
human socio-biology 94, 99, 110, 111

I

Ictalurus punctatus 193
information costs 156, 157
infrastructure 41
international organizations 2
International Sustainable Bioeconomy Working Group (ISBWG) 34

J

Japanese knotweed 219
Juglans cinerea 228

K

knowledge based bioeconomy (KBBE) 7
kudzu vine 219, 229

L

laminated veneer lumber (LVL) 70
life cycle assessment (LCA) 38, 40
Life Cycle Costing (LCC) 40
Life Cycle Impact Analysis (LCIA) 42
Life Cycle Inventory (LCI) 42
lumber 70, 71, 72
Lymantria dispar 226, 227

M

management plan 165
marine biomass 72
marine protected areas (MPAs) 176
maximum economic yield (MEY) 167
microorganisms 123
mollusks 222

N

natural phenomenon 218
noise pollution 189

O

Oncorhynchus mykiss 193
organic agriculture 123, 124, 147
organic foods 123
organic supermarket chains 123
Organization for Economic Co-operation and Development (OECD) 73

P

pharmaceuticals 3, 22, 26, 27
physiology 105
polymers 72
poverty 184, 215
product system 42
psychology 105, 113
pulp 68, 70, 72

Q

Quercus spp 228

R

racism 92
radar 159
rainbow trout 193
raw mineral phosphorus 123
renewable energy (RE) 130
renewable resources 2, 9, 21, 25
reptiles 222
resource biology 154
resource scarcity 66

S

salmon 184, 188, 189, 190, 191, 192, 193, 216
Salmo salar 193
saltwater systems 189
Salvelinus namaycush 225, 226
sexism 92
shrimp 184, 205
Social Life Cycle Assessment (S-LCA) 40
social trap condition 156

sociobiology 98, 101, 106, 107, 108, 110, 112, 113, 117, 119
Standard economic theory 116
Strongylocentrotus purpuratus 218
superorganism 94
sustainable bioeconomy 34, 35, 37, 38, 40, 48, 52, 53, 55, 59
sustainable development 65, 66, 79, 80
Sustainable Development Goals (SDGs) 62
Sustainable Europe Research Institute's (SERI) 39

T

total allowable catch (TAC) 165
toxicants 189
Tsuga canadensis 228

U

Ulmus americana 228
urban agriculture 125, 126

V

vegetation 122, 135

W

waste collection 3
waste management systems 41
water column pollution 189
water consumption 164

Y

yellow star thistle 219